AF587359

GLYCOBIOLOGY RESEARCH TRENDS

GLYCOBIOLOGY RESEARCH TRENDS

GEORGE POWELL
AND
OLIVER MCCABE
EDITORS

Nova Biomedical Books
New York

For permission to use material from this book please contact us:
Telephone 631-231-7269; Fax 631-231-8175
Web Site: http://www.novapublishers.com

LIBRARY OF CONGRESS CATALOGING-IN-PUBLICATION DATA

ISBN 978-1-60692-841-7

Available upon request

Published by Nova Science Publishers, Inc. ✝ New York

Contents

Preface		**vii**
Chapter I	Multidrug Resistance Protein 1 (MDR1) and Glycosphingolipid Biosynthesis *María Fabiana De Rosa and Clifford A. Lingwood*	**1**
Chapter II	Sophisticated Modes of Sugar Recognition by Intracellular Lectins Involved in Quality Control of Glycoproteins *Yukiko Kamiya, Daiki Kamiya, Reiko Urade, Tadashi Suzuki, and Koichi Kato*	**27**
Chapter III	Glyco-Biosensors: Properties, Materials and Applications *S. Cunningham, J. Gerlach and L Joshi*	**41**
Chapter IV	Structure and Activities of Natural Complex Polysaccharides *Nicola Volpi and Francesca Maccari*	**61**
Chapter V	CMP-N-acetylneuraminic Acid Synthetases and Sialyltransferases from Bacterial Sources *Takeshi Yamamoto*	**85**
Chapter VI	Heparan Sulfate D-glucosaminyl 3-O-sulfotransferases *Tomio Yabe and Nobuaki Maeda*	**109**
Chapter VII	Click Chemistry in Carbohydrate Based Drug Development and Glycobiology *Brendan L. Wilkinson, Laurent F. Bornaghi, Todd A. Houston and Sally-Ann Poulsen*	**127**
Chapter VIII	Functional Enzyme Complex Involved in HNK-1 Carbohydrate Biosynthesis *Yasuhiko Kizuka, Yasuhiro Tonoyama, and Shogo Oka*	**173**
Short Communication A		
	Protein-specific Glycosylation: Cis-Controlling Peptidic Elements *Franz-Georg Hanisch*	**189**

Short Communication B

Activation of Invariant Vα19-Jα33 TCR α-bearing Cells by Stimulation withcertain α-mannosylated Glycolipids **199**
Michio Shimamura

Index **211**

PREFACE

Glycobiology is the study of the structure, biosynthesis and biology of saccharides that are widely distributed in nature and are essential components of all living things. This field combines expertise in both carbohydrate biochemistry and molecular biology. The ABC drug transporter, P glycoprotein or MDR1 is one such example discussed in this book. The relationship of glycosphingolipid (GSL) biosynthesis to the Pgp-mediated multiple drug resistance phenotype, and the role of MDR1 in lipid biosynthesis in particular is evaluated. The structure and activities of natural complex polysaccharides are also analyzed. Specifically, heparin (Hep) as an anticoagulant and antithrombotic (macro)molecule and CS in the treatment of osteoarthritis is reviewed, and the GAG structure-function relationship, which has led to the discovery of novel drugs for the possible treatment of some serious diseases. The gene targeting analyses of certain enzymes, which revealed that the HNK-1 carbohydrate plays important roles in synaptic plasticity and memory function, is described as well. Recent work on the topic "protein-specific glycosylation" is highlighted also, focusing on the primary structure of a protein which directly exerts a cis-control of the glycosylation event.

Chapter I - The ABC drug transporter, P glycoprotein (Pgp) or MDR1 (ABCB1) has been shown to function as a glycosphingolipid (GSL) flippase in model systems using fluorescent GSL analogues, and in microsomes using exogenous glucosylceramide. MDR1 transfection of cells results in a marked increase in GSL synthesis, consistent with the proposed function as a glucosylceramide flippase which translocates this precursor GSL from the cytosolic to the luminal surface of the Golgi for GSL synthesis. Although this mechanism has recently been questioned, GSL biosynthesis and cellular drug resistance are intimately linked. Translocation of glucosylceramide in the Golgi for GSL biosynthesis can reduce drug elevated ceramide levels which would otherwise induce apoptosis. Inhibition of GSL biosynthesis in part, reverses drug resistance and inhibition of MDR1 prevents GSL biosynthesis. Functional cell surface MDR1 is contained within lipid rafts and is involved in cholesterol homeostasis via regulation of esterification. A raft location is required for MDR1 mediated drug efflux and MDR1 function can regulate the metabolism of both cholesterol and GSLs, the central components of such lipid microdomains. This generates a complex mechanism for the potential feedback inhibition of MDR1 function. In addition, GSLs biosynthesis and their anterograde trafficking can influence surface and subcellular protein trafficking. Studies in Lysosomal Storage Diseases (in which GSLs accumulate) show GSL and cholesterol anterograde and retrograde intracellular trafficking are linked together, implying that GSL and cholesterol levels could also affect MDR1 plasma membrane trafficking. It has been recently shown that depletion of GSLs results in the loss of cell surface MDR1 expression and selective GSL analogues function as MDR1 inhibitors. This review will consider the relationship of GSL biosynthesis to the Pgp-mediated multiple drug resistance phenotype, and the role of MDR1 in lipid homeostasis.

Chapter II - Recent advances in molecular and structural glycobiology have shown that a variety of lectins recognize partially trimmed intermediates of the high-mannose-type oligosaccharides and thereby determine the fates of their carrier proteins in cells. The carbohydrate recognition domains (CRDs) of these intracellular lectins exhibit structural similarities with those of the extracellular lectins such as leguminous lectins and galectins. This article primarily focuses on sugar recognition events in the early secretory pathway and in the ubiquitin/proteasome-mediated degradation pathway. Based on our main findings, we discuss the structural and molecular basis for the mechanisms underlying the quality control of glycoprotsins through the interplay of proteins possessing homologous CRDs in cells.

Chapter III - For almost half a century, the development of methods for the detection and quantification of biomolecules has been an intensive area of research. The field of 'biosensors' now encompasses several areas specifically geared toward the rapid and sensitive detection, identification, and quantification of target analytes. In general biosensors function through characteristic properties, ideally singular, of direct ligand/receptor interaction or associated enzyme activity. This area of research has been greatly aided by advancement brought by interdisciplinary mergers of biological, chemical and physical sciences and engineering along with miniaturizing of devices. However, relative to time and investment committed, limited commercial success has been made to date, with the notable exceptions being those biosensors for diabetes and pregnancy testing.

Although, the field of biosensors is significantly advanced for the detection of nucleic acids and proteins, it is at a relatively nascent stage for detecting glycans, their conjugates, and glycan-related enzymes. Further development and application of such 'glyco-biosensors' or 'glycosensors' would open novel analytical avenues. These avenues would be valuable for industrial bioprocesses, diagnoses of chronic and infectious diseases in clinical settings and for academic researchers. Here, we describe the current status and potential applications of glycosensors, examine methods currently being explored, and discuss the potential impact of such technologies. Recent reports of novel or refined strategies of rapid, robust and low cost analysis are also presented.

Chapter IV - Glycosaminoglycans (GAGs), hyaluronic acid or hyaluronan (HA), keratan sulfate (KS), chondroitin sulfates (CSs) and heparin (Hep)/heparan sulfate (HS), are complex ubiquitous natural polysaccharides that exhibit a wide range of biological functions by participating and regulating multiple cellular events and (patho)physiological processes. They are generally present either as free chains (HA) or as side chains of proteoglycans (PGs) (CS/dermatan sulfate (DS), Hep/HS and KS) and are most often found in cell membranes and in the extracellular matrix. The recent emergence of improved analytical tools for the study of these complex sugars has produced a virtual explosion in the field of glycomics. In particular, the well-known therapeutic applications of some of these macromolecules, in particular Hep as an anticoagulant and antithrombotic (macro)molecule and CS in the treatment of osteoarthritis (OA), and the increased understanding of GAG structure-function relationship has led to the discovery of novel drugs for the possible treatment of some serious diseases.

Chapter V - Sialic acids (Sias) are an important component of carbohydrate chains and are linked to terminal positions of the carbohydrate moiety of glycoconjugates, including glycoproteins and glycolipids. Various studies have focused on clarifying the structure–function relationship of Sias and have revealed that *N*-acetylneuraminic acid (Neu5Ac) is the major Sia component of glycoconjugates, and that the sialylated carbohydrate chains (sialosides) of glycoconjugates play significant roles in many biological processes.

Four main linkage patterns found in the sialosides—Neu5Acα2-6Gal, Neu5Acα2-3Gal, Neu5Acα2-6GalNAc, and Neu5Acα2-8Neu5Ac—are formed by specific sialyltransferases. All known mammalian, bacterial, and viral sialyltransferases use cytidine 5′-monophosphate *N*-acetylneuraminic acid (CMP-Neu5Ac), synthesized by CMP-Neu5Ac synthetase, as the common donor substrate. For this reason, the sialyltransferases and CMP-Neu5Ac synthetases are considered key enzymes in the biosynthesis and *in vitro* enzymatic synthesis of several sialosides.

The sialyltransferases and CMP-Neu5Ac synthetases have been cloned from various sources, including mammalian organs and bacteria. In general, the enzymes with a bacterial origin are more stable and productive in *Escherichia coli* protein expression systems than the mammalian-derived enzymes. In the case of sialyltransferases, the bacterial-derived enzymes show broader acceptor substrate specificity than the mammalian enzymes. These advantages highlight the capacity of bacterial enzymes as efficient tools for the *in vitro* enzymatic synthesis of sialosides.

The recent increase in research focusing on sialyltransferases from a diverse range of bacteria has led to the identification of many bacterial sialyltransferases. Several bacterial CMP-Neu5Ac synthetases have also recently been identified. This chapter describes the bacterial CMP-Neu5Ac synthetases and sialyltransferases that show promise as tools for the large-scale production of sialosides.

Chapter VI - Heparan sulfate (HS) is a polysaccharide belonging to the family of glycosaminoglycans that is ubiquitously expressed on the cell surface and as an extracellular substance. The sulfated domains of HS give specific binding affinities for ligands and regulate the biological functions by binding with various proteins, including growth factors, morphogens, and extracellular matrix molecules. These regulatory interactions of HS are considered to be structure-dependent and are determined by HS sulfotransferases. HS has a disaccharide unit comprising *N*-acetylglucosamine (GlcNAc) and glucuronic acid (GlcA). In the biosynthesis of HS, after the elongation of the skeletal chains involved in the sequential addition of alternating GlcNAc and GlcA residues, further modification by HS sulfotransferases produces the fine structures of HS. One of the HS sulfotransferases is 3-*O*-sulfotransferase (3-OST), which catalyzes the transfer of sulfate groups to the 3-*O*-position of D-glucosamine residues of HS from 3'-phosphoadenosine 5'-phosphosulfate. This is a rare, but essential HS chain modification required to produce the specific binding affinities between active HS and the target proteins. Recent studies on HS biosynthesis have indicated that 3-OST plays a pivotal role in bioactivity.

Chapter VII - The 1,3-dipolar cycloaddition reaction (1,3-DCR) of a 1,3-dipole to a dipolarophile for the synthesis of heterocycles is a ubiquitous transformation in synthetic organic chemistry. Recently, the Sharpless and Meldal groups have reported the dramatic rate enhancement (up to 10^7 times) and improved regioselectivity of the Huisgen 1,3-DCR of an organic azide to a terminal acetylene to afford, specifically a 1,4-disubstituted-1,2,3-triazole in the presence of a Cu^I catalyst. This Cu^I catalysed 1,3-DCR has successfully fulfilled the requirements of "click chemistry" as prescribed by Sharpless and within the past few years has become a premier component of the click chemistry paradigm. Click chemistry has proven to be of remarkable utility and broad scope, not only in organic synthesis, but in chemical biology and drug discovery. Click chemistry is highly modular and simplifies difficult syntheses. The biocompatibility of the reaction, tolerance towards a broad range of pH and relative inertness of acetylenes and azides within highly functionalised biological milieus has allowed click chemistry to become a viable bioconjugation strategy both for labelling biomolecules and for *in situ* lead discovery applications. More recently, click chemistry has emerged as a powerful conjugation strategy for the preparation of structurally diverse neoglycoconjugates of biomedical interest. This chapter aims to highlight recent developments appearing within the literature concerning the use of click chemistry in carbohydrate based drug discovery and glycobiology. Topics range from small molecule probes and drug leads, multivalent neoglycoconjugates acting as lectin inhibitors and potential vaccines, to bioconjugation strategies for labelling of engineered cell surface glycans. The chapter aims to be comprehensive with commentary on future perspective.

Chapter VIII - The HNK-1 carbohydrate epitope, also known as CD57, has a unique structural feature comprising a negatively charged trisaccharide (HSO_3-3GlcAβ1-3Galβ1-4GlcNAc-). This carbohydrate was first found to be an epitope on the human natural killer cell surface, but later experiments revealed that the HNK-1 carbohydrate was mainly expressed in the nervous system in various species. As key enzymes for its biosynthesis, we and others have cloned two glucuronyltransferases (GlcAT-P and GlcAT-S) and a sulfotransferase (HNK-1ST). Gene targeting analyses of these enzymes clearly revealed that the HNK-1 carbohydrate plays important roles in synaptic plasticity and memory formation. Although the key molecules related to its biosynthesis have been cloned, the detailed expression mechanism remains unclear probably due to the low expression levels of these transferases and to the complexity of the cellular glycosylation system in the ER and Golgi apparatus. During the course of a study to clarify the complicated HNK-1 biosynthesis system, we discovered a functional enzyme complex consisting of GlcAT-P (or –S) and HNK-1ST. Since the specific activity of HNK-1ST was up-regulated with the co-existence of GlcAT-P (or –S), the complex was found to raise the efficiency of HNK-1 biosynthesis. Moreover, recent research on the diverse glycan biosynthesis mechanism indicated that several enzyme complexes exist in cells to control the well-regulated glycan expression. It has been evident that to these complexes, non-transferase proteins are also related such as sugar-nucleotide transporter, glycosidase, phosphatase, and so on. These results suggest that not only the expression level of a single glycosyltransferase but also its surrounding environment serve as regulators of the expression of diverse glycoconjugates.

Short Communication A - The term "Protein-specific glycosylation" points to important functional implications of a subset of glycosylation types that are under direct control of recognition determinants on the protein. An example of this type is found in the β4-GalNAc modification of N-linked glycans on a selected panel of proteins, like carbonic anhydrase or glycodelin, which was demonstrated recently to require specific protein (sequence) determinants proximal to the glycosylation site functioning as cis-regulating elements. Another example of such a cis-regulating element was described for the control of mammalian O-mannosylation. In this case the structural features of the substrate sites (located within the mucin domain of human α-dystroglycan) are necessary, but not sufficient for determining the transfer of mannose to Ser/Thr. Evidence was provided that an upstream-located peptide was also essential. This short overview will highlight recent work on the topic "Protein-specific Glycosylation" focussing on the above-cited examples, where the primary structure of a protein directly exerts a cis-control of the glycosylation event and will not refer to other examples, where conformational epitopes serve as signal patches, as revealed for the modification of lysosomal hydrolases with mannose-6-phosphate or for the polysialylation of NCAM.

Short Communication B - A novel invariant Vα19-Jα33 TCR α chain was first found from mammalian blood cells. This TCR α chain as well as an invariant Vα14-Jα18 TCR α chain was primarily expressed by $NK1.1^+$ T cell repertoire (Vα19 NKT cell). Vα19 NKT cells produce immunoregulatory cytokines in response to TCR engagement, thus are suggested to be a potent therapeutic target. Attempts have been made to find specific antigens for Vα19 NKT cells. A series of α- and β-glycosyl ceramides were synthesized and tested whether they had potential to stimulate the cells isolated from invariant Vα19-Jα33 TCR transgenic mice (where development of Vα19 NKT cells is facilitated). Following comprehensive examinations substantial antigenic activity was found in α-ManCer that was presented by MR1, one of the MHC class Ib molecules. To determine structural requirements for natural ligands for Vα19 NKT cells, naturally occurring and synthetic α-mannosyl glycolipids were further analyzed. As a result, α-mannosyl phosphatidyl inositols (PI) such as $(\alpha\text{-Man})_2$-PI and α-Man-α-$GlcNH_2$-PI (a partial structure of mycobacterial lipoarabinomannan and GPI-anchors) as well as α-ManCer derivatives with structural modification in their sphingosine portion were found to activate Vα19 NKT cells *in vivo* and *in vitro*. Taken together, it is strongly suggested that Vα19 NKT cells are responsive to certain α-mannosylated glycolipids in the context of MR1.

In: Glycobiology Research Trends
Editors: G. Powell and O. McCabe

ISBN: 978-1-60692-841-7

Chapter I

MULTIDRUG RESISTANCE PROTEIN 1 (MDR1) AND GLYCOSPHINGOLIPID BIOSYNTHESIS

María Fabiana De Rosa[1,2], and Clifford A. Lingwood[1,2,3,*]

[1]Molecular Structure and Function, The Research Institute,
Hospital for Sick Children, Toronto, Ontario M5G 1X8, Canada
[2]Department of Laboratory Medicine & Pathobiology
[3]Department of Biochemistry, Faculty of Medicine,
University of Toronto, Toronto, Ontario M5S 1A8, Canada

ABSTRACT

The ABC drug transporter, P glycoprotein (Pgp) or MDR1 (ABCB1) has been shown to function as a glycosphingolipid (GSL) flippase in model systems using fluorescent GSL analogues, and in microsomes using exogenous glucosylceramide. MDR1 transfection of cells results in a marked increase in GSL synthesis, consistent with the proposed function as a glucosylceramide flippase which translocates this precursor GSL from the cytosolic to the luminal surface of the Golgi for GSL synthesis. Although this mechanism has recently been questioned, GSL biosynthesis and cellular drug resistance are intimately linked. Translocation of glucosylceramide in the Golgi for GSL biosynthesis can reduce drug elevated ceramide levels which would otherwise induce apoptosis. Inhibition of GSL biosynthesis in part, reverses drug resistance and inhibition of MDR1 prevents GSL biosynthesis. Functional cell surface MDR1 is contained within lipid rafts and is involved in cholesterol homeostasis via regulation of esterification. A raft location is required for MDR1 mediated drug efflux and MDR1 function can regulate the metabolism of both cholesterol and GSLs, the central components of such lipid microdomains. This generates a complex mechanism for the potential feedback inhibition of MDR1 function. In addition, GSLs biosynthesis and their anterograde trafficking can influence surface and subcellular protein trafficking. Studies in Lysosomal Storage Diseases (in which GSLs accumulate) show GSL and cholesterol anterograde and

[*] Corresponding author: Fax: (416) 813-5993; Email: cling@sickkids.ca

retrograde intracellular trafficking are linked together, implying that GSL and cholesterol levels could also affect MDR1 plasma membrane trafficking. It has been recently shown that depletion of GSLs results in the loss of cell surface MDR1 expression and selective GSL analogues function as MDR1 inhibitors. This review will consider the relationship of GSL biosynthesis to the Pgp-mediated multiple drug resistance phenotype, and the role of MDR1 in lipid homeostasis.

1. Introduction

Multidrug resistance (MDR), a drug resistance phenotype characterized by the ability of tumor cells to become insensitive to a variety of chemically unrelated chemotherapeutics, was first described by Ling and colleagues [1, 2]. Subsequent research has revealed a variety of mechanisms presumed to be responsible for the MDR phenotype. The best characterized mechanism leading to decreased intracellular drug levels, is the overexpression of energy-dependent drug efflux pump proteins such as P-glycoprotein (P-gp), the product of the MDR1-gene [3-5].

In recent years, links have been established between MDR and glycosphingolipids (GSLs). In MDR cancer cells, higher levels of specific glycosphingolipids have been reported with respect to their parental sensitive cells, such as glucosylceramide (GlcCer), sphingomyelin (SM), and galactosylceramide (GalCer), whereas lactosylceramide (LacCer) and all the more complex GSLs are present in lower amounts [6-10]. Cell transfection with the MDR1 gene also results in elevated GSLs levels [11]. These and other reports discussed later, led to the suggestion that MDR1 may be able to function as a lipid flippase in both the plasma membrane and the Golgi apparatus [11, 12]. The observation that MDR1 cells consistently overexpress GlcCer raises questions concerning the cellular localization of this increased lipid pool. Detergent-insoluble glycosphingolipid-enriched complexes (DIG) or lipid rafts, which are virtually present in all cell types, provide an interesting aspect of this field. Among other lipids such as globotriaosylceramide, gangliosides, SM and cholesterol, these membrane domains are highly enriched in GlcCer [13]. Upregulation of constituents of lipid rafts and caveolae, a specialized form of DIG, in particular caveolin 1, has been detected in MDR1 cells, and it has been suggested that caveolin 1 facilitates the transport of drugs from intracellular compartments to drug transporters in the plasma membrane [14]. MDR1 is also involved in the esterification of cholesterol, as the level decreases when MDR1 activity is modulated by certain inhibitors [15, 16]. In addition, it appears that the lipid environment can drastically influence the characteristics of the ATPase activity of MDR1 [17]. Cholesterol seems to play an exceptional role in MDR1 function; the active cholesterol redistribution across the membrane appears to be mediated by MDR1 [18]. Furthermore, ATPase activity [19] and drug binding [20] are affected by cholesterol. Some authors report two functional populations of MDR1 in the plasma membrane: the first is located in rafts and displays optimal ATPase activity, while the second is located elsewhere in the membrane and displays a lower ATPase activity. According to this report, after cholesterol depletion, MDR1 translocates to the non-raft fraction of the membrane, and after its repletion, MDR1 reinserts into rafts and recovers its optimal ATPase activity [21]. There was also evidence that glycolipid analogs may act as MDR1 inhibitors [22, 23].

The goal of this chapter is to provide an update of the interesting relationship that exists between GSLs biosynthesis and MDR1, in particular the role of MDR1 in lipid homeostasis, and to highlight the positive effects that can be obtained on MDR1 by manipulating GSLs biosynthesis, and *vice versa*.

2. MDR1 (P-GLYCOPROTEIN)

P-glycoprotein (P-gp), encoded by the *MDR1* gene, is a 170 kDa plasma membrane ATP dependent transporter that belongs to the so-called ABC transporters, and it works as a pump to extrude anticancer drugs and other compounds out of the cells [4, 24]. The MDR phenotype can be intrinsic or acquired during chemotherapy, and is characterized by the fact that cells expressing it, show cross resistance to a variety of structurally and functionally unrelated drugs [25].

Like all ABC proteins, P-gp comprises two membrane-bound domains, each made up of six transmembrane (TM) α-helices, and two nucleotide-binding (NB) domains that hydrolyze ATP to power drug transport [26]. Thus, P-gp reduces intracellular drug concentration and finally, protects cells from being killed [4].

A physiological efflux role for MDR1 has been proposed based on its expression on the apical membranes of gut epithelia, liver cells, kidney tubules and at blood tissue barriers [27]. As a consequence, MDR1 plays a key role in determining the pharmacokinetic properties of a number of pharmaceuticals, such as digestive absorption, cerebral disposition, and biliary and urinary elimination [28]. This has been subsequently confirmed in *Mdr1a/*1b knockout mice that had profound defects in drug distribution [29]. However, MDR1 is also expressed in the adrenal gland, on hematopoietic stem cells, natural killer cells, antigen-presenting dendritic cells, and T and B lymphocytes [30, 31].

In addition, P-gp expression can be regulated by some of its transport substrates [3]. P-gp is also able to handle some endogenous substrates, such as hydrophobic peptides [32] like cytokines [33], steroid metabolites [34] and even lipids [35, 36]. It has been reported that P-gp transports its various substrates after binding within the cytoplasmic leaflet of the plasma membrane [37, 38]. This may explain the high sensitivity of P-gp function on its hydrophobic environment, particularly on the state and composition of the surrounding membrane [39].

Although P-gp is generally considered to be localized in the cytoplasmic membrane and to mediate drug resistance by lowering total intracellular drug concentrations [40], some reports have suggested a more complex subcellular distribution of this protein. P-gp is also expressed in the membrane of the nucleus of MDR cell lines, suggesting its possible involvement in the extrusion of drugs from the nucleus of resistant cells [41]. Figure 1 shows localization of MDR1 and *Escherichia coli* verotoxin at the nuclear envelope in MDR1-MDCK cells, following retrograde transport of cell, surface bound verotoxin to the ER and nuclear envelope [42, 43] Other studies have localized P-gp within intracellular membranes [44, 45]. In particular, significant levels of P-gp were found in intracytoplasmic vesicles, identified as part of the Golgi apparatus, in different resistant cell lines, suggesting that P-gp functions in this region also [46-50]. Functionally active MDR1 has been found in the mitochondrial membrane of MDR-positive cells, to remove anticancer drugs from mitochondria into the cytosol; therefore, MDR1 could be involved in the protection of mitochondrial DNA from damage due to antiproliferative drugs [51].

One of the unresolved questions in the study of MDR1 is the transport mechanism by which P-gp can recognize structurally diverse compounds and couple their removal from the plasma membrane to the ATP hydrolysis. One approach to understanding this mechanism has been to identify residues and domains critical for drug transport, through structure-function analysis. This approach revealed a number of key amino acids which when changed, alter the substrate specificity of P-gp [52-56]. All of these residues are located within predicted transmembrane segments or cytoplasmic loops connecting these segments. The interface between the TM6 and TM7, which are the C-terminal and N-terminal membrane spanning sections of the first and second transmembrane domains, has been implicated as the position of the MDR1 common drug-binding site [54].

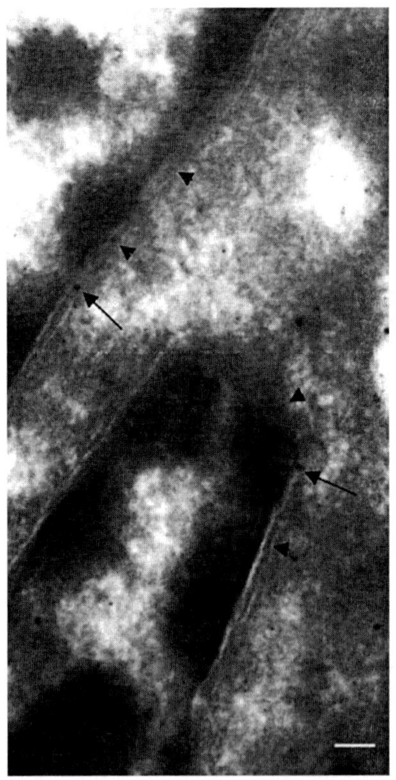

Figure 1. Nuclear envelope MDR1 and Verotoxin. Double labeling cryo-immunoelectronmicroscopy of MDR1 and Verotoxin 1B. Post-embedding cryo-double label immunoelectronmicroscopy using monoclonal antibody MRK16 andti-MDR1 and rabbit anti VT1B with 10- (arrows) and 5-nm (arrowheads) gold anti-species antibodies, respectively, was performed on MDR1-MDCK cells pretreated with VT1B at 37°C for 2hr to define relative location of antibody-bound MDR1 and globotriaosylceramide (Gb_3), VT1B receptor. Bar=500nm.

3. MDR1 AND GLYCOSPHINGOLIPIDS

Apart from the multidrug transporter function, normal physiological functions of MDR1 have also been reported [57]. MDR1 could be responsible for translocation of platelet-activating factor (PAF) across the plasma membrane, as PAF inhibited MDR1 drug transport in cancer cells [58]. High expression of MDR1 prevents stem-cell differentiation, leading to the proliferation and amplification of this cell repertoire [59], and MDR1 also plays a fundamental role in regulating programmed cell death, apoptosis [60, 61]. MDR1 might also be involved in the transport of cytokines, in particular IL-1, IL-2, IL-4, and IFN-γ, out of activated normal lymphocytes into the media [33, 62]. MDR1 is also associated with a volume-activated chloride channel [63], thus MDR1 is bifunctional with both transport and channel regulators.

MDR1 has also a close relationship with lipids, and in particular, glycosphingolipids. MDR1 can translocate both C6-NBD-PC and C(6)-NBD-PE across the apical membrane of multidrug resistant cells [64, 65]. MDR1 was also shown to regulate the translocation of sphingomyelin [66], and glucosylceramide [11], and short chain C(6)-NBD-GlcCer was found in the apical medium of MDR1 cells exclusively, and not in the basolateral membrane[12, 67]. Cells transfected with MDR1 were reported to esterify more cholesterol than their sensitive parental counterparts [68].

3.1. MDR1 and Ceramide Metabolism

It has been known for a long time that phospholipids [69], triglycerides [70] and cholesterol composition [71, 72] of MDR1 cells can differ from that of drug-sensitive cells. In addition, observations in the early 1980s, reported differences in gangliosides composition of MDR1 cells when compared to sensitive cells [73]. However, it was not until the end of the 1990s, that MDR-related differences in simple sphingolipid composition were further investigated.

Ceramide is one of the lipids constituting membrane microdomains, but is also present in internal membranes, where it is synthesized. Ceramide is known to regulate anti-proliferative responses, such apoptosis, growth arrest, differentiation and senescence in various human cancer cell lines [74]. In recent studies, metabolites of ceramide (Cer) have emerged as important MDR regulators. Cabot et al. showed that the levels of GlcCer, precursor of all higher glycosphingolipids and a simple glycosylated form of ceramide, was consistently elevated in several MDR1 overexpressing cells [6]. Thus, accumulation of glucosylceramide (GlcCer), has been shown to be a characteristic of some MDR cells [6, 7, 11], and MDR1 cells are more sensitive to depletion of GlcCer by various MDR1 reversing agents than their non-MDR counterparts [75]. Work on human ovarian carcinoma cells demonstrated that, in addition to GlcCer, also sphingomyelin and GalCer levels were significantly enhanced in MDR1 overexpressing cells, when compared to their drug sensitive counterpart [10]. However, LacCer and all higher GSLs were substantially decreased in these cells, due to altered intracellular localization of the biosynthesis enzyme LacCer synthase [10]. In contrast, in MDR1-transfected HepG2 cells, LacCer is prominently elevated compared with control cells, along with more moderate increases in GlcCer, ganglioside GM3 and SM levels. These alterations were unaffected by cyclosporin A but accompanied by upregulation of LacCer synthase [76]. Nevertheless, *MDR1*-transfected MDCK cell line showed dramatic accumulation of globotriaosylceramide (Gb_3) and globoseries GSLs, correlated with MDR1 activity [11]. Furthermore, observation of the requirement in Gb_3 expression for a cell to exhibit lectin-induced stimulation of MDR1 activity for both, PS externalization and rhodamine 123 efflux, indicates a functional relationship between MDR1 and Gb_3, even though this remains to be clarified [77]. LacCer and Gb_3 synthesis dependent on MDR1 functional activity was subsequently observed in various MDR1 cell lines, and it was proposed that MDR1 performs a translocase role for GlcCer at the level of the Golgi membranes [50]. The role of MDR1 as a lipid translocase will be discussed in detail later in this chapter. This might be consistent with a general positive role for P-gp in glycolipid biosynthesis and intracellular traffic.

On the other hand, the levels of the bioactive sphingolipids metabolites, ceramide and sphingosine, were quite comparable in resistant cells and their sensitive counterpart [10]. Furthermore, the deviations in GlcCer levels in multidrug resistant cells, are not restricted to MDR1 overexpressing cells since also MRP1-overexpressing cells, also show a 2- to 3- fold increase in this lipid [8]. Several metabolic mechanisms can be proposed to underlie the observed differences in GlcCer levels. Not much data is reported on enzymatic activities, but an increased glucosylceramide synthase (GCS) activity is probably an important factor. In addition to an increased GCS activity, enhanced GlcCer levels might be explained by a decreased GlcCer translocation into the Golgi apparatus, and this also might explain the decreased levels of LacCer and all higher GSLs [10].

It has also been described that administration of a variety of chemotherapeutic drugs leads to the production of ceramide, but this mechanism differs between agents, and possibly between cell types [78-80]. Some agents, such as daunorubicin, induces ceramide accumulation and subsequent cell death by either activation of a neutral SMase or, alternatively, by an increased of *de novo* synthesis through ceramide synthase [81, 82]. Upon administration of the MDR1 inhibitor etoposide, an increased activity of serine palmitoyltranferase, the first and rate-limiting step in the synthesis of all sphingolipids, was observed [83]. Modulators of MDR1, such as verapamil, cyclosporin A and its analog SDZ PSC833, and tamoxifen, which are well-known to inhibit the pump activity of P-gp and reverse resistance, have been shown to retard the levels of GlcCer in various human cancer cells that do not express MDR1 or other drug transporter proteins, implicating an additional mechanism of action of these agents [74].

Inhibition of the Cer glycosylation pathway has been shown to increase MDR1 cell sensitivity to cytotoxic drugs [6, 7, 84]. Moreover, overexpression of GlcCer synthase confers doxorubicin (Adriamycin) resistance in human breast cancer cells [85], whereas transfection of GlcCer synthase antisense reverses doxorubicin resistance in MDR cells [86]. P-gp inhibitors, such as PSC833 were found to facilitate Cer accumulation by stimulating Cer synthase-mediated *de novo* Cer synthesis [7, 87]. Together these results suggest that Cer and Cer metabolites may play an important role in regulating P-gp function. From this perspective, mifepristone, one of the GlcCer synthase inhibitors used in these reports [88], had been previously reported in an independent study to inhibit P-gp activity in KG1a cells

[89]. Thus, GSL synthesis may represent a ceramide sink which prevents accumulation and toxicity.

MDR1 transfected cells were observed to be sensitive to the pro-apoptotic action of the GlcCer synthase inhibitor, PDMP, whereas control cells were not, and this seems to be related to an effect on ceramide metabolism rather than a direct action on P-gp mediated cytotoxic drug transport [90]. PDMP has been described to sensitize neuroblastoma cells to paclitaxel (Taxol) and vincristine by reducing drug efflux presumably through P-gp inhibition [91]. Indeed, the pro-apoptotic effect on sensitive cells of GlcCer synthesis inhibition depends on the cell type considered, varying from moderate in MCF-7 [92], marginal in Hep-G2 [93], to null on melanoma cells [94], and even to a cell-protective effect on leukemic cells [95], whereas the effect is marked on MDR cells [96]. These observations strongly support a role for endogenous Cer in regulating MDR1 capacity.

3.2. MDR1 as a Lipid Flippase

Higgins and Gottesman have initially proposed that MDR1 might function as a drug flippase, moving hydrophobic molecules from the inner to the outer leaflet of the plasma membrane [37]. Since it was found that the P-glycoproteins encoded by the *MDR3* (*MDR2*) gene in humans and the *Mdr2* gene in mice are primarily phosphatidylcholine translocators [35, 97], and that MRP1, the only member of the MRP family known to transport lipid analogues, is the main transporter for the cellular excretion of the lipid metabolite leukotriene C4 [98], there has been increasing interest in the possibility that other ABC transporters are involved in lipid transport. Further studies in intact cells by the group of van Meer provided evidence that drug-transporting P-gps are also able to translocate short-chain lipid analogs, phospholipids and glycosphingolipids, from the inner to the outer leaflet of the plasma membrane [12, 35, 99]. The range of lipid analogues translocated is remarkably large: not only C_6-NBD-PC, but also C_6-NBD-phosphatidylethanolamine (PE) is transported. The MDR1 P-gp is able to transport C_6-NBD-SM, C_6-NBD-GlcCer, C_6-GlcCer and C_8C_8-GlcCer. Transport is inhibited by P-gp inhibitors, such as verapamil and PSC833, and by energy depletion of the kidney cells. Experiments using P-gp reconstituted in proteoliposomes also provided convincing evidence of this flippase activity [26]. Romsicki and Sharom used a fluorescence quenching technique to show that P-gp could flip a variety of NBD-labeled phospholipids and sphingomyelin [100], and this work was extended by Eckford and Sharom, who found that fluorescent analogues of the simple glycosphingolipids galactosyl- and glucosylceramide were also flipped at high rates [101]. Flipping of lactosylceramide was substantially slower, suggesting that addition of a second polar sugar residue to the headgroup presents a significant barrier to movement across the membrane. The process of lipid flipping resembles drug transport as it requires ATP hydrolysis and is inhibited by orthovanadate. Drugs and modulators are able to compete with membrane lipids for flipping, and their inhibitory potency is highly correlated with their MDR1 binding affinity [100, 101], suggesting that lipid translocation and drug transport share the same path in the protein.

The fluorescent lipids used in flippase studies in intact cells and model systems usually (but not always) have one short acyl chain. It is still unknown whether P-gp can transport natural membrane lipids, with two long acyl chains, as it is very difficult to test flipping of normal unlabelled lipids in model systems. However, P-gp was able to translocate a fluorescent derivative with two 16- or 18-carbon acyl chains. In addition, a large number of ABC transporters appear to translocate natural membrane lipids and sterols, and this may be a side activity of all the proteins in this family [102-104]. Another approach reported was the use of a microsomes system instead of intact cells, showing that lactosylceramide and globotriaosylceramide synthesis from endogenous or exogenously added liposomal glucosylceramide was inhibited by the MDR1 inhibitor cyclosporin A, consistent with a direct role for MDR1/glucosylceramide translocase activity in their synthesis [50].

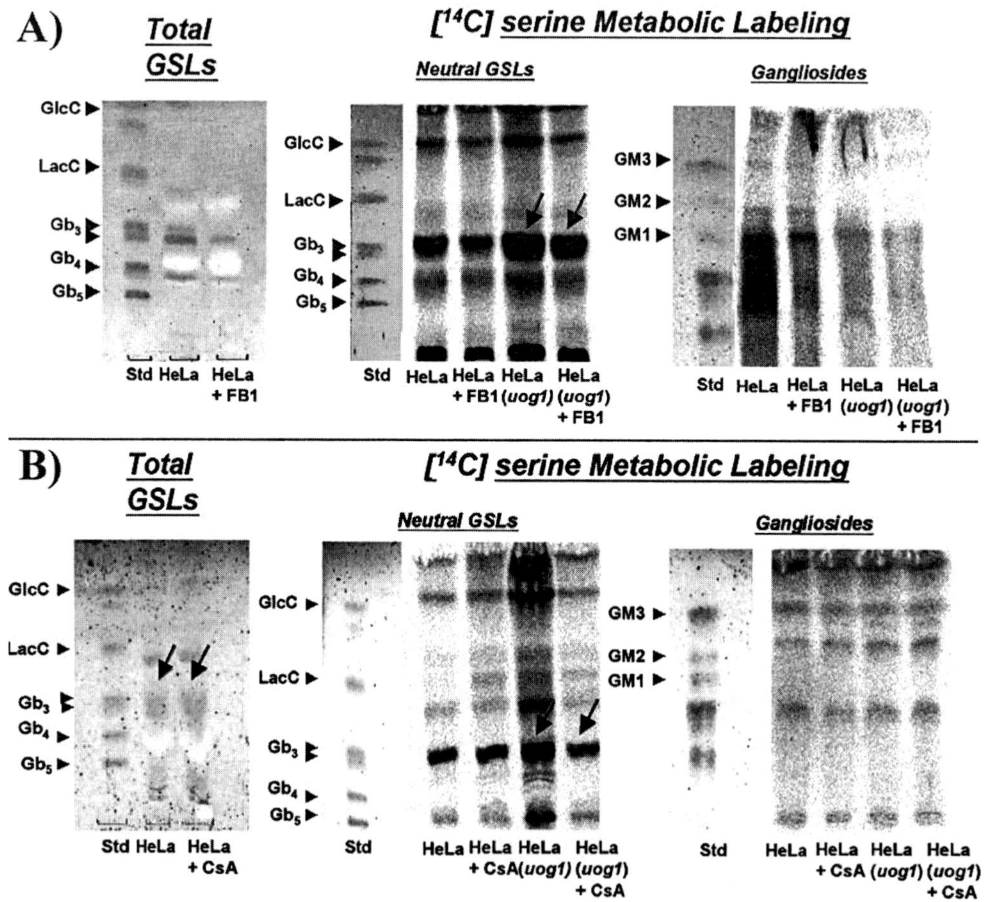

Figure 2. MDR1 dependent and independent mechanisms for GlcCer translocation into the Golgi for GSL synthesis in HeLa cells. HeLa cells were transfected with uog1 and extracted GSLs separated by TLC were detected by orcinol (total GSLs, left panel) and metabolically labeled neutral (center panel) and gangliosides (right panel) separated by ion exchange and detected by autoradiography. A) C18 ceramide synthesis (*uog1*) is channeled into neutral GSLs but not into gangliosides. B) Gb_3 synthesized via C18 ceramide synthase (*uog1*) is inhibited by CsA in HeLa cells (arrows in center panel) but total Gb_3 detected by orcinol is unaffected (arrows left panel).

Overexpression of MDR1 by retroviral transfection of MDCK cells with human MDR1 cDNA results in the accumulation of globotriaosylceramide (Gb_3), the receptor for the *Escherichia coli*-derived verotoxin, and increased sensitivity of cells (about 10^6-fold) to verotoxin (VT) [11]. These data show that P-gp plays a role in the synthesis of glycolipids, and also support that verotoxin might be a potential anticancer agent for the treatment of MDR1-expressing drug resistant cells. MDR1 inhibitors, *e.g.* ketoconazole or cyclosporin A (CsA), prevented the increased Gb_3 and VT sensitivity in various cell lines. In contrast, cellular ganglioside synthesis in the same cells, was unaffected by MDR1 inhibition, suggesting neutral and acid GSLs are synthesized from distinct precursor GSL pools [50]. Metabolic labeling in wild type and knockout (MDR*1a, 1b,* MRP1) mouse fibroblasts showed the same loss of neutral glycosphingolipids (glucosylceramide and lactosylceramide) but not ganglioside GM3 synthesis, confirming the proposed role for MDR1 translocase activity for GlcCer [50]. Gb_3 synthase was not significantly elevated in this MDR1-MDCK transfected cell line, and was not affected by CsA. Instead, synthesis of fluorescent analogs of LacCer (NBD-LacCer) and NBD-Gb_3 via NBD-GlcCer from exogenous NBD-C^6-ceramide was prevented by CsA [11]. Immunoelectron microscopy showed that MDR1 drug efflux pump is, in part, Golgi associated, and as a consequence, it is proposed MDR1 mediated flipping of GlcCer within the Golgi as a primary mechanism by which GlcCer is provided as a substrate for the various luminal glucosyltransferases involved in neutral GSL biosynthesis [50].

3.2.1. Alternative Mechanisms for GlcCer Transport into the Golgi

The fact that GlcCer is synthesized on the cytosolic surface of the Golgi but converted to LacCer in the lumen [105, 106] forced questions as whether and how GlcCer crosses the Golgi membrane. This appeared to have been solved according to the previous reports already considered, which can be summarized: short-chain GlcCer analogues were able to cross the Golgi membrane [105-107], the identification of the multidrug transporters ABCB1 and –C1 as *floppases* for these molecules [104] and a correlation between MDR1 activity and complex GSLs synthesis in living cells [50].

It has been reported that MDR1 inhibitors, CsA or ketoconazole, inhibit GSL biosynthesis in various cell lines studied but not in HeLa cells, as they do not express MDR1 [50]. However, when these cells were transfected with *uog1* gene, a *lass* (longevity assurance gene family) gene encoding C18 ceramide synthase, subsequent C18 GSL synthesis (Gb_3 is the major species) was found to be fumonisin B resistant as previously reported [108] but CsA sensitive, indicating that both MDR1 dependent and independent mechanisms for GlcCer translocation are present in these cells (Figure 2 A, B). Endogenous HeLa cell Gb_3 synthesis is CsA resistant but C18 *uog* mediated Gb_3 synthesis is MDR1 mediated (CsA sensitive).

The recent discoveries of two sphingolipid transfer proteins, CERT and FAPP2, have introduced potential alternative pathways in GSL anabolism. According to new reports, glucosylceramide synthase (GCS) is concentrated in the trans-Golgi, not in the cis-Golgi, and its activity was inhibited more than twofold upon CERT knockdown [109, 110]. Natural GlcCer did not flop efficiently across the Golgi membrane and GlcCer translocation was not affected in mouse MDR1 knockout TOK cells [110], despite the fact that our studies showed

GlcCer and LacCer depletion in these same cells [50]. Instead newly synthesized GlcCer reached the outer leaflet of the plasma membrane bilayer by a non-vesicular transport pathway and was translocated within the Golgi by a mechanism that involves a proton gradient. Finally, most GlcCer reached LacCer synthase in the Golgi lumen via the ER, with a role from the trans-Golgi glycolipid-binding protein, FAPP2 in shuttling Golgi cytosolic membrane GlcCer to the ER [110]. CERT transfers ceramide from the endoplasmic reticulum (ER) to the Golgi apparatus, a crucial step for the synthesis of sphingomyelin [109, 111, 112], and FAPP2 transfers GlcCer to appropriate sites for the synthesis of complex GSLs [110, 113]. These observations indicate that lipid transfer proteins, CERT and FAPP2, spatially regulate lipid metabolism on the cytosolic side [114] of the Golgi/ER.

Interestingly, most of the work on FAPP2 has been done using HeLa cells [113], where MDR1 is not the predominant mechanism by which GlcCer is translocated into the Golgi, and in other cell lines where the MDR phenotype has not been studied. FAPP2 also has several splicing variants and some of them lack the C terminus half of the GLPT domain, necessary to interact with GlcCer for its transport [113]. Tissue distribution and their effects on cellular functions are not established yet. Further studies are required for general understanding of GlcCer translocation into the Golgi.

3.3. MDR1 and Lipid Rafts

MDR1 has a special relationship with its membrane environment since it recognizes its substrates within the lipid bilayer [115]. The transport substrates gain access to the multiple drug binding sites which are located inside the cytoplasmic leaflet after partitioning into the lipid phase of the membrane [37]. Strong evidence indicates, that the lipid composition of biological membranes is closely related to MDR1 function [116]. Surrounding lipids modulates the ATPase activity of MDR1 [117] and its interaction with its substrates [20].

Glucosylceramide and other glycosphingolipids are important constituents of detergent-insoluble membrane domains termed DIGs [118] that are enriched also in sphingomyelin and cholesterol [119]. DIGs or lipid rafts are related in their lipid composition and their insolubility in cold non-ionic detergents to nonclathrin-coated, plasma membrane vesicular invaginations termed caveolae [120]. Caveolin-1, a 21-kDa integral membrane protein, is a major caveolar coat protein [121] that has the ability to engage in complex interactions with other caveolin molecules, as well as other proteins [122]. Lavie *et al.* first reported that in MDR1 cells there was a dramatic increase in the number of caveolae and in the level of caveolin-1 [14]. These findings may be related to the fact that a significant fraction of cellular P-gp is associated with caveolin-rich membrane domains [19]. In contrast, Hinrichs *et al.* determined that MDR1 was localized in the non-caveolar fraction of the plasma membrane [123], while flow cytometry and confocal microscopy showed that a substantial fraction of MDR1 was associated with lipid rafts and the cytoskeleton in human colon carcinoma cells [124]. In 2005, it was reported that MDR1 does not interact directly with caveolin-1, and is localized in intermediate-density domains distinct from classical lipid rafts and caveolae, which can be isolated using Brij-96 [125]. Barakat *et al.* reported the existence of two functional populations of P-gp: the first one localized in rafts, that displays optimal ATPase activity almost completely inhibited by orthovanadate and activated by verapamil; and the second one, located elsewhere in the membrane, displays a lower ATPase activity, less sensitive to orthovanadate and inhibited by verapamil [21]. Finally, after an exhaustive literature analysis, Orlowski *et al.* concludes that P-gp exists in rafts and non-rafts domains, depending on the cell considered, the experimental conditions and the method used; that when P-gp is present in rafts, interaction with protein partners regulates its activity; and that P-gp handles the raft-constituting lipids with particular efficiency, and it also influences membrane trafficking in the cell [126].

Cholesterol seems to play an exceptional role in MDR1 function; the active cholesterol redistribution across the membrane appears to be mediated by MDR1 [19]. MDR1 has been proposed to play a role in cholesterol esterification [15, 16]. MDR1 has also been reported to regulate the translocation of sphingomyelin (SM) [127], and it is possible that these two functions are interrelated. Depletion of plasma membrane SM can induce cholesterol esterification by releasing cholesterol from the plasma membrane into the intracellular pool [128]. However, it cannot be discounted that MDR1 somehow plays a direct role in trafficking cholesterol from the inner plasma membrane to the endoplasmic reticulum [129]. MDR1 may also be involved in the relocation of cholesterol from the cytosolic to the exoplasmic leaflet of the plasma membrane and in stabilization of the cholesterol-rich microdomains, rafts [18, 126]. The ATPase activity of Pgp measured in the presence and absence (basal ATPase activity) of Pgp substrates and modulators exhibit a strong dependence on the amount of cholesterol incorporated into native membrane vesicles or proteoliposomes [18, 65, 130]. Cholesterol depletion by increasing concentrations of methyl-cyclodextrin decreased its ATPase activity [18]. Acute cholesterol depletion or saturation in different cell lines inhibited transport activity [124, 131, 132], and in certain cell types cholesterol saturation enhanced active drug efflux [133, 134]. Thus, modulation of membrane cholesterol content can significantly alter Pgp function, but the relationship between the cholesterol content and Pgp function may be complex. Enrichment of membrane cholesterol changed lipid raft distribution but not the localization of P-gp, so MDR1 capacity depends on accurate lipid raft properties [131].

It is well known that GSLs interact specifically with cholesterol, as demonstrated in both artificial and cellular membranes [135, 136]. Pagano *et al.* have thoroughly studied this interaction by using BODIPY-GSLs in normal and lysosomal storage diseases (LSD) cell lines, in particular Niemann-Pick type-A and type-C [137-141]. In normal cells, BIODIPY-LacCer is transported to the Golgi after endocytosis, while in multiple LSD cell types, it accumulates in endosomal structures [137-139, 141]. He concluded that GSL accumulation significantly alters cholesterol distribution in LSD cells, and that cholesterol plays a major role in modulating the intracellular targeting of GSLs. He also identified rab7 and rab 9 as the main proteins responsible in GSL transport and this suggests another possible therapeutic approach to LSD treatment [141]. These findings that showed GSL and cholesterol anterograde and retrograde intracellular trafficking linked together, imply that GSL and cholesterol levels could also affect MDR1 plasma membrane trafficking, and could explain why depletion of GSLs with PPMP results in the loss of cell surface MDR1 expression and why selective GSL analogs function as MDR1 inhibitors [23].

3.4. MDR1 Inhibition and Lipid Analogs

Various attempts have been made to reverse multidrug resistance with MDR1 inhibitors that interact with MDR1 and block drug efflux. Three generations of drugs have already been tested with this purpose. While some of them were relatively non specific [142, 143], others were weak inhibitors that were also substrates, or even their deleterious toxicities with their use at concentrations required to inhibit MDR1, limited their clinical use [144-147]. The development of new MDR1 inhibitors with higher selectivity and stronger potency remains a major goal for this field of research.

Veldman *et al.* were the first to report MDR1 inhibition by neutral hexanoyl-glycosphingolipids [22]. When MDR1 ovarian carcinoma cells were incubated in medium supplemented with C_6-SM, C_6-GlcCer or C_6-GalCer, they regained their sensitivity to doxorubicin. In contrast, hexanoylceramide and gangliosides GM_3 and GM_2 had no effect on MDR1 activity, whereas sphingosine had a stimulating effect. They concluded that exogenous applied hexanoyl-glycosphingolipids modulated MDR1 activity through interactions with domains of the protein embedded in the inner leaflet of the plasma membrane, depending on the lipid headgroup structure [22].

In addition, MDR1 is in significant part Golgi associated [11] and able to flip exogenous GlcCer inside the Golgi to enhance neutral GSL biosynthesis [50]. Co-localization of MDR1 with Gb_3 at the cell surface was also reported [11, 23]. Furthermore, inhibition of GSL biosynthesis results in the loss of drug resistance and cell surface MDR1 expression [23], suggesting that GSLs could be involved in MDR1 cell surface transit. GSL involvement in protein trafficking has been previously reported [148]. Moreover, all MDR1 inhibitors tested resulted in the loss of cell surface MDR1 expression despite increased total MDR1. This could result from MDR1 inhibition preventing GSL synthesis to restrict MDR1 surface access. AdamantylGb_3 (AdaGb_3), a semi-synthetic analog of Gb_3, in which the fatty acid is replaced with a rigid globular hydrocarbon frame, with marked increased water solubility compared to Gb_3 [149], was able to reverse MDR1 cell drug resistance, to deplete cell surface MDR1, significantly reduced rhodamine 123 efflux in MDR1-MDCK and SK VLB cells (MDR variant of the parental ovarian carcinoma SKOV3 cell line), and significantly increased vinblastine apical-to basal transport, when added to the culture media of polarized C2BBe1 gastrointestinal epithelial cells. Disulfide cross-linking of mutant MDR1s showed no binding of AdaGb_3 to the MDR1 verapamil/cyclosporin-binding site between surface proximal helices of transmembrane segments (TM) 6 and TM7, but rather to an adjacent site nearer the center of TM6 and the TM7 extracellular face, *i.e.* closer to the bilayer leaflet interface. Verotoxin-mediated Gb_3 endocytosis also up-regulated total MDR1 and inhibited MDR1-dependant rhodamine 123 drug efflux. Thus, AdaGb_3 is an inhibitor of MDR1 [23]. In contrast, adamantyl galactosyl ceramide had no effect on drug resistance [23]. Since C_6-GalCer reversed doxorubicin resistance [22], the lipid and the carbohydrate must both play key roles in MDR1 inhibition.

3.5. MDR1 and Lysosomal Storage Diseases

The balance of synthesis and degradation of GSLs is completely regulated in the cell. If a glycosylhydrolase is lacking due to a genetic deficiency, a GSL accumulates in the lysosomes and causes a serious disease. The lysosomal storage diseases are genetic deficiencies in glycoconjugate metabolism, each due to a lack of a specific lysosomal sugar hydrolase or its activator protein [150]. To date, around 40 genetically distinct forms of them have been described. They are inherited as either autosomal recessive or X-linked traits. They are often associated with severe neurodegenerative pathology, caused by the intracellular accumulation of the enzyme substrate, and are termed glycosphingolipidoses [151]. Some examples of glycosphingolipidoses include Fabry disease, Gaucher disease, Sandhoff disease, Tay Sachs disease, and GM1 gangliosidosis. The symptoms of each disease depend on the enzyme, age of onset and residual enzyme activity [150]. Because approximately only 10% residual enzyme activity may be sufficient to develop clinical symptoms, exogenous enzyme-replacement therapy (ERT) has been developed, particularly in the two neutral GSL storage diseases, Fabry and Gaucher [152-154]

Fabry disease is characterized by the accumulation of galabiosylceramide and Gb_3, (as well as the blood group B and B1 antigenic GSLs to a minor extent), as a consequence of the deficiency of the lysosomal enzyme α-galactosidase A [155]. Gb_3 accumulates in the affected tissues of Fabry disease, including the kidney, vascular endothelium, peripheral nerves, and heart. The major clinical manifestations include renal failure, cerebral vascular disease, myocardial infarction, neuropathic pain, and skin lesions. Recombinant α-galactosidase administered to Fabry patients is able to reduce serum levels by 50%, liver Gb_3, and kidney Gb_3 levels [156, 157].

Gaucher disease (GD) is a storage disease in which macrophage sphingolipidosis accumulation occurs. This progressive disease results from deficiency of glucocerebrosidase (acid-β-glucosidase) in lysosomes. This enzyme is responsible for cleaving β-glycoside into β-glucose and ceramide subunits [158]. GD is inherited in an autosomal recessive fashion. It has 2 major forms: non-neuropathic (type 1, most commonly observed type in adulthood) and neuropathic (types 2 and 3). The disease is characterized by massive splenomegaly due to excessive accumulation of glucosylceramide in splenic macrophages. Other than spleen, so called "Gaucher cells" are the lipid-laden macrophages and can be observed in liver and bone marrow. Therefore GD causes organ damage where macrophages are densely present. Clinical alterations in bone, liver and spleen resulting in splenomegaly, hepatomegaly, hematological changes and orthopedic complications are the most predominant ones. Rarely kidney, skin, heart and central nervous system may be involved. Recent advances in genetic technology have made it possible to manage the GD patients with enzyme replacement treatment (ERT).

Despite its clinical success, the extraordinary cost of ERT has limited patient access and promoted the development of alternative strategies. Gene therapy is a candidate strategy for Fabry [159] and Gaucher [160] which showed promising results for the future. The third approach - substrate reduction therapy (SRT)- has been to use inhibitors of glucosylceramide synthase, the first enzyme required for the synthesis of most GSLs. In this line of research, the glucosylceramide synthase substrate mimic, PDMP and its derivatives have been tested on one side [161-163], and on the other, imino sugar-based GlcCer synthase inhibitors. The latter have proven effective in animal storage disease models [164], and in clinical trials for Gaucher disease [165, 166]. But such imino sugars also inhibit glucosidase processing of N-linked high mannose oligosaccharides [167] and glycogen breakdown [168].

The fact that MDR1 can function as a glycolipid flippase [35, 101], being responsible for the translocation of GlcCer into the Golgi in the majority of cultured cells[11, 50], and that MDR1-translocated GlcCer is used only for neutral GSL synthesis because inhibition of MDR1 does not affect cellular ganglioside synthesis [50], provide a degree of selectivity not available in the other approaches to substrate reduction therapy as a clinical management for Gaucher and Fabry diseases. Reported evidence for this is that MDR1 inhibitor, cyclosporin A, can deplete Gaucher lymphoid cell lines of accumulated glucosylceramide and Fabry cell lines of Gb_3, by preventing *de novo* synthesis [169]. Adult Fabry mice treated with ERT and CsA, showed after 9 weeks post-ERT, that their serum Gb_3 recovered to lower levels, and that the level of Gb_3 (but not gangliosides) in the liver was significantly depleted by CsA treatment. CsA treatment also reduced VT1 liver staining of hepatic venule endothelial and Kupffer cells that accumulated Gb_3 in the untreated Fabry mice [169]. This has been confirmed with additional, more selective MDR1 inhibitors (in progress). Thus MDR1 inhibition offers a potential alternative approach to SRT of such storage diseases.

4. Conclusion

Several direct molecular interactions between GSLs and MDR1 have been described. In summary, MDR1 expressing cell lines showed elevated levels of GlcCer and SM [6, 7, 10], inhibitors of GlcCer synthase promote drug toxicity in MDR1 cells [75], transfection of MDR1 also results in elevated GSLs levels [11], MDR1 expression can be down-regulated by GCS antisense transfection or chemical inhibition of GCS [170], cell surface MDR1 can flip glucosylceramide from the cytosolic Golgi membrane to the external plasma membrane bilayer [171], and inhibition of GSL biosynthesis results in the loss of MDR1 phenotype and cell surface MDR1 [23]. MDR1 has also been implicated in the transport and esterification of plasma cholesterol. It was recently reported that cholesterol acts on MDR1 transport activity by structuring the plasma membrane and structural organization of lipid rafts. Other aspects of this interesting relationship remain to be elucidated such as the interaction of MDR1 and lipid transport proteins.

The interplay between glycosphingolipids and MDR1 could provide a new spectrum of therapeutic options:

a) Verotoxin may provide a new approach to cancer treatment [172-174] as it was observed that drug resistant tumor cells are hypersensitive to verotoxin, based on altered expression of its glycolipid receptor, Gb_3, on multidrug resistant cells [42, 175].
b) The strong MDR1 reversal effects of adaGb_3, as well as its favourable *in vivo* features make it a possible choice for inhibition of MDR1 to increase bioavailability

of drugs across the intestinal epithelium [23]. Thus, specific GSL analogs provide a new approach to MDR reversal.

c) Cyclosporin A treatment has been found to reduce the recovery of serum Gb_3 levels in Fabry mice following α-galactosidase treatment. In such mice, the expression of Gb_3 within the liver is also reduced. Thus, MDR1 inhibition represents a potential novel treatment of neutral GSL storage diseases [169].

5. Acknowledgments

We would like to thank Dr Anthony Futerman, Weizmann Institute for his gift of the *uogl* plasmid. Studies from our laboratory have been supported by CIHR grant no 13073.

6. References

[1] Ling V, Kartner N, Sudo T, Siminovitch L, Riordan JR. Multidrug-resistance phenotype in Chinese hamster ovary cells. *Cancer Treat Rep* 1983;67(10):869-74.

[2] Biedler JL. Drug resistance: genotype versus phenotype--thirty-second G. H. A. Clowes Memorial Award Lecture. *Cancer Res* 1994;54(3):666-78.

[3] Bosch I, Croop J. P-glycoprotein multidrug resistance and cancer. *Biochem Biophys Acta* 1996;1288:F37-F54.

[4] Gottesman MM, Pastan I. Biochemistry of multidrug resistance mediated by the multidrug transporter. *Annu Rev Biochem* 1993;62:385-427.

[5] Gros P, Croop J, Housman D. Mammalian multidrug resistance gene: complete cDNA sequence indicates strong homology to bacterial transport proteins. *Cell* 1986;47(3):371-80.

[6] Lavie Y, Cao H, Bursten SL, Giuliano AE, Cabot MC. Accumulation of glucosylceramides in multidrug-resistant cancer cells. *J Biol Chem* 1996;271 (32): 19530-19536.

[7] Lucci A, Cho W, Han T, Giuliano A, Morton D, Cabot M. Glucosyl ceramide: a marker for multiple-drug resistant cancers. *Anticancer Res* 1998;18:475-480.

[8] Kok JW, Veldman RJ, Klappe K, Koning H, Filipeanu CM, Muller M. Differential expression of sphingolipids in MRP1 overexpressing HT29 cells. *Int J Cancer* 2000;87(2):172-8.

[9] Morjani H, Aouali N, Belhoussine R, Veldman RJ, Levade T, Manfait M. Elevation of glucosylceramide in multidrug-resistant cancer cells and accumulation in cytoplasmic droplets. *Int J Cancer* 2001;94(2):157-65.

[10] Veldman J, Klappe K, Hinrichs J, Hummel I, van der Schaaf G, Sietsma H, et al. Altered sphingolipid metabolism in multidrug-resistant ovarian cancer cells is due to uncoupling of glycolipid biosynthesis in the Golgi apparatus. *FASEB J* 2002;10:1096.

[11] Lala P, Ito S, Lingwood CA. Transfection of MDCK cells with the MDR1 gene results in a major increase in globotriaosyl ceramide and cell sensitivity to verocytotoxin: role of P-gp in glycolipid biosynthesis. *J Biol Chem* 2000;275(9):6246-6251.

[12] van Meer G, Sillence D, Sprong H, Kalin N, Raggers R. Transport of (glyco)sphingolipids in and between cellular membranes; multidrug transporters and lateral domains. *Biosci Rep* 1999;19(4):327-33.

[13] Kurzchalia T, Parton R. Membrane microdomains and caveolae. *Curt Op Cell Biol* 1999;11:424-431.

[14] Lavie Y, Fiucci G, Lisovitch M. Up-regulation of calveolae and calveolar constituents in Multidrug-resistant cancer cells. *J Biol Chem* 1998;273:32380-32383.

[15] Debry P, Nash EA, Neklason DW, Metherall JE. Role of multidrug resistance P-glycoproteins in cholesterol esterification. *J Biol Chem* 1997;272(2):1026-31.

[16] Luker GD, Nilsson KR, Covey DF, Piwnica-Worms D. Multidrug resistance (MDR1) P-glycoprotein enhances esterification of plasma membrane cholesterol. *J Biol Chem* 1999;274:6979-91.

[17] Urbatsch IL, Senior AE. Effects of lipids on ATPase activity of purified Chinese hamster P-glycoprotein. *Arch Biochem Biophys* 1995;316(1):135-40.

[18] Garrigues A, Escargueil AE, Orlowski S. The multidrug transporter, P-glycoprotein, actively mediates cholesterol redistribution in the cell membrane. *Proc Natl Acad Sci* U S A 2002;99(16):10347-52.

[19] Luker G, Pica C, Kumar A, Covey D, Piwnica-Worms D. Effects of cholesterol and enantiomeric cholesterol on P-glycoprotein localization and function in low-density membrane domains. *Biochem* 2000;39:7651-61.

[20] Romsicki Y, Sharom FJ. The membrane lipid environment modulates drug interactions with the P-glycoprotein multidrug transporter. *Biochem* 1999;38:6887-6896.

[21] Barakat S, Gayet L, Dayan G, Labialle S, Lazar A, Oleinikov V, et al. Multidrug-resistant cancer cells contain two populations of P-glycoprotein with differently stimulated P-gp ATPase activities: evidence from atomic force microscopy and biochemical analysis. *Biochem J* 2005;388(Pt 2):563-71.

[22] Veldman R, Sietsma H, Klappe K, Hoekstra D, Kok J. Inhibition of P-glycoprotein activity and chemosensitization of multidrug-resistant ovarian carcinoma 2780AD cells by hexanoylglucosyl ceramide. *Biochim Biophys Res Comm* 1999;266:492-496.

[23] De Rosa MF, Ackerley C, Wang B, Ito S, Clarke DM, Lingwood C. Inhibition of multidrug resistance by adamantylgb3, a globotriaosylceramide analog. *J Biol Chem* 2008;283(8):4501-11.

[24] Germann UA. P-glycoprotein--a mediator of multidrug resistance in tumour cells. *Eur J Cancer* 1996;32A(6):927-44.

[25] Breier A, Barancik M, Sulova Z, Uhrik B. P-glycoprotein--implications of metabolism of neoplastic cells and cancer therapy. *Curr Cancer Drug Targets* 2005;5(6):457-68.

[26] Sharom FJ, Lugo MR, Eckford PD. New insights into the drug binding, transport and lipid flippase activities of the p-glycoprotein multidrug transporter. *J Bioenerg Biomembr* 2005;37(6):481-7.

[27] Cordon-Cardo C, O'Brien JP, Boccia J, Casals D, Bertino JR, Melamed MR. Expression of the multidrug resistance gene product (P-glycoprotein) in human normal and tumor tissues. J Histochem Cytochem 1990;38(9):1277-87.

[28] Schinkel AH. The physiological function of drug-transporting P-glycoproteins. *Semin Cancer Biol* 1997;8(3):161-70.

[29] Schinkel AH, Mayer U, Wagenaar E, Mol CA, van Deemter L, Smit JJ, et al. Normal viability and altered pharmacokinetics in mice lacking mdr1-type (drug-transporting) P-glycoproteins. *Proc Natl Acad Sci* U S A 1997;94(8):4028-33.

[30] Klimecki WT, Futscher BW, Grogan TM, Dalton WS. P-glycoprotein expression and function in circulating blood cells from normal volunteers. *Blood* 1994;83(9):2451-8.

[31] Randolph GJ, Beaulieu S, Pope M, Sugawara I, Hoffman L, Steinman RM, et al. A physiologic function for p-glycoprotein (MDR-1) during the migration of dendritic cells from skin via afferent lymphatic vessels. *Proc Natl Acad Sci* U S A 1998;95(12):6924-9.

[32] Sarkadi B, Muller M, Homolya L, Hollo Z, Seprodi J, Germann UA, et al. Interaction of bioactive hydrophobic peptides with the human multidrug transporter. *Faseb J* 1994;8(10):766-70.

[33] Drach J, Gsur A, Hamilton G, Zhao S, Angerler J, Fiegl M, et al. Involvement of P-glycoprotein in the transmembrane transport of interleukin-2 (IL-2), IL-4, and interferon-gamma in normal human T lymphocytes. *Blood* 1996;88(5):1747-54.

[34] Barnes KM, Dickstein B, Cutler GB, Jr., Fojo T, Bates SE. Steroid treatment, accumulation, and antagonism of P-glycoprotein in multidrug-resistant cells. *Biochemistry* 1996;35(15):4820-7.

[35] van Helvoort A, Smith A, Sprong H, Fritzsche I, Schinkel A, Borst P, et al. MDR1 P-Glycoprotein is a lipid translocase of broad specificity, while MDR3 P-glycoprotein specifically translocates phosphatidyl choline. *Cell* 1996;87:507-517.

[36] Bosch I, Dunussi-Joannopoulos K, Wu R, Furlong S, Croop J. Phosphatidylcholine and phosphatidylethanolamine behave as substrates of the human MDR1 P-glycoprotein. *Biochem* 1997;36:5685-94.

[37] Higgins C, Gottesman M. Is the multidrug transporter a flippase? *Trends Biochem Sci* 1992;17:18-21.

[38] Sharom FJ. The P-glycoprotein efflux pump: How does it transport drugs? *J Membrane Biol* 1997;160:161-175.

[39] Ferte J. Analysis of the tangled relationships between P-glycoprotein-mediated multidrug resistance and the lipid phase of the cell membrane. *Eur J Biochem* 2000;267(2):277-94.

[40] Stride BD, Cole SP, Deeley RG. Localization of a substrate specificity domain in the multidrug resistance protein. *J Biol Chem* 1999;274(32):22877-83.

[41] Maraldi NM, Zini N, Santi S, Scotlandi K, Serra M, Baldini N. P-glycoprotein subcellular localization and cell morphotype in MDR1 gene-transfected human osteosarcoma cells. *Biol Cell* 1999;91(1):17-28.

[42] Arab S, Lingwood C. Intracellular targeting of the endoplasmic reticulum/nuclear envelope by retrograde transport may determine cell hypersensitivity to Verotoxin: sodium butyrate or selection of drug resistance may induce nuclear toxin targeting via globotriaosyl ceramide fatty acid isoform traffic. *J Cell Physiol* 1998;177:646-660.

[43] Tam P, Mahfoud R, Nutikka A, Khine AA, Binnington B, Paroutis P, et al. Differential intracellular transport and binding of verotoxin 1 and verotoxin 2 to globotriaosylceramide-containing lipid assemblies. *J Cell Physiol* 2008;216(3):750-63.

[44] Willingham MC, Rutherford AV, Cheng SY. Immunohistochemical localization of a thyroid hormone-binding protein (p55) in human tissues. *J Histochem Cytochem* 1987;35(10):1043-6.

[45] Gong Y, Wang Y, Chen F, Han J, Miao J, Shao N, et al. Identification of the subcellular localization of daunorubicin in multidrug-resistant K562 cell line. *Leuk Res* 2000;24(9):769-74.

[46] Gervasoni JE, Jr., Fields SZ, Krishna S, Baker MA, Rosado M, Thuraisamy K, et al. Subcellular distribution of daunorubicin in P-glycoprotein-positive and -negative drug-resistant cell lines using laser-assisted confocal microscopy. *Cancer Res* 1991;51(18):4955-63.

[47] Rutherford AV, Willingham MC. Ultrastructural localization of daunomycin in multidrug-resistant cultured cells with modulation of the multidrug transporter. *J Histochem Cytochem* 1993;41(10):1573-7.

[48] Molinari A, Cianfriglia M, Meschini S, Calcabrini A, Arancia G. P-Glycoprotein expression in the Golgi apparatus of multidrug-resistant cells. *Int J. Cancer* 1994;59:789-795.

[49] Meschini S, Calcabrini A, Monti E, Del Bufalo D, Stringaro A, Dolfini E, et al. Intracellular P-glycoprotein expression is associated with the intrinsic multidrug resistance phenotype in human colon adenocarcinoma cells. *Int J Cancer* 2000;87(5):615-28.

[50] De Rosa MF, Sillence D, Ackerley C, Lingwood C. Role of Multiple Drug Resistance Protein 1 in neutral but not acidic glycosphingolipid biosynthesis. *J. Biol. Chem.* 2004;279:7867-7876.

[51] Solazzo M, Fantappie O, Lasagna N, Sassoli C, Nosi D, Mazzanti R. P-gp localization in mitochondria and its functional characterization in multiple drug-resistant cell lines. *Exp Cell Res* 2006;312(20):4070-8.

[52] Loo TW, Clarke DM. Functional consequences of phenylalanine mutations in the predicted transmembrane domain of P-glycoprotein. *J Biol Chem* 1993;268(27):19965-72.

[53] Loo TW, Clarke DM. Functional consequences of proline mutations in the predicted transmembrane domain of P-glycoprotein. *J Biol Chem* 1993;268(5):3143-9.

[54] Loo TW, Clarke DM. Functional consequences of glycine mutations in the predicted cytoplasmic loops of P-glycoprotein. *J Biol Chem* 1994;269(10):7243-8.

[55] Currier SJ, Kane SE, Willingham MC, Cardarelli CO, Pastan I, Gottesman MM. Identification of residues in the first cytoplasmic loop of P-glycoprotein involved in the function of chimeric human MDR1-MDR2 transporters. *J Biol Chem* 1992;267(35):25153-9.

[56] Kajiji S, Dreslin JA, Grizzuti K, Gros P. Structurally distinct MDR modulators show specific patterns of reversal against P-glycoproteins bearing unique mutations at serine939/941. *Biochemistry* 1994;33(17):5041-8.

[57] Mizutani T, Masuda M, Nakai E, Furumiya K, Togawa H, Nakamura Y, et al. Genuine functions of P-glycoprotein (ABCB1). *Curr Drug Metab* 2008;9(2):167-74.

[58] Raggers R, Vogels I, van Meer G. Multidrug-resistance P-glycoprotein (MDR1) secretes platelet-activating factor. *Biochem J.* 2001;357:859-65.

[59] Bunting KD, Zhou S, Lu T, Sorrentino BP. Enforced P-glycoprotein pump function in murine bone marrow cells results in expansion of side population stem cells in vitro and repopulating cells in vivo. *Blood* 2000;96(3):902-9.

[60] Smyth MJ, Krasovskis E, Sutton VR, Johnstone RW. The drug efflux protein, P-glycoprotein, additionally protects drug-resistant tumor cells from multiple forms of caspase-dependent apoptosis. *Proc Natl Acad Sci* U S A 1998;95(12):7024-9.

[61] Mantovani I, Cappellini A, Tazzari PL, Papa V, Cocco L, Martelli AM. Caspase-dependent cleavage of 170-kDa P-glycoprotein during apoptosis of human T-lymphoblastoid CEM cells. *J Cell Physiol* 2006;207(3):836-44.

[62] Pawlik A, Baskiewicz-Masiuk M, Machalinski B, Safranow K, Gawronska-Szklarz B. Involvement of P-glycoprotein in the release of cytokines from peripheral blood mononuclear cells treated with methotrexate and dexamethasone. *J Pharm Pharmacol* 2005;57(11):1421-5.

[63] Marin M, Poret A, Maillet G, Leboulenger F, Le Foll F. Regulation of volume-sensitive Cl- channels in multi-drug resistant MCF7 cells. *Biochem Biophys Res Commun* 2005;334(4):1266-78.

[64] Abulrob AG, Gumbleton M. Transport of phosphatidylcholine in MDR3-negative epithelial cell lines via drug-induced MDR1 P-glycoprotein. *Biochem Biophys Res Commun* 1999;262(1):121-6.

[65] Rothnie A, Theron D, Soceneantu L, Martin C, Traikia M, Berridge G, et al. The importance of cholesterol in maintenance of P-glycoprotein activity and its membrane perturbing influence. *Eur Biophys* J 2001;30(6):430-42.

[66] Come MG, Bettaieb A, Skladanowski A, Larsen AK, Laurent G. Alteration of the daunorubicin-triggered sphingomyelin-ceramide pathway and apoptosis in MDR cells: influence of drug transport abnormalities. *Int J Cancer* 1999;81(4):580-7.

[67] van IJzendoorn SC, Zegers MM, Kok JW, Hoekstra D. Segregation of glucosylceramide and sphingomyelin occurs in the apical to basolateral transcytotic route in HepG2 cells. *J Cell Biol* 1997;137(2):347-57.

[68] Gayet L, Dayan G, Barakat S, Labialle S, Michaud M, Cogne S, et al. Control of P-glycoprotein activity by membrane cholesterol amounts and their relation to multidrug resistance in human CEM leukemia cells. *Biochemistry* 2005;44(11):4499-509.

[69] May GL, Wright LC, Dyne M, Mackinnon WB, Fox RM, Mountford CE. Plasma membrane lipid composition of vinblastine sensitive and resistant human leukaemic lymphoblasts. *Int J Cancer* 1988;42(5):728-33.

[70] Ramu A, Glaubiger D, Weintraub H. Differences in lipid composition of doxorubicin-sensitive and -resistant P388 cells. *Cancer Treat Rep* 1984;68(4):637-41.

[71] Mazzoni A, Trave F. Cytoplasmic membrane cholesterol and doxorubicin cytotoxicity in drug-sensitive and multidrug-resistant human ovarian cancer cells. *Oncol Res* 1993;5(2):75-82.

[72] Mountford CE, Wright LC. Organization of lipids in the plasma membranes of malignant and stimulated cells: a new model. *Trends Biochem Sci* 1988;13(5):172-7.

[73] Gascoyne N, Van Heyningen WE. Binding of cholera toxin by various tissues. *Infect Immun* 1975;12(3):466-9.

[74] Ogretmen B, Hannun YA. Updates on functions of ceramide in chemotherapy-induced cell death and in multidrug resistance. *Drug Resist Updat* 2001;4(6):368-77.

[75] Nicholson K, Quinn D, Kellett G, Warr J. Preferential killing of multidrug resistant KB cells by inhibitors of glucosylceramide synthase. *Br J Cancer* 1999;81:423-430.

[76] Hummel I, Klappe K, Kok JW. Up-regulation of lactosylceramide synthase in MDR1 overexpressing human liver tumour cells. *FEBS Lett* 2005;579(16):3381-4.

[77] Sugawara S, Hosono M, Ogawa Y, Takayanagi M, Nitta K. Catfish egg lectin causes rapid activation of multidrug resistance 1 P-glycoprotein as a lipid translocase. *Biol Pharm Bull* 2005;28(3):434-41.

[78] Strum JC, Small GW, Pauig SB, Daniel LW. 1-beta-D-Arabinofuranosylcytosine stimulates ceramide and diglyceride formation in HL-60 cells. *J Biol Chem* 1994;269(22):15493-7.

[79] Suzuki A, Iwasaki M, Kato M, Wagai N. Sequential operation of ceramide synthesis and ICE cascade in CPT-11-initiated apoptotic death signaling. *Exp Cell Res* 1997;233(1):41-7.

[80] Maurer BJ, Metelitsa LS, Seeger RC, Cabot MC, Reynolds CP. Increase of ceramide and induction of mixed apoptosis/necrosis by N-(4-hydroxyphenyl)- retinamide in neuroblastoma cell lines. *J Natl Cancer Inst* 1999;91(13):1138-46.

[81] Bose R, Vereheij M, Haimovitz-Friedman A, Scotto K, Fuks Z, Kolesnick R. Ceramide synthase mediates daunorubicin-induced apoptosis: an alternative mechanism for generating death signals. *Cell* 1995;82:405-414.

[82] Jaffrezou JP, Levade T, Bettaieb A, Andrieu N, Bezombes C, Maestre N, et al. Daunorubicin-induced apoptosis: triggering of ceramide generation through sphingomyelin hydrolysis. *Embo J* 1996;15(10):2417-24.

[83] Perry DK, Carton J, Shah AK, Meredith F, Uhlinger DJ, Hannun YA. Serine palmitoyltransferase regulates de novo ceramide generation during etoposide-induced apoptosis. *J Biol Chem* 2000;275(12):9078-84.

[84] Lavie Y, Cao H, Volner A, Lucci A, Han T-Y, Geffen V, et al. Agents that reverse multidrug resistance, tamoxifen, verapamil, and cyclosporin A, block glycosphingolipid metabolism by inhibiting ceramide glycosylation in human cancer cells. *J Biol Chem* 1997;272(3):1682-1687.

[85] Liu Y-Y, Han T-Y, Giuliano AE, Cabot MC. Expression of glucosylceramide synthase, converting ceramide to glucosylceramide, confers adriamycin resistance in human breast cancer cells. *J Biol Chem* 1999;274(21):1140-1146.

[86] Liu Y, Han, TY, Giuliano, AE, Hansen, N, Cabot, MC. Uncoupling ceramide glycosylation by transfection of glucosylceramide synthase antisense reverses adriamycin resistance. *J Biol Chem* 2000;275:7138-43.

[87] Cabot MC, Han T-Y, Giuliano AE. The multidrug resistance modulator SDZ PSC 833 is a potent activator of cellular ceramide formation. *FEBS Lett* 1998;431:185-188.

[88] Lucci A, Han TY, Liu YY, Giuliano AE, Cabot MC. Multidrug resistance modulators and doxorubicin synergize to elevate ceramide levels and elicit apoptosis in drug-resistant cancer cells. *Cancer* 1999;86(2):300-11.

[89] Fardel O, Courtois A, Drenou B, Lamy T, Lecureur V, le Prise PY, et al. Inhibition of P-glycoprotein activity in human leukemic cells by mifepristone. *Anticancer Drugs* 1996;7(6):671-7.

[90] Shabbits JA, Mayer LD. P-glycoprotein modulates ceramide-mediated sensitivity of human breast cancer cells to tubulin-binding anticancer drugs. *Mol Cancer Ther* 2002;1(3):205-13.

[91] Sietsma H, Veldman RJ, Kok JW. The involvement of sphingolipids in multidrug resistance. *J Membr Biol* 2001;181(3):153-62.

[92] Liu YY, Han TY, Yu JY, Bitterman A, Le A, Giuliano AE, et al. Oligonucleotides blocking glucosylceramide synthase expression selectively reverse drug resistance in cancer cells. *J Lipid Res* 2004;45(5):933-40.

[93] di Bartolomeo S, Spinedi A. Differential chemosensitizing effect of two glucosylceramide synthase inhibitors in hepatoma cells. *Biochem Biophys Res Commun* 2001;288(1):269-74.

[94] Veldman RJ, Mita A, Cuvillier O, Garcia V, Klappe K, Medin JA, et al. The absence of functional glucosylceramide synthase does not sensitize melanoma cells for anticancer drugs. *Faseb J* 2003;17(9):1144-6.

[95] Grazide S, Terrisse AD, Lerouge S, Laurent G, Jaffrezou JP. Cytoprotective effect of glucosylceramide synthase inhibition against daunorubicin-induced apoptosis in human leukemic cell lines. *J Biol Chem* 2004;279(18):18256-61.

[96] Liu Y, Han, TY, Giuliano, AE, Cabot, MC. Ceramide glycosylation potentiates cellular multidrug resistance. *FASEB J* 2001;15:719-30.

[97] Ruetz S, Gros P. Phosphatidylcholine translocase: a physiological role for the mdr2 gene. *Cell* 1994;77:1071-1081.

[98] Wijnholds J, Evers R, van Leusden MR, Mol CA, Zaman GJ, Mayer U, et al. Increased sensitivity to anticancer drugs and decreased inflammatory response in mice lacking the multidrug resistance-associated protein. *Nat Med 1*997;3(11):1275-9.

[99] van Helvoort A, Giudici ML, Thielemans M, van Meer G. Transport of sphingomyelin to the cell surface is inhibited by brefeldin A and in mitosis, where C6-NBD-sphingomyelin is translocated across the plasma membrane by a multidrug transporter activity. *J Cell Sci* 1997;110 (Pt 1):75-83.

[100] Romsicki Y, Sharom, FJ. Phospholipid flippase activity of the reconstituted P-glycoprotein multidrug transporter. *Biochemistry* 2001;40:6937-47.

[101] Eckford PD, Sharom FJ. The reconstituted P-glycoprotein multidrug transporter is a flippase for glucosylceramide and other simple glycosphingolipids. *Biochem J* 2005;389(Pt 2):517-26.

[102] Borst P, Zelcer N, van Helvoort A. ABC transporters in lipid transport. *Biochim Biophys Acta* 2000;1486(1):128-44.

[103] Kalin N, Fernandes J, Hrafnsdottir S, van Meer G. Natural phosphatidylcholine is actively translocated across the plasma membrane to the surface of mammalian cells. *J Biol Chem* 2004;279(32):33228-36.

[104] van Meer G, Halter D, Sprong H, Somerharju P, Egmond MR. ABC lipid transporters: extruders, flippases, or flopless activators? *FEBS Lett* 2006;580(4):1171-7.

[105] Lannert H, Bunning C, Jeckel D, Wieland F. Lactosyl ceramide is synthesized in the lumen of the Golgi apparatus. *FEBS Lett* 1994;342:91-96.

[106] Lannert H, Gorgas K, Meißner I, Wieland FT, Jeckel D. Functional organization of the Golgi apparatus in glycosphingolipid biosynthesis. Lactosylceramide and subsequent glycosphingolipids are formed in the lumen of the late Golgi. *J Biol Chem* 1998;273(5):2939-2946.

[107] Burger K, van der Bijl P, van Meer G. Topology of sphingolipid galactosyl transferase in ER and Golgi: transbilayer movement of monohexyl sphingolipids is required for higher glycosphingolipid biosynthesis. *J Cell Biol* 1996;133:15-28.

[108] Venkataraman K, Riebeling C, Bodennec J, Riezman H, Allegood JC, Sullards MC, et al. Upstream of growth and differentiation factor 1 (uog1), a mammalian homolog of the yeast longevity assurance gene 1 (LAG1), regulates N- stearoyl-sphinganine (C18-(dihydro)ceramide) synthesis in a fumonisin B1-independent manner in mammalian cells. *J Biol Chem* 2002;277(38):35642-9.

[109] Hanada K, Kumagai K, Yasuda S, Miura Y, Kawano M, Fukasawa M, et al. Molecular machinery for non-vesicular trafficking of ceramide. Nature 2003;426(6968):803-9.

[110] Halter D, Neumann S, van Dijk SM, Wolthoorn J, de Maziere AM, Vieira OV, et al. Pre- and post-Golgi translocation of glucosylceramide in glycosphingolipid synthesis. *J Cell Biol* 2007;179(1):101-15.

[111] Hanada K, Hara T, Fukasawa M, Yamaji A, Umeda M, Nishijima M. Mammalian cell mutants resistant to a sphingomyelin-directed cytolysin. Genetic and biochemical evidence for complex formation of the LCB1 protein with the LCB2 protein for serine palmitoyltransferase. *J Biol Chem* 1998;273(50):33787-94.

[112] Fukasawa M, Nishijima M, Hanada K. Genetic evidence for ATP-dependent endoplasmic reticulum-to-Golgi apparatus trafficking of ceramide for sphingomyelin synthesis in Chinese hamster ovary cells. *J Cell Biol* 1999;144(4):673-85.

[113] D'Angelo G, Polishchuk E, Di Tullio G, Santoro M, Di Campli A, Godi A, et al. Glycosphingolipid synthesis requires FAPP2 transfer of glucosylceramide. *Nature* 2007;449(7158):62-7.

[114] Yamaji T, Kumagai K, Tomishige N, Hanada K. Two sphingolipid transfer proteins, CERT and FAPP2: their roles in sphingolipid metabolism. *IUBMB Life* 2008;60(8):511-8.

[115] Homolya L, Hollo Z, Germann UA, Pastan I, Gottesman MM, Sarkadi B. Fluorescent cellular indicators are extruded by the multidrug resistance protein. *J Biol Chem* 1993;268(29):21493-6.

[116] Modok S, Heyward C, Callaghan R. P-glycoprotein retains function when reconstituted into a sphingolipid- and cholesterol-rich environment. *J Lipid Res* 2004;45(10):1910-8.

[117] Romsicki Y, Sharom FJ. The ATPase and ATP-binding functions of P-glycoprotein--modulation by interaction with defined phospholipids. *Eur J Biochem* 1998;256(1):170-8.

[118] Parton RG, Simons K. Digging into caveolae. *Science* 1995;269(5229):1398-9.

[119] Harder T, Simons K. Caveolae, DIGs, and the dynamics of sphingolipid-cholesterol microdomains. *Curr Opin Cell Biol* 1997;9(4):534-42.

[120] Parton RG. Caveolae and caveolins. *Curr Opin Cell Biol* 1996;8(4):542-8.

[121] Rothberg KG, Heuser JE, Donzell WC, Ying Y, Glenney JR, Anderson RGW. Caveolin, a protein component of caveolae membrane coats. *Cell* 1992;68:673-682.

[122] Okamoto T, Schlegel A, Scherer PE, Lisanti MP. Caveolins, a family of scaffolding proteins for organizing "preassembled signaling complexes" at the plasma membrane. *J Biol Chem* 1998;273(10):5419-22.

[123] Hinrichs JW, Klappe K, Hummel I, Kok JW. ATP-binding cassette transporters are enriched in non-caveolar detergent-insoluble glycosphingolipid-enriched membrane domains (DIGs) in human multidrug-resistant cancer cells. *J Biol Chem* 2004;279(7):5734-8.

[124] Bacso Z, Nagy H, Goda K, Bene L, Fenyvesi F, Matko J, et al. Raft and cytoskeleton associations of an ABC transporter: P-glycoprotein. *Cytometry A* 2004;61(2):105-16.

[125] Radeva G, Perabo J, Sharom FJ. P-Glycoprotein is localized in intermediate-density membrane microdomains distinct from classical lipid rafts and caveolar domains. *Febs J* 2005;272(19):4924-37.

[126] Orlowski S, Martin S, Escargueil A. P-glycoprotein and 'lipid rafts': some ambiguous mutual relationships (floating on them, building them or meeting them by chance?). *Cell Mol Life Sci* 2006;63(9):1038-59.

[127] Bezombes C, Maestre N, Laurent G, Levade T, Bettaieb A, Jaffrezou JP. Restoration of TNF-alpha-induced ceramide generation and apoptosis in resistant human leukemia KG1a cells by the P-glycoprotein blocker PSC833. *Faseb J* 1998;12(1):101-9.

[128] Slotte JP, Bierman EL. Depletion of plasma-membrane sphingomyelin rapidly alters the distribution of cholesterol between plasma membranes and intracellular cholesterol pools in cultured fibroblasts. *Biochem J* 1988;250(3):653-8.

[129] Johnstone RW, Ruefli AA, Smyth MJ. Multiple physiological functions for multidrug transporter P-glycoprotein. *Trends Biochem Sci* 2000;25(1):1-6.

[130] Kimura Y, Kioka N, Kato H, Matsuo M, Ueda K. Modulation of drug-stimulated ATPase activity of human MDR1/P-glycoprotein by cholesterol. *Biochem J* 2007;401(2):597-605.

[131] Dos Santos SM, Weber CC, Franke C, Muller WE, Eckert GP. Cholesterol: Coupling between membrane microenvironment and ABC transporter activity. *Biochem Biophys Res Commun* 2007;354(1):216-21.

[132] Cai C, Zhu H, Chen J. Overexpression of caveolin-1 increases plasma membrane fluidity and reduces P-glycoprotein function in Hs578T/Dox. *Biochem Biophys Res Commun* 2004;320(3):868-74.

[133] Troost J, Lindenmaier H, Haefeli WE, Weiss J. Modulation of cellular cholesterol alters P-glycoprotein activity in multidrug-resistant cells. *Mol Pharmacol* 2004;66(5):1332-9.

[134] Troost J, Albermann N, Emil Haefeli W, Weiss J. Cholesterol modulates P-glycoprotein activity in human peripheral blood mononuclear cells. *Biochem Biophys Res Commun* 2004;316(3):705-11.

[135] Brown RE. Sphingolipid organization in biomembranes: what physical studies of model membranes reveal. *J Cell Sci* 1998;111 (Pt 1):1-9.

[136] Brown DA, London E. Structure and function of sphingolipid- and cholesterol-rich membrane rafts. *J Biol Chem* 2000;275(23):17221-4.

[137] Chen CS, Bach G, Pagano RE. Abnormal transport along the lysosomal pathway in mucolipidosis, type IV disease. *Proc Natl Acad Sci* U S A 1998;95(11):6373-8.

[138] Chen C, Patterson M, Wheatley C, O'Brien J, Pagano R. Broad screening test for sphingolipid-storage diseases. *Lancet.* 1999;354:901-5.

[139] Puri V, Watanabe R, Dominguez M, Sun X, Wheatley CL, Marks DL, et al. Cholesterol modulates membrane traffic along the endocytic pathway in sphingolipid-storage diseases. *Nat Cell Biol* 1999;1:386-388.

[140] Puri V, Watanabe R, Singh RD, Dominguez M, Brown JC, Wheatley CL, et al. Clathrin-dependent and -independent internalization of plasma membrane sphingolipids initiates two Golgi targeting pathways. *J Cell Biol* 2001;154(3):535-47.

[141] Choudhury A, Dominguez M, Puri V, Sharma DK, Narita K, Wheatley CL, et al. Rab proteins mediate Golgi transport of caveola-internalized glycosphingolipids and correct lipid trafficking in Niemann-Pick C cells. *J Clin Invest* 2002;109(12):1541-50.

[142] Ford JM, Yang JM, Hait WN. P-glycoprotein-mediated multidrug resistance: experimental and clinical strategies for its reversal. *Cancer Treat Res* 1996;87:3-38.

[143] Sikic BI, Fisher GA, Lum BL, Halsey J, Beketic-Oreskovic L, Chen G. Modulation and prevention of multidrug resistance by inhibitors of P-glycoprotein. *Cancer Chemother Pharmacol* 1997;40 Suppl:S13-9.

[144] Hyafil F, Vergely C, Du Vignaud P, Grand-Perret T. In vitro and in vivo reversal of multidrug resistance by GF120918, an acridonecarboxamide derivative. *Cancer Res* 1993;53(19):4595-602.

[145] Fisher GA, Lum BL, Hausdorff J, Sikic BI. Pharmacological considerations in the modulation of multidrug resistance. *Eur J Cancer* 1996;32A(6):1082-8.

[146] Dantzig AH, Shepard RL, Cao J, Law KL, Ehlhardt WJ, Baughman TM, et al. Reversal of P-glycoprotein-mediated multidrug resistance by a potent cyclopropyldibenzosuberane modulator, LY335979. *Cancer Res* 1996;56(18):4171-9.

[147] Rowinsky EK, Smith L, Wang YM, Chaturvedi P, Villalona M, Campbell E, et al. Phase I and pharmacokinetic study of paclitaxel in combination with biricodar, a novel agent that reverses multidrug resistance conferred by overexpression of both MDR1 and MRP. *J Clin Oncol* 1998;16(9):2964-76.

[148] Sillence DJ, Puri V, Marks DL, Butters TD, Dwek RA, Pagano RE, et al. Glucosylceramide modulates membrane traffic along the endocytic pathway. *J Lipid Res* 2002;43(11):1837-1845.

[149] Mylvaganam M, Lingwood C. Adamantyl globotriaosyl ceramide- a monovalent soluble glycolipid mimic which inhibits verotoxin binding to its glycolipid receptor. *Biochem. Biophys. Res. Commun.* 1999;257:391-394.

[150] Schuette CG, Doering T, Kolter T, Sandhoff K. The glycosphingolipidoses-from disease to basic principles of metabolism. *Biol Chem* 1999;380(7-8):759-66.

[151] Sillence DJ, Platt FM. Storage diseases: new insights into sphingolipid functions. *Trends Cell Biol* 2003;13(4):195-203.

[152] Brady RO. Gaucher and Fabry diseases: from understanding pathophysiology to rational therapies. *Acta Paediatr Suppl* 2003;92(443):19-24.

[153] Heukamp LC, Schroder DW, Plassmann D, Homann J, Buttner R. Marked clinical and histologic improvement in a patient with type-1 Gaucher's disease following long-term

glucocerebroside substitution. A case report and review of current diagnosis and management. *Pathol Res Pract* 2003;199(3):159-63.

[154] Wilcox WR, Banikazemi M, Guffon N, Waldek S, Lee P, Linthorst GE, et al. Long-term safety and efficacy of enzyme replacement therapy for Fabry disease. *Am J Hum Genet* 2004;75(1):65-74.

[155] Brady RO. Enzymatic abnormalities in diseases of sphingolipid metabolism. *Clin Chem* 1967;13(7):565-77.

[156] Schiffmann R, Murray GJ, Treco D, Daniel P, Sellos-Moura M, Myers M, et al. Infusion of a-galactosidase A reduces tissue globotriaosylceramide storage in patients with Fabry disease. *Proc Nat Acad Sci,* USA 2000;97(1):365-370.

[157] Schiffmann R, Kopp JB, Austin HA, 3rd, Sabnis S, Moore DF, Weibel T, et al. Enzyme replacement therapy in Fabry disease: a randomized controlled trial. *Jama* 2001;285(21):2743-9.

[158] Desnick RJ. Gaucher disease: a century of delineation and understanding. *Prog Clin Biol Res* 1982;95:1-30.

[159] Yoshimitsu M, Sato T, Tao K, Walia JS, Rasaiah VI, Sleep GT, et al. Bioluminescent imaging of a marking transgene and correction of Fabry mice by neonatal injection of recombinant lentiviral vectors. *Proc Natl Acad Sci* U S A 2004;101(48):16909-14.

[160] Enquist IB, Nilsson E, Ooka A, Mansson JE, Olsson K, Ehinger M, et al. Effective cell and gene therapy in a murine model of Gaucher disease. *Proc Natl Acad Sci* U S A 2006;103(37):13819-24.

[161] Lee L, Abe A, Shayman JA. Improved inhibitors of glucosylceramide synthase. *J Biol Chem* 1999;274(21):14662-14669.

[162] Abe A, Arend LJ, Lee L, Lingwood CA, Brady RO, Shayman JA. Glycosphingolipid depletion in Fabry disease lymphoblasts with potent inhibitors of glucosylceramide synthase. *Kid Intl* 2000;57(2):446-454.

[163] Abe A, Wild SR, Lee WL, Shayman JA. Agents for the treatment of glycosphingolipid storage disorders. *Curr Drug Metab* 2001;2(3):331-8.

[164] Platt FM, Jeyakumar M, Andersson U, Heare T, Dwek RA, Butters TD. Substrate reduction therapy in mouse models of the glycosphingolipidoses. *Philos Trans R Soc Lond B Biol Sci* 2003;358(1433):947-54.

[165] Zimran A, Elstein D. Gaucher disease and the clinical experience with substrate reduction therapy. *Philos Trans R Soc Lond B Biol Sci* 2003;358(1433):961-6.

[166] Futerman AH, Sussman JL, Horowitz M, Silman I, Zimran A. New directions in the treatment of Gaucher disease. *Trends Pharmacol Sci* 2004;25(3):147-51.

[167] Tian G, Wilcockson D, Perry VH, Rudd PM, Dwek RA, Platt FM, et al. Inhibition of alpha-glucosidases I and II increases the cell surface expression of functional class A macrophage scavenger receptor (SR-A) by extending its half-life. *J Biol Chem* 2004;279(38):39303-9.

[168] Andersson U, Reinkensmeier G, Butters TD, Dwek RA, Platt FM. Inhibition of glycogen breakdown by imino sugars in vitro and in vivo. *Biochem Pharmacol* 2004;67(4):697-705.

[169] Mattocks M, Bagovich M, De Rosa M, Bond S, Binnington B, Rasaiah V, et al. Treatment of neutral glycosphingolipid storage disease via inhibition of the ABC Drug

Transporter, MDR1: Cyclosporin A can lower serum and some tissue globotriaosyl ceramide levels in the Fabry's mouse model. *FASEB J* 2006;273:2064-2075.

[170] Gouaze V, Liu YY, Prickett CS, Yu JY, Giuliano AE, Cabot MC. Glucosylceramide synthase blockade down-regulates P-glycoprotein and resensitizes multidrug-resistant breast cancer cells to anticancer drugs. *Cancer Res* 2005;65(9):3861-7.

[171] Raggers RJ, Pomorski T, Holthuis JC, Kalin N, van Meer G. Lipid traffic: the ABC of transbilayer movement. *Traffic* 2000;1(3):226-34.

[172] Arab S, Russel E, Chapman W, Rosen B, Lingwood C. Expression of the Verotoxin receptor glycolipid, globotriaosylceramide, in Ovarian Hyperplasias. *Oncol Res* 1997;9:553-563.

[173] Arab S, Murakami M, Dirks P, Boyd B, Hubbard S, Lingwood C, et al. Verotoxins inhibit the growth of and induce apoptosis in human astrocytoma cells. *J. Neuro. Oncol.* 1998;40:137-150.

[174] Arab S, Rutka J, Lingwood C. Verotoxin induces apoptosis and the complete, rapid, long-term elimination of human astrocytoma xenografts in nude mice. *Oncol Res* 1999;11:33-39.

[175] Farkas-Himsley H, Rosen B, Hill R, Arab S, Lingwood CA. Bacterial colicin active against tumour cells in vitro and in vivo is verotoxin 1. *Proc Natl Acad Sci* 1995;92:6996-7000.

In: Glycobiology Research Trends
Editors: G. Powell and O. McCabe
ISBN: 978-1-60692-841-7

Chapter II

SOPHISTICATED MODES OF SUGAR RECOGNITION BY INTRACELLULAR LECTINS INVOLVED IN QUALITY CONTROL OF GLYCOPROTEINS

Yukiko Kamiya[1,2], Daiki Kamiya[2], Reiko Urade[3], Tadashi Suzuki[4], and Koichi Kato[1,2,5,6]

[1]Okazaki Institute for Integrative Bioscience and Institute for Molecular Science, National Institutes of Natural Sciences, 5-1 Higashiyama, Myodaiji, Okazaki, Aichi 444-8787, Japan,

[2]Graduate School of Pharmaceutical Sciences, Nagoya City University, 3-1 Tanabe-dori, Mizuho-ku, Nagoya 467-8603, Japan,

[3]Graduate School of Agriculture, Kyoto University, Uji, Kyoto 611-0011, Japan,

[4]Glycometabolome Team, Systems Glycobiology Research Group, RIKEN Advanced Science Institute, 2-1 Hirosawa, Wako-shi, Saitama 351-0198, Japan,

[5]The Glycoscience Institute, Ochanomizu University, 2-1-1 Ohtsuka, Bunkyo-ku, Tokyo 112-8610, Japan, and

[6]GLYENCE Co., Ltd., 2-22-8 Chikusa, Chikusa-ku, Nagoya 464-0858, Japan

ABSTRACT

Recent advances in molecular and structural glycobiology have shown that a variety of lectins recognize partially trimmed intermediates of the high-mannose-type oligosaccharides and thereby determine the fates of their carrier proteins in cells. The carbohydrate recognition domains (CRDs) of these intracellular lectins exhibit structural similarities with those of the extracellular lectins such as leguminous lectins and galectins. This article primarily focuses on sugar recognition events in the early secretory pathway and in the ubiquitin/proteasome-mediated degradation pathway. Based on our main findings, we discuss the structural and molecular basis for the mechanisms underlying the quality control of glycoprotsins through the interplay of proteins possessing homologous CRDs in cells.

Introduction

Biological functions of glycoconjugates are expressed, in many cases, through the interactions of carbohydrate moieties with lectins operating as *decoders of glycocodes*. The carbohydrate-lectin interactions have frequently been characterized in comparison with antigen-antibody interactions, underscoring fuzzy specificities for glycotopes and weak affinities amplified by multivalency. The molecular basis of the sugar recognition by lectins has long been investigated in the context of extracellular events such as host-defense, viral infection, and cell-cell communication. However, the past decade has seen growing evidence of the intracellular roles of lectins primarily in quality control of glycoproteins [1,2]. The intracellular lectins recognize *N*-linked glycans as *quality tags* of the polypeptide and thereby determine the fates of the glycoproteins in cells, i.e. folding, translocation, and degradation (Figure 1). For instance, the endoplasmic reticulum (ER) chaperones calreticulin (CRT) and calnexin (CNX) assist the folding of nascent glycoproteins by recognizing a specific glycoform displayed thereon. Whilst correctly folded glycoproteins are sorted into the secretory pathway through interactions with type I membrane proteins ERGIC-53, VIP36, and VIPL, those failing to achieve correct folding are transported in retrograde from the ER to the cytosol, where they are subjected to ubiquitination, deglycosylation, and consequently proteasomal degradation. Carbohydrate recognition domains (CRDs) are found in those type I membrane proteins along with the ubiquitin ligases and peptide:*N*-glycanase (PNGase) working in the cytosol.

In this article, we will focus on these lectins and provide molecular and structural basis of their functions in the quality control of glycoproteins. While lectins have traditionally been classified based on taxonomy and saccharide-specificity, the accumulation of molecular and structural biology data has prompted us to categorize lectins on the basis of homology of amino acid sequence and/or three-dimensional structure. In this viewpoint, the term 'intracellular lectins' may sound broad and superficial, but the authors preferentially use it to highlight their similarity and difference from extracellular lectins.

Processing of *N*-Glycans in Cells

The common precursor of *N*-linked glycan is $Glc_3Man_9GlcNAc_2$, which possesses branches designated as D1, D2, and D3 arms (Figure 2A) [3]. After attaching to an asparagine residue in the consensus sequence Asn-X-Ser/Thr, this high-mannose-type oligosaccharide is trimmed by the actions of a series of glycosidases in the ER. The outermost glucose residue of the D1 arm is removed by the glucosidase I [4]. Subsequently, the second and third glucose residues are trimmed by glucosidase II [3]. While re-glucosylation at the D1 arm is catalyzed by UDP-glucose: glycoprotein glucosyltransferase (UGGT) reproducing the monoglucosylated glycoform [3,5-8], the non-reducing terminal mannose residues are irreversibly removed by the ER α1-2 mannosidase [3,9,10]. In the Golgi complex, the *N*-glycans are further trimmed by mannosidases and thereafter grow to mature glycoforms through a sequence of reactions catalyzed by glycosyltransferases [11].

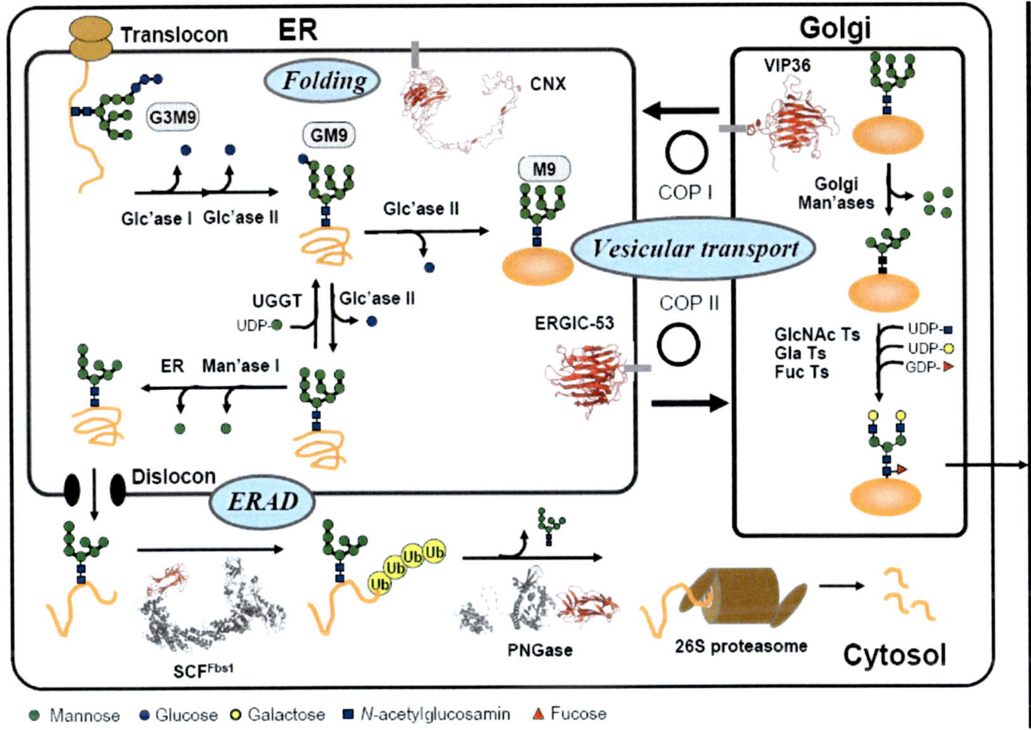

Figure 1. Scheme of glycoprotein-fate determination in cells through interactions with intracellular lectins. CRD of each protein or protein complex is colored red. Symbols representing the sugar residues are used according to the glycans nomenclature adopted by the CFG (http://www.functionalglycomics.org/static/consortium/). Notations for the high-mannose-type glycoforms are indicated as G3M9, GM9 and M9. The abbreviations: Glc'ase, glucosidase; Man'ase, mannosidase; UGGT, UDP-glucose:glycoprotein glucosyltransferase; GlcNAc T; *N*-acetylglucosaminyltransferase; Gal T, galactosyltransferase; Fuc T, fucosyltransferase; CRT, calreticulin; CNX, calnexin; PNGase, peptide:*N*-glycanase; ERAD, ER-associated degradation; ERGIC-53, 53 kDa membrane protein of the ER-Golgi intermediate compartment; VIP36, vesicular integral protein of 36 kDa; COPI and COPII, coat protein complexes I and II; Ub, ubiquitin.

Thus, during these processes, a variety of high-mannose-type glycoforms are generated, which are thought to be the targets of the intracellular lectins. By frontal affinity chromatography (FAC) using a pyridylaminated sugar library [12-14], the authors have performed comprehensive and quantitative analyses of the sugar-binding of intracellular lectins (Figure 2B). As a key example, CRT and CNX specifically interact with monoglucosylated high-mannose-type oligosaccharides and thereby assist the folding of carrier polypeptides [15]. Therefore, the molecular actions of these ER chaperones are governed by the trimming of the non-reducing terminal glucose at the D1 branch by glucosidase II and the re-glucosylation by UGGT. Through this deglucosylation/reglucosylation cycle, the correctly folded glycoproteins and the terminally misfolded glycoproteins are doomed to secretory and degradation pathways, respectively. The FAC data also revealed that removal of the outer mannose residues of the carbohydrate chains compromises the binding to CRT and, though moderately, to CNX, suggesting that the irreversible mannose trimming urge the glycoproteins to exit from the CNX/CRT cycle.

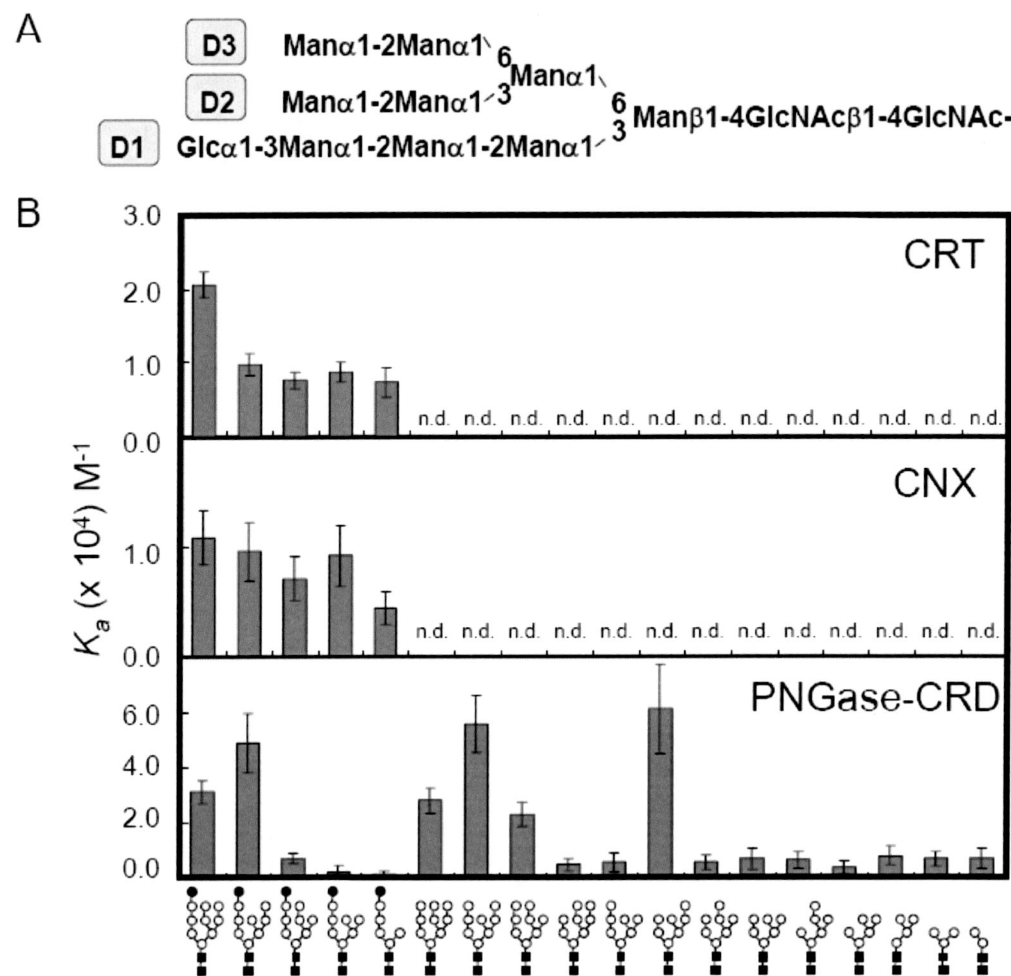

Figure 2. (A) Schematic representation of $GlcMan_9GlcNAc_2$ (GM9) structure showing the nomenclature of the sugar residues and blanches. (B) K_a values for CRT, CNX, and the CRD of mouse PNGase determined by FAC. Human CRT was expressed with an N-terminal GST moiety as described previously [54]. The DNA fragment encoding residues 1-462 of human CNX and 471-651 of mouse PNGase were inserted into the pET30 and pET28a plasmid vector, respectively, with an N-terminal hexahistidine moiety [55,56]. The protein was expressed in *Escherichia coli* BL21(DE3) strain. GST- and hexahistidine-fusion proteins were purified from cell lysates on glutathione-sepharose column and Ni^{2+}-sepharose high performance column affinity column, respectively. The CRT and the luminal region of CNX were immobilized on a HiTrap NHS-activated column after cleavage of the tag moiety, whilst the CRD of PNGase was on a Ni^{2+} sepharose high perormance column. The FAC analyses using a pyridylaminated sugar library were conducted as described previously [13,14]. n.d., not detected.

L-Type Lectins in the Early Secretory Pathway

The correctly folded glycoproteins are sorted as cargos and transported from the ER *via* the ERGIC (ER-Golgi intermiediate compartment) to the Golgi complex. In mammals, p58/ERGIC-53, VIP36, and VIPL contribute to this early secretory pathway by recognizing the sugar chains of the cargos. These type I transmembrane proteins are classified as L-type lectins because they conserve a luminal CRD that is homologous to leguminous lectins [16-18]. Indeed, the crystallographic studies showed that the CRDs of ERGIC-53 and VIP36 assume a jelly roll-like fold with structural resemblance to each other and to concanavalin A (Figure 3A-3C) [19-21]. In ERGIC-53, the CRD is connected to a coiled-coil domain, followed by transmembrane region, along with the cytosolic region that contains both ER exit and retrieval motifs [22], whereas VIP36 and VIPL have a stalk region instead of the coiled-coil domain [17,18,23]. ERGIC-53 forms homohexamer and cycles between the ER and the ERGIC. VIP36 cycles between the Golgi complex and the ER, while VIPL is a resident of the ER. The different localizations of these L-type lectins suggest their distinct roles in the secretory pathway in spite of the structural similarities of their CRDs. The overexpression of VIPL causes redistribution of ERGIC-53, suggesting the interaction and functional interplay between these lectins in the ER [17,18]. ERGIC-53 contributes to ER export of some glycoproteins, including the lysosomal glycoproteins cathepsin C and cathepsin Z [24,25]. This lectin interacts with MCFD2 (multi coagulation factor deficiency 2), a 16-kDa protein having two EF-hand domain [26-28], forming a complex for vesicular transport of blood coagulation factors V and VIII [29,30]. On the other hand, VIP36 has been reported to form a complex with a molecular chaperone BiP [31].

The sugar-binding properties of the three L-type lectins were characterized in detail by FAC analyses [13,14]. The FAC data revealed that both VIP36 and VIPL preferentially bind high-mannose-type oligosaccharides that retain the deglucosylated trimannosyl structure, i.e. Manα1-2Manα1-2Man, in the D1 branch, while ERGIC-53 exhibits a broader specificity and lower affinity in binding to high-mannose-type oligosaccharides, irrespective of the presence or absence of the terminal glucose residue. Interactions of a mannose residue with asparagine and aspartate in a pocket neighboring the Ca^{2+}-binding site are widely conserved among leguminous lectins [32,33]. The crystal structures of VIP36-CRD complexed with mannosyl ligands indicate that, in addition to these canonical interactions, Ser96 and Asp261 interact with Man-D1, putatively leading to the steric crush between the non-reducing terminal glucose residue and Glu98 (Figure 3D) [20]. While VIPL conserved these residues, the sugar-binding pocket of ERGIC-53 is significantly shallower because of the lack of the protrusive Asp261 side chain. Our mutational analysis demonstrated that abolishing the structural constraint by the Asp261 side chain enables VIP36-CRD to accept the monoglucosylated oligosaccharides, providing the structural basis of the deglucosylation sensing by VIP36 and VIPL [14].

The sugar-binding affinities of the L-type lectins are pH- and Ca^{2+}-dependent [13,14,34-36]. Biochemical and NMR studies showed that ERGIC-53 effectively binds glycans under neutral conditions but not at lower pH [34,36] (Kamiya Y. *et al.*, unpublished data). This can be attributed to the ionization of His178 (in human) destabilizes the nearby Ca^{2+} coordination [36]. The crystal structural data of the apo-form of rat ERGIC-53-CRD shows no interpretable electron density for a putative sugar-binding loop [21]. While VIPL, like ERGIC-53, shows stronger binding with M9 at higher pH, VIP36 exhibits a bell-shaped pH dependence of binding with a lower pH optimum (pH 6.5) [13], notwithstanding the fact that the corresponding histidine residue is conserved in both lectins, i.e. His190 in VIP36 and His188 in VIPL. Our mutational study revealed that the observed difference in the pH dependence between VIP36 and VIPL is attributed to the insertion of the lysine residue in a loop close to the sugar-binding site, i.e. Lys166 in VIPL [14]. Thus, the L-type lectins are equipped with distinct glucosylation- and/or pH-sensing sites in their homologous CRDs, which are conceivably relevant to their functions (Figure 3D).

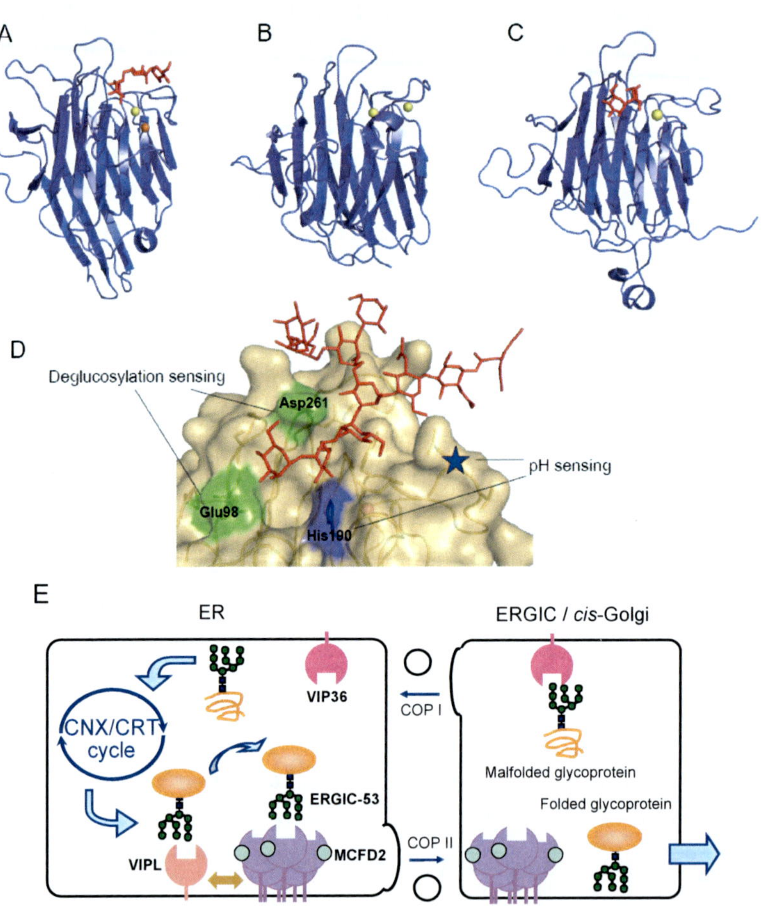

Figure 3. Crystal structures of (A) concanavalin A, (B) p58/ERGIC-53, and (C) VIP36. Manganese and calcium ions are shown as orange and yellow spheres, respectively. (D) The model of VIP36-CRD complexed with $Man_8GlcNAc_2$-Asn made based on the crystal structures (PDB accession codes: 1BSI, 1LTE, 1H4P, 2DUR, and 2E6V) [14,20], displaying the amino acid residues involved in deglucosylation-sensing and pH-sensing amino acid residues in green and blue, respectively. The blue star indicates the corresponding site to the Lys166 insertion in VIPL. The ligand oligosaccharides are shown in stick model in red. The graphics were generated with PyMOL (http://www.pymol.org). (E) Scheme of the interplay of the L-type lectins in the sorting and transport of glycoproteins. PDB accession codes: (A) 1CVN, (B) 1R1Z, and (C) 2DUR.

By taking into account their sugar-binding properties and subcellular localization together with the organellar pH, we have proposed a working model of the functions of the three L-type lectins in the sorting and transport of glycoproteins as illustrated (Figure 3E). In the ER, VIPL captures the deglucosylated glycoproteins immediately after their exit from the CNX/CRT cycle and thereby protect them from demannosylation leading to ERAD. Subsequently, the glycoproteins are handed over as cargos from VIPL to ERGIC-53. The binding of the glycoproteins to the ERGIC-53 hexamer can be enhanced by avidity and also by direct or MCFD2-mediated protein-protein interactions. In the absence of the protein-protein interactions, ERGIC-53 facilitates anterograde transport with only low efficiency *via* its lectin function with broad specificity. Under ER stress condition, however, the upregulation of ERGIC-53 [37,38] may contribute, with its broad specificity, to efficient and non-selective removal of glycoproteins from the ER in the case of emergency. In the Golgi, VIP36 in cooperation with BiP captures malfolded glycoproteins that have inadvertently escaped quality control in the ER and retrieves them back to the ER, protecting their D1 trimannosyl branch from the attack of mannosidases for re-challenging to the CNX/CRT cycle. Thus, the three L-type lectins ERGIC-53, VIP36, and VIPL possess the homologous CRDs, which, however, exhibit different sugar-binding specificities and affinities finely tuned by sensing pH, reflecting their distinct roles in transport and ER and post-ER quality control of glycoproteins.

Sugar Recognition Modes in the Ubiquitin-Proteasome Pathway

The terminally misfolded glycoproteins are sorted into the cytosol for the ubiquitin/proteasome-mediated degradation pathway. In this ER-associated degradation (ERAD) system, the malfolded glycoproteins are thought to be recognized by the ER lectins having a mannose-6-phosphate receptor homology (MRH) domain such as OS-9 [39]. Although the detailed sugar-binding properties of these ERAD-associated lectins remain to be elucidated, the mannose trimming in the ER has been suggested to be a crucial factor in the condemnation of the terminally misfolded glycoproteins.

In the cytosol, the malfolded glycoproteins are captured by an intracellular lectin Fbs1 [40]. This lectin operates as a chaperone by protecting glycoproteins from aggregation [41] and also functions as a substrate-binding subunit of SCF^{Fbs1}, an Skp1-Cullin1-F-box (SCF) ubiquitin ligase that recognizes glycoproteins [40]. The CRD of Fbs1 consists of ten-stranded antiparallel β-sandwich with two α-helices (Figure 4B) [42,43]. Although the β-sandwich fold shares structural similarity with the CRDs of galectins [44], their sugar-binding modes are obviously different: The Fbs1-CRD contacts the glycosylated aspragine side-chain and the reducing terminal portion of the carbohydrate moiety through the loops connecting the β-strands [42,43], whereas the CRDs of galectins accommodate the non-reducing terminal sugar residues in the concave of a β-sheet (Figure 4A) [44]. In many cases, the innermost portion of the carbohydrate chains is shielded by intra-molecular interactions with the surrounding amino acid residues in native glycoproteins. Hence, Fbs1 recognizes the exposed carbohydrate-polypeptide junction as a hallmark of malfolded glycoproteins for ubiquitination.

The ubiquitinated glycoproteins have to be subjected to deglycosylation prior to proteasomal degradation [45], because the bulky carbohydrate moieties hinder the accessibility of the substrates to the active sites for proteolysis, which are located within a narrow channel of the proteasome [46]. The deglycosylation is catalyzed by PNGase, which cleaves the β-aspartyl-glucosamine bond of the glycoproteins and thereby facilitates glycoprotein degradation in the ERAD system [45,47]. In mammalian PNGase, the catalytic domain is flanked by the N-terminal PUB [peptide:*N*-glycanase/Ub-associated or Ubiquitin regulatory X-containing proteins] [48] and the C-terminal CRD domains [49]. While the PUB domain is engaged in interactions with Derlin-1 and p97, components of the ERAD machinery on the ER membrane [50,51], the CRD domain captures the carbohydrate chain of the ERAD substrate allowing its sugar-polypeptide junction to orient toward the active site of the catalytic domain for an efficient cleavage of the glycan.

Intriguingly, this CRD domain also exhibits a β-sandwich architecture with a close structural similarity to Fbs1-CRD, despite low sequence identity (Figure 4C) [49]. However, the crystal structures also revealed that the sugar-binding site of PNGase-CRD exhibit a more open conformation than Fbs1-CRD, accommodating therein the α3, α6-mannotriose part of a mannopentaose ligand.

The FAC analysis provided the profile of the affinities of PNGase-CRD for a series of high-mannose-type oligosaccharides (Figure 2B). These data clearly demonstrate that this CRD selectively binds larger carbohydrate chains, especially with those retaining the intact D3 arm but lack the terminal mannose residue in the D2 arm. The present FAC data along with the crystallographic data indicate that PNGase-CRD recognizes the outer branches of the high-mannose-type oligosaccharides. The preferential capturing of the bulky carbohydrate moieties, appears suitable for removing the steric hindrance in the proteasomal degradation.

Thus, despite the structural similarities, the CRDs of Fbs1 and PNGase recognize the different portions of the common target glycans, reflecting their distinct roles in the glycoprotein degradation pathway. Rather, Fbs1-CRD is directly competitive with the catalytic domain of PNGase because the sugar-polypeptide junction is their common target [52]. This raises the possibility that Fbs1 captures the non-native glycoproteins in the cytosol, protecting them from the attack of PNGase during the chaperoning or ubiquitinating operation (Figure 4D). In this scenario, the Fbs1-bound glycoproteins are handed over PNGase after sensing an adequate elongation of the polyubiquitin chain by an as yet elucidated mechanism.

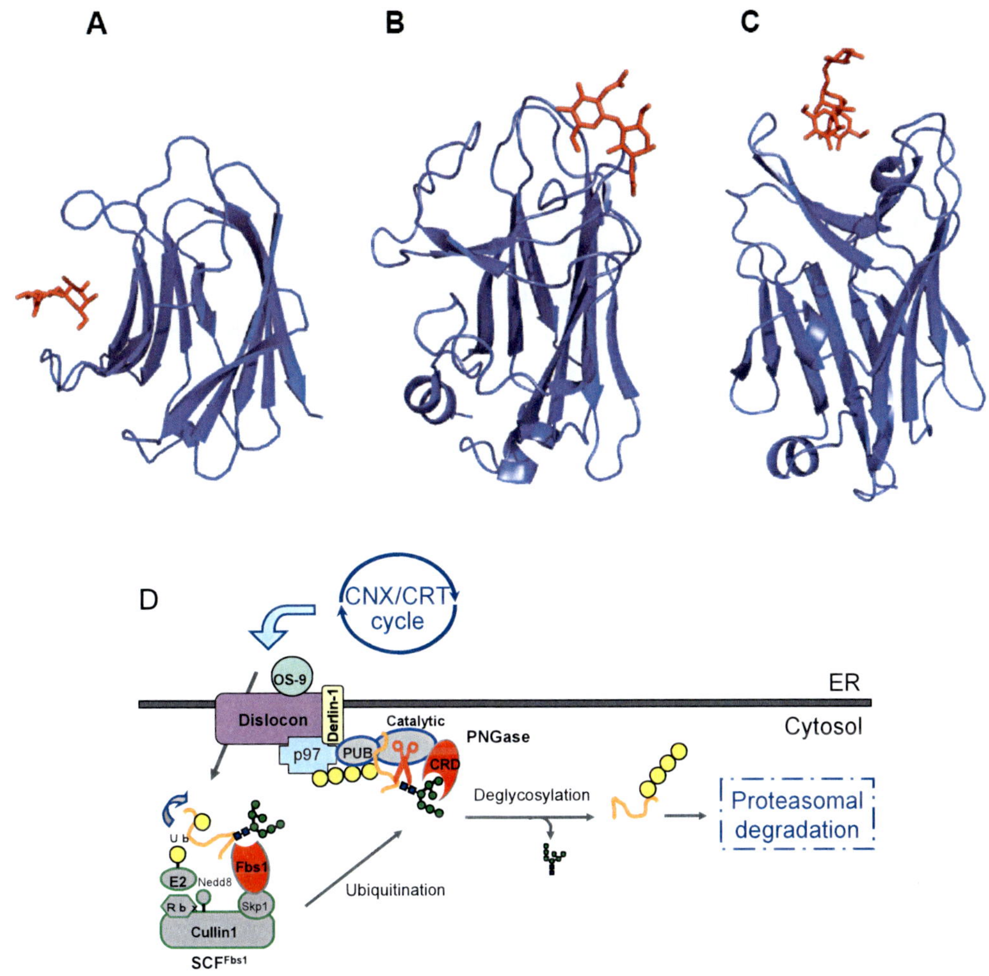

Figure 4. Crystal structures of (A) galectin-3, (B) Fbs1-CRD and (C) PNGase-CRD. The ligand oligosaccharides are shown in stick model in red. (D) Scheme of the interplay of Fbs1 and PNGase in the ubiquitin/proteasome-mediated glycoprotein degradation pathway. PDB accession codes: (A) 1A3K, (B) 2E31, (C) 2I74.

Conclusion

Although the intracellular lectins share structural similarities in their CRDs with the well studied extracellular lectins such as leguminous lectins and galectins, their way of recognizing targets seems to be more sophisticated than those involved in the extracellular events. The fineness of the sugar recognition by the intracellular lectins can be exemplified by distinguishing the isomeric structures of the processing intermediates of the high-mannose-type glycans, sensing of the folding state of the glycoprotein polypeptide chains, and catching-and-releasing of glycoproteins upon sensing the organeller pH. Accumulating evidence further underscores the conservativeness of CRD structures among intracellular lectins. For example, malectin, a newly identified membrane-anchored ER lectin, specifically binds diglucosylated high-mannose-type glycans produced by glucosidase I through its luminal CRD domain [53]. Interestingly, this CRD assumes a β-sandwich fold similar to Fbs1-CRD, despite their different sugar-binding modes. A common usage of CRD fold is also found in glucosidase II and OS-9, both of which adopt an MRH domain putatively for recognizing *N*-glycans, although their sugar-binding properties remain to be fully understood. The quality control of glycoproteins is temporally and spatially regulated through a network of the intracellular lectins and the glycan-processing enzymes. In order to better understand the underlying mechanisms of this network operation, we need to further extend our knowledge of the structural basis for the functional interplay of these proteins.

Acknowledgements

We thank Drs. Shunji Natori and Shunji Natsuka for providing the expression construct of CRT. The part of this study was performed jointly with Prof. Hans-Peter Hauri, University of Basel, Switzerland, Prof. Kazuo Yamamoto, The University of Tokyo, Japan, and Dr. Keiji Tanaka, Tokyo Metropolitan Insitute of Medical Science, Japan. Financial support by CREST (Core Research for Evolution Science and Technology) project from the Japan Science and Technology Agency and by Grants-in-Aid from the Ministry of Education, Culture, Sports, Science and Technology of Japan are fully acknowledged.

References

[1] Kamiya Y, Kato K. (2006). Sugar recognition by intracellular lectins that determine the fates of glycoproteins. *Trends in Glycosci. Glycotech. 18*: 231-44.

[2] Kato K, Kamiya Y. (2007). Structural views of glycoprotein-fate determination in cells. *Glycobiology 17*: 1031-44.

[3] Spiro RG. (2004). Role of *N*-linked polymannose oligosaccharides in targeting glycoproteins for endoplasmic reticulum-associated degradation. *Cell Mol. Life Sci. 61*: 1025-41.

[4] Grinna LS, Robbins PW. (1980). Substrate specificities of rat liver microsomal glucosidases which process glycoproteins. *J. Biol. Chem. .255*: 2255-8.

[5] Taylor SC, Ferguson AD, Bergeron JJ, Thomas DY. (2004). The ER protein folding sensor UDP-glucose glycoprotein-glucosyltransferase modifies substrates distant to local changes in glycoprotein conformation. *Nat. Struct. Mol. Biol. 11*: 128-34.

[6] Parodi AJ. (2000). Protein glucosylation and its role in protein folding. *Annu. Rev. Biochem 69*: 69-93.

[7] Guerin M, Parodi AJ. (2003). The UDP-glucose:glycoprotein glucosyltransferase is organized in at least two tightly bound domains from yeast to mammals. *J. Biol. Chem. 278*: 20540-6.

[8] Totani K, Ihara Y, Matsuo I, Koshino H, Ito Y. (2005). Synthetic substrates for an endoplasmic reticulum protein-folding sensor, UDP-glucose: glycoprotein glucosyltransferase. *Angew Chem. Int. Ed. Engl 44*: 7950-4.
[9] Weng S, Spiro RG. (1993). Demonstration that a kifunensine-resistant alpha-mannosidase with a unique processing action on *N*-linked oligosaccharides occurs in rat liver endoplasmic reticulum and various cultured cells. *J. Biol. Chem. 268*: 25656-63.
[10] Herscovics A. (2001). Structure and function of Class I alpha 1,2-mannosidases involved in glycoprotein synthesis and endoplasmic reticulum quality control. *Biochimie 83*: 757-62.
[11] Kornfeld R, Kornfeld S. (1985). Assembly of asparagine-linked oligosaccharides. *Annu. Rev. Biochem. 54*: 631-64.
[12] Kasai K, Oda Y, Nishikawa M, Ishii S. (1986). Frontal affinity chromatoraphy: theory for its application to studies on specofoc interactions of biomolecules. *J. Chromatogr. 376*: 33-47.
[13] Kamiya Y, Yamaguchi Y, Takahashi N, Arata Y, Kasai KI, Ihara Y, Matsuo I, Ito Y, Yamamoto K, Kato K. (2005). Sugar-binding properties of VIP36, an intracellular animal lectin operating as a cargo receptor. *J. Biol. Chem. 280*: 37178-82.
[14] Kamiya Y, Kamiya D, Yamamoto K, Nyfeler B, Hauri HP, Kato K. (2008). Molecular basis of sugar recognition by the human L-type lectins ERGIC-53, VIPL, and VIP36. *J. Biol. Chem. 283*: 1857-61.
[15] Williams DB. (2006). Beyond lectins: the calnexin/calreticulin chaperone system of the endoplasmic reticulum. *J. Cell Sci. 119*: 615-23.
[16] Fiedler K, Simons K. (1994). A putative novel class of animal lectins in the secretory pathway homologous to leguminous lectins. *Cell 77*: 625-6.
[17] Neve EP, Svensson K, Fuxe J, Pettersson RF. (2003). VIPL, a VIP36-like membrane protein with a putative function in the export of glycoproteins from the endoplasmic reticulum. *Exp. Cell Res. 288*: 70-83.
[18] Nufer O, Mitrovic S, Hauri HP. (2003). Profile-based data base scanning for animal L-type lectins and characterization of VIPL, a novel VIP36-like endoplasmic reticulum protein. *J. Biol. Chem. 278*: 15886-96.
[19] Velloso LM, Svensson K, Pettersson RF, Lindqvist Y. (2003). The crystal structure of the carbohydrate-recognition domain of the glycoprotein sorting receptor p58/ERGIC-53 reveals an unpredicted metal-binding site and conformational changes associated with calcium ion binding. *J. Mol. Biol .334*: 845-51.
[20] Satoh T, Cowieson NP, Hakamata W, Ideo H, Fukushima K, Kurihara M, Kato R, Yamashita K, Wakatsuki S. (2007). Structural basis for recognition of high mannose type glycoproteins by mammalian transport lectin VIP36. *J. Biol. Chem. 282*: 28246-55.
[21] Velloso LM, Svensson K, Schneider G, Pettersson RF, Lindqvist Y. (2002). Crystal structure of the carbohydrate recognition domain of p58/ERGIC-53, a protein involved in glycoprotein export from the endoplasmic reticulum. *J. Biol. Chem. 277*: 15979-84.
[22] Hauri HP, Kappeler F, Andersson H, Appenzeller C. (2000). ERGIC-53 and traffic in the secretory pathway. *J. Cell Sci. 113 (Pt 4)*: 587-96.

[23] Hauri HP, Appenzeller C, Kuhn F, Nufer O. (2000). Lectins and traffic in the secretory pathway. *FEBS Lett. 476*: 32-7.

[24] Vollenweider F, Kappeler F, Itin C, Hauri HP. (1998). Mistargeting of the lectin ERGIC-53 to the endoplasmic reticulum of HeLa cells impairs the secretion of a lysosomal enzyme. *J. Cell Biol. 142*: 377-89.

[25] Appenzeller C, Andersson H, Kappeler F, Hauri HP. (1999). The lectin ERGIC-53 is a cargo transport receptor for glycoproteins. *Nat. Cell Biol. 1*: 330-4.

[26] Zhang B, Cunningham MA, Nichols WC, Bernat JA, Seligsohn U, Pipe SW, McVey JH, Schulte-Overberg U, de Bosch NB, Ruiz-Saez A, White GC, Tuddenham EG, Kaufman RJ, Ginsburg D. (2003). Bleeding due to disruption of a cargo-specific ER-to-Golgi transport complex. *Nat. Genet. 34*: 220-5.

[27] Nyfeler B, Michnick SW, Hauri HP. (2005). Capturing protein interactions in the secretory pathway of living cells. *Proc. Natl. Acad. Sci. U S A 102*: 6350-5.

[28] Kawasaki N, Ichikawa Y, Matsuo I, Totani K, Matsumoto N, Ito Y, Yamamoto K. (2008). The sugar-binding ability of ERGIC-53 is enhanced by its interaction with MCFD2. *Blood 111*: 1972-9.

[29] Zhang B, Kaufman RJ, Ginsburg D. (2005). LMAN1 and MCFD2 form a cargo receptor complex and interact with coagulation factor VIII in the early secretory pathway. *J. Biol. Chem. 280*: 25881-6.

[30] Nyfeler B, Kamiya Y, Boehlen F, Yamamoto K, Kato K, de Moerloose P, Hauri HP, Neerman-Arbez M. (2008). Deletion of 3 residues from the C-terminus of MCFD2 affects binding to ERGIC-53 and causes combined factor V and factor VIII deficiency. *Blood 111*: 1299-301.

[31] Nawa D, Shimada O, Kawasaki N, Matsumoto N, Yamamoto K. (2007). Stable interaction of the cargo receptor VIP36 with molecular chaperone BiP. *Glycobiology 17*: 913-21.

[32] Bouckaert J, Dewallef Y, Poortmans F, Wyns L, Loris R. (2000). The structural features of concanavalin A governing non-proline peptide isomerization. *J. Biol. Chem. 275*: 19778-87.

[33] Lescar J, Loris R, Mitchell E, Gautier C, Chazalet V, Cox V, Wyns L, Perez S, Breton C, Imberty A. (2002). Isolectins I-A and I-B of Griffonia (Bandeiraea) simplicifolia. Crystal structure of metal-free GS I-B(4) and molecular basis for metal binding and monosaccharide specificity. *J. Biol. Chem. 277*: 6608-14.

[34] Itin C, Roche AC, Monsigny M, Hauri HP. (1996). ERGIC-53 is a functional mannose-selective and calcium-dependent human homologue of leguminous lectins. *Mo.l Biol. Cell 7*: 483-93.

[35] Hara-Kuge S, Ohkura T, Seko A, Yamashita K. (1999). Vesicular-integral membrane protein, VIP36, recognizes high-mannose type glycans containing alpha1-->2 mannosyl residues in MDCK cells. *Glycobiology 9*: 833-9.

[36] Appenzeller-Herzog C, Roche AC, Nufer O, Hauri HP. (2004). pH-induced conversion of the transport lectin ERGIC-53 triggers glycoprotein release. *J. Biol. Chem. 279*: 12943-50.

[37] Nyfeler B, Nufer O, Matsui T, Mori K, Hauri HP. (2003). The cargo receptor ERGIC-53 is a target of the unfolded protein response. *Biochem. Biophys. Res. Commun. 304*: 599-604.

[38] Renna M, Caporaso MG, Bonatti S, Kaufman RJ, Remondelli P. (2007). Regulation of ERGIC-53 gene transcription in response to endoplasmic reticulum stress. *J. Biol. Chem. 282*: 22499-512.

[39] Bernasconi R, Pertel T, Luban J, Molinari M. (2008). A dual task for the Xbp1-responsive OS-9 variants in the mammalian endoplasmic reticulum: inhibiting secretion of misfolded protein conformers and enhancing their disposal. *J. Biol. Chem. 283*: 16446-54.

[40] Yoshida Y, Chiba T, Tokunaga F, Kawasaki H, Iwai K, Suzuki T, Ito Y, Matsuoka K, Yoshida M, Tanaka K, Tai T. (2002). E3 ubiquitin ligase that recognizes sugar chains. *Nature 418*: 438-42.

[41] Yoshida Y, Murakami A, Iwai K, Tanaka K. (2007). A Neural-specific F-box Protein Fbs1 Functions as a Chaperone Suppressing Glycoprotein Aggregation. *J. Biol. Chem. 282*: 7137-44.

[42] Mizushima T, Hirao T, Yoshida Y, Lee SJ, Chiba T, Iwai K, Yamaguchi Y, Kato K, Tsukihara T, Tanaka K. (2004). Structural basis of sugar-recognizing ubiquitin ligase. *Nat. Struct. Mol. Biol. 11*: 365-70.

[43] Mizushima T, Yoshida Y, Kumanomidou T, Hasegawa Y, Suzuki A, Yamane T, Tanaka K. (2007). Structural basis for the selection of glycosylated substrates by SCF^{Fbs1} ubiquitin ligase. *Proc. Natl. Acad. Sci. U S A 104*: 5777-81.

[44] Seetharaman J, Kanigsberg A, Slaaby R, Leffler H, Barondes SH, Rini JM. (1998). X-ray crystal structure of the human galectin-3 carbohydrate recognition domain at 2.1-A resolution. *J. Biol. Chem. 273*: 13047-52.

[45] Hirsch C, Blom D, Ploegh HL. (2003). A role for *N*-glycanase in the cytosolic turnover of glycoproteins. *Embo. J. 22*: 1036-46.

[46] Unno M, Mizushima T, Morimoto Y, Tomisugi Y, Tanaka K, Yasuoka N, Tsukihara T. (2002). The structure of the mammalian 20S proteasome at 2.75 A resolution. *Structure 10*: 609-18.

[47] Kim I, Ahn J, Liu C, Tanabe K, Apodaca J, Suzuki T, Rao H. (2006). The Png1-Rad23 complex regulates glycoprotein turnover. *J. Cell Biol. 172*: 211-9.

[48] Suzuki T. (2007). Cytoplasmic peptide:*N*-glycanase and catabolic pathway for free *N*-glycans in the cytosol. *Semin Cell Dev. Biol .18*: 762-9.

[49] Zhou X, Zhao G, Truglio JJ, Wang L, Li G, Lennarz WJ, Schindelin H. (2006). Structural and biochemical studies of the C-terminal domain of mouse peptide-N-glycanase identify it as a mannose-binding module. *Proc. Natl. Acad. Sci. USA 103*: 17214-9.

[50] Katiyar S, Joshi S, Lennarz WJ. (2005). The retrotranslocation protein Derlin-1 binds peptide:*N*-glycanase to the endoplasmic reticulum. *Mol. Biol. Cell 16*: 4584-94.

[51] Li G, Zhao G, Zhou X, Schindelin H, Lennarz WJ. (2006). The AAA ATPase p97 links peptide *N*-glycanase to the endoplasmic reticulum-associated E3 ligase autocrine motility factor receptor. *Proc. Natl. Acad. Sci .U S A 103*: 8348-53.

[52] Yamaguchi Y, Hirao T, Sakata E, Kamiya Y, Kurimoto E, Yoshida Y, Suzuki T, Tanaka K, Kato K. (2007). Fbs1 protects the malfolded glycoproteins from the attack of peptide:*N*-glycanase. *Biochem. Biophys Res. Commun. 362*: 712-6.

[53] Schallus T, Jaeckh C, Feher K, Palma AS, Liu Y, Simpson JC, Mackeen M, Stier G, Gibson TJ, Feizi T, Pieler T, Muhle-Goll C. (2008). Malectin: a novel carbohydrate-binding protein of the endoplasmic reticulum and a candidate player in the early steps of protein *N*-glycosylation. *Mol. Biol. Cell 19*: 3404-14.

[54] Matsuo I, Wada M, Manabe S, Yamaguchi Y, Otake K, Kato K, Ito Y. (2003). Synthesis of monoglucosylated high-mannose-type dodecasaccharide, a putative ligand for molecular chaperone, calnexin, and calreticulin. *J. Am. Chem. Soc. 125*: 3402-3.

[55] Park H, Suzuki T, Lennarz WJ. (2001). Identification of proteins that interact with mammalian peptide:*N*-glycanase and implicate this hydrolase in the proteasome-dependent pathway for protein degradation. *Proc. Natl. Acad. Sci. U S A 98*: 11163-8.

[56] Urade R, Okudo H, Kato H, Moriyama T, Arakaki Y. (2004). ER-60 domains responsible for interaction with calnexin and calreticulin. *Biochemistry 43*: 8858-68.

In: Glycobiology Research Trends
Editors: G. Powell and O. McCabe
ISBN: 978-1-60692-841-7

Chapter III

GLYCO-BIOSENSORS: PROPERTIES, MATERIALS AND APPLICATIONS

S. Cunningham., J. Gerlach and L Joshi
Glycoscience and Glycotechnology Group,
National Centre for Biomedical Engineering Science and Martin Ryan Institute
National University of Ireland, Galway

ABSTRACT

For almost half a century, the development of methods for the detection and quantification of biomolecules has been an intensive area of research. The field of 'biosensors' now encompasses several areas specifically geared toward the rapid and sensitive detection, identification, and quantification of target analytes. In general biosensors function through characteristic properties, ideally singular, of direct ligand/receptor interaction or associated enzyme activity. This area of research has been greatly aided by advancement brought by interdisciplinary mergers of biological, chemical and physical sciences and engineering along with miniaturizing of devices. However, relative to time and investment committed, limited commercial success has been made to date, with the notable exceptions being those biosensors for diabetes and pregnancy testing.

Although, the field of biosensors is significantly advanced for the detection of nucleic acids and proteins, it is at a relatively nascent stage for detecting glycans, their conjugates, and glycan-related enzymes. Further development and application of such 'glyco-biosensors' or 'glycosensors' would open novel analytical avenues. These avenues would be valuable for industrial bioprocesses, diagnoses of chronic and infectious diseases in clinical settings and for academic researchers. Here, we describe the current status and potential applications of glycosensors, examine methods currently being explored, and discuss the potential impact of such technologies. Recent reports of novel or refined strategies of rapid, robust and low cost analysis are also presented.

INTRODUCTION

In addition to nucleic acids and proteins, carbohydrate chains (glycans) play critical roles in determining biological functions and affect many physiological processes (Turnbull and Field 2007; Alavi and Axford 2008). Thus, the study and characterization of glycans and glycoconjugates have become increasingly important. Cellular and molecular biology of glycans and their roles in evolution, development, physiology and pathology are being further elucidated through accelerated inquiry (Turnbull and Field 2007; Alavi and Axford 2008). However, research in this field has been hampered by the wide diversity of glycan structures which may be attached to glycoprotein, glycolipid or proteoglycan, and by the relatively small amounts of materials available for analysis (Raman et al. 2005). This challenge is amplified by the lack of suitable analytical tools for rapid, sensitive and high throughput (HTP) measurements. The design of HTP technologies for the characterization of glycan/protein interactions is therefore emerging as an attractive field for researchers and clinicians (Hirabayashi 2004). In particular, these approaches are necessary for monitoring glycan and lectin information with relation to recombinant protein production, disease diagnostic and prognostic applications. This review aims to provide an overview of recent developments in the area of glycosensor and HTP monitoring of glycans. In addition, it provides a brief introduction into the role and functions of glycans. We have omitted the field of glucose sensing which has been reported in-depth elsewhere (Newman and Turner 2005; Staiano et al. 2005).

THE SIGNIFICANCE OF GLYCOSYLATION:

Of more than 200 different types of post-translational modifications (PTM) that can be found on proteins (Schweppe et al. 2003; Walsh et al. 2005; Reinders and Sickmann 2007) glycosylation is one of the most frequent and important events (Dalpathado and Desaire 2008). Glycosylation is the covalent attachment of carbohydrates, usually in oligomeric chains to specific amino acids within polypeptide chains of proteins. The biological roles of glycosylation are diverse, including fertilization, development, differentiation, inflammation, cell migration/adhesion, cancer metastasis and host-pathogen/parasite interactions (Varki 1993; Moens and Vanderleyden 1997; Dwek et al. 2001; Daniels et al. 2002; Gabius et al. 2004). At least 50% of all mammalian proteins, (and approximately 80% of membrane proteins) are estimated to be glycosylated (Apweiler et al. 1999; Butler 2006). Because many of these proteins are bioactive molecules, directly stimulating or stabilizing systems within organisms, they are also of great clinical and industrial interest.

Glycosylation reactions are catalyzed by the action of glycosyltransferases and glycosidases, which add and remove sugar chains at various glycoconjugates, (Muramatsu 2000; Spiro 2002). A large number of glycosyltransferases (products of approximately 170 genes) have been cloned (Varki 1993; Narimatsu 2006), and some of their important functions have been clarified (Dwek et al. 1995; Saxon and Bertozzi 2001). These enzymes lead to the synthesis of a wide array of structurally related but distinct glycans (Weiss and Iyer 2007) and their attachment affects many properties of glycoconjugates, including structural characteristics, secretion and biomolecular interactions. Surface glycans serve as identification molecules to the surrounding world (Gornik et al. 2006). Glycans are the markers for self and non-self proteins and cells, and in doing so they are intrinsically linked to the immune system. However, the same communication mechanism permits cancerous cells and pathogens such as viruses and bacteria to use glycans to evade the immune system.

Unlike DNA replication, RNA transcription and protein translation processes, glycosylation is not a template dependent process. Glycosylation takes place in the endomembrane system (the endoplasmic reticulum and Golgi complex) and is only indirectly controlled by the genome. On the other hand, glycosylation is more susceptible to biochemical environment as well as the physiological state of the cell, i.e. the glycosylation patterns differ with the growth stage and with diseases states (Patwa et al. 2006; Qiu et al. 2008). For example, cancer cells frequently display glycans at different levels or with fundamentally different structures than those observed on normal cells (Warren et al. 1978; Brockhausen 2006; Qiu et al. 2008). Advances in genomics, proteomics and mass spectrometry have enabled the association of specific glycan structures with some disease states and, occasionally, the functional significance of disease-associated changes in glycosylation has been revealed (Itzkowitz et al. 1995).

The characterization of changes in glycosylation has to date been a challenging undertaking requiring profound analytical expertise, considerable time and expensive equipment. In an analytical setting, glycan structures can be cleaved chemically and/or enzymatically and characterised by using established instrumentation such as liquid chromatography (LC), mass spectrometry (MS) and nuclear magnetic resonance (NMR), but these methods have restrictive limitations in point-of-care settings with respect to time, throughput, sample size and also require robust technical expertise. Use of lectins (sugar binding proteins), glycosidases and a variety of chemical approaches have enhanced analysis and permitted the development of improved resources for glycosylation analysis and profiling. Combining these materials and methods with novel technologies and sensor strategies will in turn permit structural and functional characterization of the glycome and greatly improve biopharmaceutical production and clinical profiling.

Biological Roles of Glycans and Receptors

Glycans are displayed on macromolecules and the surface of every cell, where the information they encode is read by glycan binding proteins (GBP) in numerous processes such as the antigen recognition machinery, bacterial and viral adhesion to host cells and evasion from host immune system, protein folding, protein stability and trafficking (Crocker and Feizi 1996; Karlsson 1998; Feizi 2000; Feizi 2000; Helenius and Aebi 2001).

Immune System: Cellular and humoral immune responses depend on interactions between glycans and GBP. Binding GBP can be in the form of (i) lectins involved in cell–cell interactions (McEver 1997) (ii) lectins of the innate immune system (Feizi 2000) or (iii) antibodies recognizing sugar-antigens on surfaces of pathogens (Willison and Yuki 2002). These antibodies have been demonstrated to be a substantial fraction of total serum IgG and IgM, and implicated in a number of autoimmune conditions especially in neuropathies like

Multiple Sclerosis and Guillain-Barre syndrome (Dotan et al. 2006) Anti-glycan antibodies of all classes (IgG, IgM, IgA and IgE) have been detected within serum, though roles and disease correlations for some of these remain as yet unclear. Anti-glycan antibodies in serum are potential candidates for diagnositic or prognostic biomarkers in inflammatory and autoimmune disease. Indeed they may be able to serve in disease management, such as in Multiple Sclerosis.

Cancer: Surface glycans present an approach for diagnosis, prevention and treatment of cancer based on differences between cancer and non-cancer cells (Galonic and Gin 2007). Glycans and glycoforms of proteins have been demonstrated to be potential candidate markers for breast cancer (Abd Hamid et al. 2008). Tumor associated cell surface glycans include Tn (GalNAc-α-1-Ser/Thr), T (Gal-β-(1→3)-GalNAc-α-1- Ser/Thr) and sialylated variants. Reports of lectins exerting different effects *in vitro* studying from breast cancer cell lines, provides evidence that some dietary lectins could inhibit growth of breast cancer cells (Valentiner et al. 2003). A number of lectins have been demonstrated to be mitogenic, coupled with anti-cancer drugs such lectins may serve to synergistically potentiate drug effectiveness (Petrossian et al. 2007). Such treatments would be dependant upon glycan differences between healthy cells and diseased cells, these differences are more likely to be conformational presentation than presence or absence of specific glycans.

Bacterial Infection: Susceptibility and resistance to gastrointestinal (GI) tract infections is often associated with presence or absence of receptors appropriately glycosylated (Mouricout 1997; Yamauchi et al. 2006). The tract is covered by a protective mucosal layer (mucins) that must fulfill many requirements in order to be an effective first line barrier against infecting microbes and viruses while at the same time permit nutrients and other smaller molecules to traverse it. The mucin layer consists of high molecular weight glycoproteins and glycolipids (forming the glycocalyx associated with the brush border membrane). Mucin glycoproteins and glycolipids act not only as a site for binding of large biomolecules (e.g. microbial toxins, surface proteins, hormones, antibodies) but also as nutrients for the commensal intestinal microbiota (Dai et al. 2000; Walker 2000). The microbial composition of the intestinal microbiota may be responsible for initiating *in vivo* synthesis of host cellular glycoconjugates required for microbes to live in the large and small intestinal tracts (Pusztai et al. 1995; Sharma and Schumacher 1995; Sharma et al. 1995; Umesaki et al. 1995; Bry et al. 1996). On-going research is being performed to develop suitable approaches to study the effect of microbiota upon the host health. An example of glycan interaction can be demonstrated by *Helicobacter pylori*, a gram negative bacteria which colonises the stomach of humans and has been associated with chronic gastritis, duodenal ulcer disease and increased risk of gastric cancer (Yamaoka et al. 2006). Interactions with carbohydrates present in the gastric supramucosal gel and epithelium are crucial for *H. pylori* colonization (Yamaoka et al. 2006).

Viral Infection: Alterations in the overall glycosylation profile of viral glycoproteins have demonstrated to be advantageous to virus survival and virulence (Vigerust and Shepherd 2007; Vigerust et al. 2007). A number of human viral pathogens depend upon specific oligosaccharides to evade detection and elimination by the host immune system. Viruses such as HIV (Scanlan et al. 2007), Hendra (Carter et al. 2005), SARS-CoV, influenza (Reading et al. 2007) and hepatitis glycosylation for critical functions such as entry into host cells, proteolytic processing and protein trafficking. One of the main difficulties encountered in the development of vaccines against HIV is that the viral surface is heavily coated with glycans which mask potential vaccine targets (Scanlan et al. 2007). These glycans on the viral surface are produced by the host cell, making the virus appear as 'self'; rendering them undetectable to the host immune system. Based upon this, the development of glycan binders (lectin and antibody) are being used for both cancer and HIV detection and vaccination (Galonic and Gin 2007). The generation of binding molecules has not yet led to clinically effective molecules, although there is clear evidence that antibodies that strongly bind to glycan antigens have been generated (Slovin et al. 2003).

Glycoprotein Therapeutics: Recombinant therapeutic glycoproteins (Jenkins et al. 1996; Butler 2006) are an increasingly important sector of medicine. These therapeutic proteins require glycans for robust biological activity and/or longer serum half-life. Examples include the drugs tissue plasminogen activator (t-PA, Werner et al. 2007) and human erythropoietin (rhEPO, Ngantung et al. 2006), each of which is greatly enhanced by glycosylation. Global market value of biologic therapeutics is expected to exceed $70 billion by 2010 (Walsh 2006). Approximately 30% of new drugs likely to achieve approval in the next decade will be recombinant proteins such as antibodies or antibody fragments, and the number may run even higher as engineering technologies improve.

Within the biotechnology and biopharmaceutical industry, the importance of glycosylation has resulted in significant emphasis on the characterization of major carbohydrate structures of recombinant protein pharmaceuticals (Butler 2006; Ngantung et al. 2006). The FDA requires accurate characterization of glycosylation and demonstration of consistency for recombinant proteins in replicate manufacturing batches before these drugs are approved for therapeutic use.

Glycoprotein therapeutics to date, have been produced using various expression systems including transgenic cultured cell lines, bacterial systems, animals, plants and yeast (Jefferis 2005; Dingermann 2008). However, the majority of glycoprotein drugs are produced by Chinese Hamster Ovary (CHO) suspension cultures within large bioreactors. The glycosylation profile of these drugs can determine serum half-life, pharmacokinetic, pharmakodynamic, and immunogenic properties in the patients. Regulatory agencies such as FDA have stringent requirements (http://www.fda.gov/cder/info/industry.htm) for glycosylation monitoring (Dingermann 2008). Simultaneously, because glycosylation is subject to several environmental factors including cell culture conditions, growth phase of the cells and the biochemical events in the cells, obtaining consistent glycosylation patterns is a formidable challenge for biopharmaceutical industries (Butler 2006). Up to 50% weight variation among isoforms of recombinant protein may be attributable to glycans alone (Apweiler et al. 1999). To meet the needs of the biopharmaceutical industry, simple, low cost, at-production line methods to ensure batch-to-batch consistency in glycosylation of protein is greatly needed.

GLYCOBIOSENSING: BIOSENSOR BACKGROUND

Research into the fabrication and application of biosensors has now been performed for nearly half a century, with the first reported use in 1962 for continuous cardiac monitoring (Clark and Lyons 1962). Recently, scientific and technological advances have demonstrated the tremendous potential which biosensors hold for applications in areas of genomics, proteomics, glycomics (carbohydrate analysis), clinical diagnostics and prognostics, environmental monitoring, food analysis, agriculture applications and in security (detection of bio-agent release). At present, the global market for medical biosensors is estimated to be $7 billion (Luong et al. 2008), with home-use health monitoring devices (e.g. glucose biosensors and pregnancy test strips) accounting for the largest percentage of this market (Luong et al. 2008). With the advances in biotechnology and nanotechnology, this industry is poised to grow exponentially. Current commercialization of sensors has recently been reviewed Luong, Male and Glennon (2008).

Biosensors can be defined as analytical tools that use either a naturally occurring or artificially-engineered biological recognition system to target a molecule or macromolecule. Such biosensors can be applied to detect small amounts of a biomolecule, and/or to report physiological and biochemical characteristics of a sample. The major advantages of using natural biological components in sensors are specificity, sensitivity, and portability. Moreover, biosensors can be specialized to avoid the use of, large and expensive instruments which reduces cost and further contribute to portability. Biosensors can be classified on the bases of the biological signaling mechanism with which they operate (i.e. antibody:antigen (Kubik et al. 2005), enzyme (Vo-Dinh and Cullum 2000; Monk and Walt 2004), nucleic acid (Vo-Dinh and Cullum 2000; Kubik et al. 2005), cell/virus interaction (Suh et al. 2004; Suh et al. 2006) or biomimetic materials (Kubik et al. 2005) or alternatively on the bases by which they measure an analyte/target biomolecule: (i) optical detection biosensors (absorption, fluorescence, phosphorescence, Raman, SERS, refraction, and dispersion spectrometry)(Rich and Myszka 2006; Borisov and Klimant 2008; Fan et al. 2008), (ii) resonant sensors (Homola 2008; Lange et al. 2008; Shiau et al. 2008) (iii) electrochemical biosensors (Bakker et al. 2006; Wang 2006; Merkoci 2007), (iv) mass-sensitive measurement (Hansen and Thundat 2005; Waggoner and Craighead 2007) and (v) thermal biosensors (Lammers and Scheper 1999; Ramanathan and Danielsson 2001).

APPLICATION OF BIOSENSORS IN GLYCOSYLATION ANALYSIS

Glycomics' most pressing issue is innovation at the analytical end (Shin et al. 2005; Raman et al. 2006; Pilobello and Mahal 2007; Turnbull and Field 2007). While there are many long-established methods available to the glycoscientist to perform *in vitro* chemical analysis, a great proportion of these methods are tedious and resource intensive. In depth, real-time glycobiological research conducted *in vivo* is nearly non-existent at the current date. In addition, state-of-the-art deconvolution of multi-unit sugar structures is still dependent upon the combination of many technologies; there is no one technology to date that will allow the detection, isolation, purification, and sequencing of a complex sugar structure (Pilobello and Mahal 2007).

The ability to rapidly detect physiologically relevant glycosylation events (including aberrant glycan motifs on proteins and lipids) with high sensitivity and specificity using biosensors offers a powerful tool for early-stage diagnosis and treatment of diseases and for quality control of recombinant protein engineering. In order to develop suitable sensors, the details of the interaction must first be known, permitting measurable interaction between analyte and the sensor. The most common approaches for glycosensors are built upon the affinities for lectins and antibodies toward glycoconjugates.

Lectins, which are defined as carbohydrate-binding proteins, exist as diverse families of proteins that bind with high specificity (Sharon and Goldstein 1998; Sharon and Lis 2004) and their ubiquitous occurrence has been established in viruses, bacteria, fungi, plants, and animals including humans. There is accumulating evidence that they play regulatory roles in diverse biological processes, such as microbe infection, immune response, cell differentiation, and tumor-cell metastasis through specific binding to their glycan ligands (Robinson et al. 2006; Crocker et al. 2007; van Vliet et al. 2007). Various lectins, which are capable of binding a diverse array of glycan structures, are a major means by which glycosignatures of various molecules are decoded and the dynamics of carbohydrate structural variations in diseases interpreted (Dabelsteen 1996; Durand and Seta 2000; Kumar et al. 2005; LaBelle et al. 2007).

Lectins comprise the most widespread class of tools for the identification of glycans on the surface of whole cells, and isolated glycoproteins and glycolipids. Lectins are often described in terms of their affinity and avidity for a carbohydrate structure. Affinity is the molecular bond strength of a single association between ligand and receptor molecule, while avidity is the combined strength of multiple bonds as in the presentation of multiple ligand epitopes to a multivalent receptor. When used in traditional assays (e.g. ELISA) for identification of carbohydrates, lectin-glycan interaction is typically observed by an endpoint method requiring secondary enzymes, substrates and a change in conditions at the time of observation (e.g. raising the pH to an alkaline level to observe *p*-nitrophenol). More recent developments in lectin-based technologies have included printed lectin microarrays and biosensors for uses such as differentiating terminal glycans in clinical environments (Jelinek and Kolusheva 2004; Dai et al. 2006; LaBelle et al. 2007). Two specific examples of plant lectins are peanut agglutinin (PNA) and Sambucus nigra agglutinin 1 (SNA-1). PNA is a homotetrameric lectin that preferentially recognizes the carbohydrate portion of the tumor-associated T-antigenic determinant (also called TF-antigen, Gal-β-(1→3)-GalNAc), but has many times less affinity for other galactose-terminated oligosaccharides such as type II N-acetyllactosamine (LacNAc, Gal-β-(1→4)GlcNAc). The extensive analysis of the carbohydrate specificity of PNA has revealed that it is unable to recognize sialylated derivatives of galactose, which has contributed to its usefulness as a tool in cancer histochemical staining. Alternatively, SNA-1 is a homotetrameric lectin that is known to preferentially bind terminal α2,6-sialylated glycoproteins such as fetuin, orosomucoid, and bovine submaxillary mucin (Shibuya et al. 1987; Shibuya et al. 1987).

The characteristics of lectin binding with mammalian glyco-conjugates have been well-established and the information utilized by researchers to obtain further details on animal glycan structure (Angeloni et al. 2005; Sharon 2007). Charge, hydrophobicity and van der Waals forces, among other factors, all play roles in the binding of lectins to carbohydrate ligands (Ambrosi et al. 2005). Recognition of individual carbohydrates by proteins is characterized by relatively weak affinity binding of a single sugar structure and receptor, but it can be strongly amplified to produce a robust bond by multiple presentations of carbohydrates (Lundquist and Toone 2002; Dam et al. 2005; Dam et al. 2005).

Glycosensors – Technology and Current Status

Biosensors for detection of glycoconjugates based on lectin/carbohydrate recognition comprise the most common class of sensor technology used for such detection. Lectin arrays have effectively been used to profile glycoproteins (Angeloni et al. 2005; Zheng et al. 2005; Hsu et al. 2008). These arrays have traditionally been constructed with natural (non recombinant) lectins from various sources; the majority of these have been from plants. The lectins are either passively adsorbed to the surface or covalently linked. Animal lectins, such as galectins and siglecs have also been used, but these have limitations in biosensor utility due to limited availability, poor stability and specificity. Hsu, Gildersleeve and Mahal (2008) have taken a different approach to the lectin carbohydrate array. A small panel of lectins selected specifically for profiling glycoproteins and cell surface carbohydrates on micelles was produced through recombinant methods. This approach was aimed at overcoming some of the reported difficulties encountered with respect to variations of purity and availability of the natural lectins with the added benefit of producing lectins free from their own glycosylation which has been reported to complicate interpretation of binding data (Hsu et al. 2008). Hsu et al used specific combinations of lectins which reduced the overall size of the array and allowed a suitable number of replicants to accompany each experiment. The recombinant lectin approach was used to profile Cy5 labeled glycoproteins from mammalian sources (porcine mucin and RNaseB) and Cy5 labeled micelles created from tumor cells. In these experiments, the potential to detect glycosylation profile differences between renal carcinoma cells and melanoma cells was demonstrated. The ability to produce a variety of lectins from recombinant sources may allow for rapid development of application specific sensor platforms with lectin production that is done entirely in-house.

Carbohydrate arrays, in which glycans are deposited onto a surface for screening with a sugar-binding analyte, have been developed primarily to discover and profile lectins and to profile glycans which bind to them (Feizi et al. 2003; Palma et al. 2006). However, another potential application of carbohydrate array technology is the exploration of carbohydrate-carbohydrate interaction (de la Fuente and Penadés 2004). The dynamics of carbohydrate interaction are believed to play roles in cell adhesion and communication.

The most desirable glycobiosensors are able to detect without the need to label analytes. However, examples of techniques where label-free detection is possible are as yet few. Some notable proof-of-concept reports have recently explored new ways to detect and identify glycoconjugates without the need for chemical modification of samples. One novel technique to work around sample labelling is a hybrid of a non-covalently assembled lectin array technology and competitive binding called bimolecular fluorescence quenching and recovery (BFQR) (Koshi et al. 2006). Fluorescein-labeled lectins are deposited onto the surface of an array in hydrogel matrices. Suitable ligands, coupled to dabcylic acid, which acts as a fluorescence quencher for fluorescein, are first allowed to bind with their respective lectins. For carbohydrate identification assays, the test sample is incubated with the quenched lectin array to displace the quenching molecules. Positive results are measured as recovered fluorescence at each lectin position. This technique could be used with a small number of lectins displaying a distinctive profile for a target, greatly simplifying the approach while enhancing its use in portable glycan detection.

The use of surface plasmon resonance (SPR) for monitoring biological interaction was first described in 1983 by (Liedberg et al. 1995). SPR has been repeatedly visited as a means for detecting glycoconjugates (Duverger et al. 2003; Beccati et al. 2005). Refractory changes in reflected polarized light can be measured after the addition of molecules to the surface of a thin metal film (as low as 2.5 nm thick) mirror (Duverger et al. 2003). Refractive index changes may be quantified, providing data on selectivity and affinity. SPR is generally praised for high sensitivity. However, traditional SPR has limitations with respect to the lower molecular mass which can be effectively detected (Wang and Zhou 2008). Lectin-based SPR, in which the gold surface of the reflective chip has been covalently modified with carbohydrate-receptor proteins, has been shown to be effective for detection of glycoproteins and glycoparticles above a method-dictated size threshold. Released glycans must usually be coupled to larger masses to enable plasmon resonance shift large enough for reliable quantification.

To increase the useable sensitivity of SPR to a level sufficient to observe binding of unlabeled disaccharides, Foley et al (2008) used a high-resolution method of differential SPR scanning (Zhang et al. 2003) to detect 383 Da disaccharides binding to lectins surface-deposited onto AuS-tethered anti-lectin antibodies. The reported result was distinct selective recognition of T-antigen at concentrations down to ~5 x 10^{-8} M (50 nM) (Foley et al. 2008). Furthermore, the differential SPR system showed no significant influence from non-specific binding of the disaccharide control, LacNAc (Gal-β-(1→4)-GlcNAc) up to concentrations exceeding 5 x 10^{-7} M (500 nM). Resolution of the system was sufficient to allow kinetic parameters to be established. The dissociation constant (K_d) calculation was 2-6 μM for LacNAc and 15-29 nM for the target disaccharide T-antigen demonstrating the very high affinity of PNA for the correct target disaccharide.

Microcantilever deflection, based on that used in atomic force microscopy, has previously been used to detect the HIV 1 envelope glycoprotein gp120 with an average lower sensitivity limit of 8 μg/mL (Lam et al. 2006). The gp120 was trapped on the microcantilever in a sandwich between a surface coat of mAb A32 and a second mAb, 17b which specifically binds to an A32- binding-induced epitope on the surface of gp120. T8, a mAb that binds to the glycoprotein gp120 but does not induce a conformational change in the protein, was used to coat the microcantilever surface in a second set of experiments. 17b did not bind the T8-gp120 complex due to the lack of an exposed epitope. Conceivably, this experimental apparatus could also be used to monitor the specific glycosylation of a captured glycoprotein by means of incubation with a suitable lectin in place of the second mAb (Dill and Bearden 1996).

Detection of bacterial content in water and food supplies is an area where simple, inexpensive and rapid methods would be especially welcome. The quartz crystal microbalance (QCM) has been widely used to study the interaction of various molecules (Corry et al. 2003; Pei et al. 2005; Lebed et al. 2006; Shen et al. 2007). Shen and colleagues (2007) used the QCM method to detect bacteria by means of binding to specific glycans at the surface of the lipopolysaccharide outer layer of *E. coli.* The authors report an experimental detection limit of 7.5 x 10^2 cells per mL, a claimed four-fold increase in sensitivity (Shen et al. 2007).

Electrochemical sensor methods, such as impedimetric measurement, the change in resistance to an alternating current, and stripping voltammetry, the measurement of electrical current brought on by the voltage-assisted release of metals at an electrode are being developed with a variety of proposed biological applications (Bakker 2004; Pejcic and de Marco 2006). Electrochemical impedance spectroscopy (EIS) has been used to detect lectin-glycan interactions between neo-glycoconjugates and glycoproteins (LaBelle et al. 2007). A chip based biosensor was designed with a three-electrode surface pattern consisting of a reference (plated silver/silver chloride), working (gold) and counter electrode (gold). Lectins deposited onto the gold surface by EDAC/NHS SAM technique were used to trap carbohydrate ligands covalently attached to gold nanoparticles previously functionalized with polyethylene glycol (PEG) linkers. An alternating current was passed through a redox probe solution and the difference in impedance between electrodes initially without binding ligands and then with binding ligands was monitored. Through impedimetric measurement, lectins SNA-1 and PNA were demonstrated to selectively bind to the both sialyl-T and T-antigen coated nanoparticles as well as to sialyl and asialo forms of bovine fetuin, respectively (LaBelle et al. 2007). The combination of this analysis technique and selectivity of carbohydrate-binding molecules presents one feasible way to miniaturize and modernize rapid glycoconjugate identification.

Nanoparticles composed of metals have been used in conjunction with stripping voltametry to produce sensors for oligosaccharides (Dai et al. 2006). Functionalized gold surfaces were coupled to lectins by EDAC/NHS self-assembled monolayer (SAM) technique. Afterward, carbohydrate recognition domains of the lectins were occupied by glyconanoparticle ligands during a preparatory incubation step prior to competitive release of the nanoparticle glyconjugates during incubation with a test glyconjugate sample. Remaining glyconanoparticles on the electrode were quantified by stripping voltametry in a three-electrode (Ag^+/AgCl reference, platinum wire counter, glassy-carbon working) cell containing 10 ppm Hg^{2+}. The representative sample under investigation as well as a sugarless control and two less preferential sugar ligands, potential was applied the resulting electron peak current from the released nanoparticle electrons was measured. Dai et al (2006) used a gold electrode covalently modified with PNA. The authors were able to show a discernable current reduction corresponding to the increased displacement of nanoparticle-labeled sugars by the preferential ligand for PNA with a detectable lower limit down to 2.77 x 10^{-7} M.

Proper sialylation is one of the most important benchmarks for effective recombinant protein drugs (Ngantung et al. 2006). Rapid estimation of the level of sialylation present during bioreactor-culture production of biopharmaceuticals has been hampered by the lack of available methods readily adaptable to the bioreactor environment. A prototype amperometric biosensor for measuring sialic acid based on the co-immobilization of two enzymes, a sialic acid aldolase and pyruvate oxidase, at a modified platinum working electrode (Marzouk et al. 2007). Amperometric detection was used to monitor production of hydrogen peroxide at the working electrode (anode). The authors report a detection of limit of 10 μM with linear response up to 3.5 mM. This method could be adapted to the bioreactor environment through an automated sampling technique allowing near-real time monitoring of free sialic acid or combined with a second hydrolase step to adapt the technique to product monitoring.

FUTURE DIRECTIONS - NOVEL SOURCE OF GLYCAN RECOGNITION

The above mentioned methods of glycosensing rely on antibodies and lectins that are currently limited in their availability and specificity. Additionally, current methods of glycosensing are in their early stages of development and lack consistency, reliability, robustness and in some cases sensitivity – the factors that are needed for wider applications of the sensing technologies.

One of the areas critical for the success of glycosensors is the availability of glycan binding proteins, their mimics, desired glycan structures and their mimics, in large quantities with optimal specificity and sensitivity to target molecule. Therefore, molecular and synthetic mimetics are required to further study carbohydrate–protein interactions and construct applicable glycomic profiles for diagnostic, prognostic and engineering applications. Such designer molecules ideally mimic the natural complexity and heterogeneity. The development of alternative and diverse approaches to glycan recognition will permit the elucidation of interactions and permit the construction of novel sensors based upon existing sensor technology. So far, promising approaches, phage display and the application of aptamer libraries are at an early stage. Such discovered glyco mimetics will be further applied in areas such as glyconanoparticles (de la Fuente et al. 2001; Rojo et al. 2004) glycoarray technology (Ratner et al. 2004) and particular self-assembled monolayers (Kleinert et al. 2008).

Phage display technology is considered the best available strategy for the discovery of glycan specific peptides and to produce antibodies directed against carbohydrate moieties. In particular, the use of phage display-derived peptide libraries has become a powerful method for selecting peptides with engineered binding properties (Kelly et al. 2006). To date, this technology has been used mostly to generate antibodies against proteins, whereas its use for carbohydrate antigens has been limited by technical constraints (primarily immobilization of carbohydrate antigens).

Aptamers are oligonucleotide ligands which bind to target molecules including proteins with high affinity and specificity (Navani and Li 2006; Shangguan et al. 2008). Since they are constructed from nucleic acids, they can be easily synthesized and chemical modification for optimal hybridization is possible. It is on the bases of the ease of synthesis, controllable modification, and tremendous possible diversity that aptamers have been proposed as a powerful tool for biomolecule detection, analysis and as potential ligands comparable to antibodies (Burgstaller et al. 2002; Deisingh 2006). A purified single target is typically used in the selection of aptamers but screening using complex targets, such as crude solution, cells, or tissue, has advantages since aptamers can be selected without the purification of target proteins (Morris et al. 1998; Bianchini et al. 2001). There have been to date several attempts for screening aptamers to proteins in complex targets (Mairal et al. 2008). Shangguan and colleagues (2007) have demonstrated a cell-based SELEX (systematic evolution of ligands by exponential enrichment) strategy to generate aptamers as molecular probes to recognize neoplastic cells. This work shows the future potential of cell-based aptamer selection to be used to generate aptamer probes to obtain molecular signatures of cancer cells in patient (Shangguan et al. 2007; Tang et al. 2007). Indeed, the ability to custom synthesize and chemically modify oligonucelotides to function as aptamers lends them to be used in a broad range of sensor platforms (Navani and Li 2006).

CONCLUSIONS

Advancing current technology platforms, creation of new platforms, along with automation and miniaturization, are promising to expand the future development of this area. As increasing biological information is presented the ability to tailor specific sensors to their end applications will be possible. For biosensors to reach a personalized, point-of-care market, the key issues remaining to be addressed are their reproducibility, simplicity, and speed. There is a definite need for developing effective new analytical tools for glycobiology and biosensors that are sensitive, rapid, simple, reliable, and cost-effective is paramount. Cost reduction will undoubtedly follow as these technologies are refined and adopted for a variety of settings. HTP biosensors would also enable the development of routine health checks at home, thereby detecting abnormalities at stages early enough to allow more effective treatment.

Although the glycobiologically-directed biosensor approaches reviewed here have been shown to be successful only from a proof-of-concept perspective, glycomics would greatly benefit from additional bio-analytical tools with the capacity to scrutinize glycosignatures rapidly and, preferably, in a label-free fashion. This would accelerate the development of novel HTP technologies including glycosensors for comprehensive studies of the diversity of carbohydrate-mediated biological interactions.

REFERENCES

Abd Hamid UM, Royle L, Saldova R, Radcliffe CM, Harvey DJ, Storr SJ, Pardo M, Antrobus R, Chapman CJ, Zitzmann N, Robertson JF, Dwek RA, Rudd PM (2008) A strategy to reveal potential glycan markers from serum glycoproteins associated with breast cancer progression. *Glycobiology:*cwn095

Alavi A, Axford JS (2008) Sweet and sour: the impact of sugars on disease. Rheumatology (Oxford) 47:760-70

Ambrosi M, Cameron NR, Davis BG, Stolnik S (2005) Investigation of the interaction between peanut agglutinin and synthetic glycopolymeric multivalent ligands. *Org. Biomol. Chem.* 3:1476-80

Angeloni S, Ridet JL, Kusy N, Gao H, Crevoisier F, Guinchard S, Kochhar S, Sigrist H, Sprenger N (2005) Glycoprofiling with micro-arrays of glycoconjugates and lectins. *Glycobiology* 15:31-41

Apweiler R, Hermjakob H, Sharon N (1999) On the frequency of protein glycosylation, as deduced from analysis of the SWISS-PROT database. Biochim Biophys Acta 1473:4-8

Bakker E (2004) Electrochemical sensors. *Anal. Chem.* 76:3285-98

Bakker E, Bühlmann P, Pretsch E (2006) Electrochemical sensors: A report on the International Conference on Electrochemical Sensors, "Mátrafüred 05", held at Mátrafüred, Hungary, 13-18 November 2005. *TrAC Trends in Analytical Chemistry* 25:93-95

Beccati D, Halkes KM, Batema GD, Guillena G, Carvalho de Souza A, van Koten G, Kamerling JP (2005) SPR studies of carbohydrate-protein interactions: signal *enhancement of low-molecular-mass analytes by organoplatinum(II)-labeling. Chembiochem* 6:1196-203

Bianchini M, Radrizzani M, Brocardo MG, Reyes GB, Gonzalez Solveyra C, Santa-Coloma TA (2001) Specific oligobodies against ERK-2 that recognize both the native and the denatured state of the protein. *J. Immunol. Methods* 252:191-7

Borisov SM, Klimant I (2008) Optical nanosensors--smart tools in bioanalytics. *Analyst* 133:1302-7

Brockhausen I (2006) Mucin-type O-glycans in human colon and breast cancer: glycodynamics and functions. *EMBO Rep.* 7:599-604

Bry L, Falk PG, Midtvedt T, Gordon JI (1996) A model of host-microbial interactions in an open mammalian ecosystem. *Science* 273:1380-3

Burgstaller P, Jenne A, Blind M (2002) Aptamers and aptazymes: accelerating small molecule drug discovery. *Curr. Opin. Drug. Discov. Devel.* 5:690-700

Butler M (2006) Optimisation of the Cellular Metabolism of Glycosylation for Recombinant Proteins Produced by Mammalian Cell Systems. *Cytotechnology* 50:57-76

Carter JR, Pager CT, Fowler SD, Dutch RE (2005) Role of N-linked glycosylation of the Hendra virus fusion protein. *J. Virol.* 79:7922-5

Clark LC, Jr., Lyons C (1962) Electrode systems for continuous monitoring in cardiovascular surgery. *Ann. N. Y. Acad. Sci.* 102:29-45

Corry B, Uilk J, Crawley C (2003) Probing direct binding affinity in electrochemical antibody-based sensors. *Analytica Chimica Acta* 496:103-116

Crocker PR, Feizi T (1996) Carbohydrate recognition systems: functional triads in cell-cell interactions. *Curr. Opin. Struct .Biol.* 6:679-91

Crocker PR, Paulson JC, Varki A (2007) Siglecs and their roles in the immune system. *Nat. Rev Immunol.* 7:255-66

Dabelsteen E (1996) REVIEW ARTICLE. CELL SURFACE CARBOHYDRATES AS PROGNOSTIC MARKERS IN HUMAN CARCINOMAS. *The Journal of Pathology* 179:358-369

Dai D, Nanthkumar NN, Newburg DS, Walker WA (2000) Role of oligosaccharides and glycoconjugates in intestinal host defense. J Pediatr Gastroenterol Nutr 30 Suppl 2:S23-33

Dai Z, Kawde AN, Xiang Y, LaBelle JT, Gerlach J, Bhavanandan VP, Joshi L, Wang J (2006) Nanoparticle-Based Sensing of Glycan-Lectin Interactions. *J. Am. Chem. Soc.* 128:10018-10019

Dalpathado DS, Desaire H (2008) Glycopeptide analysis by mass spectrometry. *Analyst* 133:731-8

Dam TK, Gabius HJ, Andre S, Kaltner H, Lensch M, Brewer CF (2005) Galectins Bind to the Multivalent Glycoprotein Asialofetuin with Enhanced Affinities and a Gradient of Decreasing Binding Constants. *Biochemistry* 44:12564-12571

Dam TK, Oscarson S, Roy R, Das SK, Page D, Macaluso F, Brewer CF (2005) Thermodynamic, Kinetic, and Electron Microscopy Studies of Concanavalin A and Dioclea grandiflora Lectin Cross-linked with Synthetic Divalent Carbohydrates. *J. Biol. Chem.* 280:8640-8646

Daniels MA, Hogquist KA, Jameson SC (2002) Sweet 'n' sour: the impact of differential glycosylation on T cell responses. *Nat Immunol* 3:903-10

de la Fuente J, Barrientos AG, Rojas TC, Rojo J, Cañada J, Fernández A, Penadés S (2001) Gold Glyconanoparticles as Water-Soluble Polyvalent Models To Study Carbohydrate Interactions. *Angewandte Chemie International Edition* 40:2257-2261

de la Fuente J, Penadés S (2004) Understanding carbohydrate-carbohydrate Interactions by means of glyconanotechnology. Glycoconjugate Journal 21:149-163

Deisingh AK (2006) Aptamer-based biosensors: biomedical applications. *Handb. Exp. Pharmacol*:341-57

Dill K, Bearden D (1996) Detection of human asialo-α1-acid glycoprotein using a heterosandwich immunoassay in conjunction with the light addressable potentiometric sensor. *Glycoconjugate Journal* 13:637-641

Dingermann T (2008) Recombinant therapeutic proteins: production platforms and challenges. *Biotechnol. J.* 3:90-7

Dotan N, Altstock RT, Schwarz M, Dukler A (2006) Anti-glycan antibodies as biomarkers for diagnosis and prognosis. *Lupus* 15:442-50

Durand G, Seta N (2000) Protein Glycosylation and Diseases: Blood and Urinary Oligosaccharides as Markers for Diagnosis and Therapeutic Monitoring. *Clin. Chem.* 46:795-805

Duverger E, Frison N, Roche A-C, Monsigny M (2003) Carbohydrate-lectin interactions assessed by surface plasmon resonance. *Biochimie* 85:167-179

Dwek MV, Ross HA, Leathem AJ (2001) Proteome and glycosylation mapping identifies post-translational modifications associated with aggressive breast cancer. Proteomics 1:756-62

Dwek RA, Lellouch AC, Wormald MR (1995) Glycobiology: 'the function of sugar in the IgG molecule'. *J. Anat* .187 (Pt 2):279-92

Fan X, White IM, Shopova SI, Zhu H, Suter JD, Sun Y (2008) Sensitive optical biosensors for unlabeled targets: a review. *Anal. Chim. Acta* 620:8-26

Feizi T (2000) Carbohydrate-mediated recognition systems in innate immunity. *Immunol. Rev.* 173:79-88

Feizi T (2000) Progress in deciphering the information content of the 'glycome'--a crescendo in the closing years of the millennium. *Glycoconj.* J 17:553-65

Feizi T, Fazio F, Chai W, Wong C-H (2003) Carbohydrate microarrays -- a new set of technologies at the frontiers of glycomics. *Current Opinion in Structural Biology* 13:637-645

Foley KJ, Forzani ES, Joshi L, Tao N (2008) Detection of lectin-glycan interaction using high resolution surface plasmon resonance. *The Analyst* 133:744-746

Gabius HJ, Siebert HC, Andre S, Jimenez-Barbero J, Rudiger H (2004) Chemical biology of the sugar code. *Chembiochem.* 5:740-64

Galonic DP, Gin DY (2007) Chemical glycosylation in the synthesis of glycoconjugate antitumour vaccines. *Nature* 446:1000-7

Gornik O, Dumic J, Flogel M, Lauc G (2006) Glycoscience -- a new frontier in rational drug design. *Acta Pharm.* 56:19-30

Hansen KM, Thundat T (2005) Microcantilever biosensors. *Methods* 37:57-64

Helenius A, Aebi M (2001) Intracellular functions of N-linked glycans. Science 291:2364-9

Hirabayashi J (2004) Lectin-based structural glycomics: glycoproteomics and glycan profiling. *Glycoconj.* J 21:35-40

Homola J (2008) Surface plasmon resonance sensors for detection of chemical and biological species. *Chem. Rev.* 108:462-93

Hsu KL, Gildersleeve JC, Mahal LK (2008) A simple strategy for the creation of a recombinant lectin microarray. *Mol. Biosyst* 4:654-62

Itzkowitz SH, Marshall A, Kornbluth A, Harpaz N, McHugh JB, Ahnen D, Sachar DB (1995) Sialosyl-Tn antigen: initial report of a new marker of malignant progression in long-standing ulcerative colitis. *Gastroenterology* 109:490-7

Jefferis R (2005) Glycosylation of recombinant antibody therapeutics. Biotechnol Prog 21:11-6

Jelinek R, Kolusheva S (2004) Carbohydrate Biosensors. *Chem. Rev.* 104:5987-6016

Jenkins N, Parekh RB, James DC (1996) Getting the glycosylation right: implications for the biotechnology industry. *Nat Biotechnol* 14:975-81

Karlsson KA (1998) Meaning and therapeutic potential of microbial recognition of host glycoconjugates. *Mol. Microbiol.* 29:1-11

Kelly KA, Clemons PA, Yu AM, Weissleder R (2006) High-throughput identification of phage-derived imaging agents. *Mol. Imaging.* 5:24-30

Kleinert M, Winkler T, Terfort A, Lindhorst TK (2008) A modular approach for the construction and modification of glyco-SAMs utilizing 1,3-dipolar cycloaddition. *Org. Biomol. Chem.* 6:2118-32

Koshi Y, Nakata E, Yamane H, Hamachi I (2006) A Fluorescent Lectin Array Using Supramolecular Hydrogel for Simple Detection and Pattern Profiling for Various Glycoconjugates. *J. Am. Chem. Soc.* 128:10413-10422

Kubik T, Bogunia-Kubik K, Sugisaka M (2005) Nanotechnology on duty in medical applications. *Curr. Pharm. Biotechnol.* 6:17-33

Kumar SR, Sauter ER, Quinn TP, Deutscher SL (2005) Thomsen-Friedenreich and Tn Antigens in Nipple Fluid: Carbohydrate Biomarkers for Breast Cancer Detection. *Clin. Cancer Res*.11:6868-6871

LaBelle JT, Gerlach JQ, Svarovsky S, Joshi L (2007) Label-Free Impedimetric Detection of Glycan-Lectin Interactions. *Anal. Chem.* 79:6959-6964

Lam Y, Abu-Lail NI, Alam MS, Zauscher S (2006) Using microcantilever deflection to detect HIV-1 envelope glycoprotein gp120. Nanomedicine: Nanotechnology, *Biology and Medicine* 2:222-229

Lammers F, Scheper T (1999) Thermal biosensors in biotechnology. Adv Biochem Eng Biotechnol 64:35-67

Lange K, Rapp BE, Rapp M (2008) Surface acoustic wave biosensors: a review. *Anal Bioanal Chem* 391:1509-19

Lebed K, Kulik AJ, Forró L, Lekka M (2006) Lectin-carbohydrate affinity measured using a quartz crystal microbalance. *Journal of Colloid and Interface Science* 299:41-48

Liedberg B, Nylander C, Lundstrom I (1995) Biosensing with surface plasmon resonance--how it all started. *Biosens Bioelectron* 10:i-ix

Lundquist JJ, Toone EJ (2002) The Cluster Glycoside Effect. Chem. Rev. 102:555-578

Luong JH, Male KB, Glennon JD (2008) Biosensor technology: technology push versus market pull. *Biotechnol Adv.* 26:492-500

Mairal T, Ozalp VC, Lozano Sanchez P, Mir M, Katakis I, O'Sullivan CK (2008) Aptamers: molecular tools for analytical applications. *Anal Bioanal. Chem.* 390:989-1007

Marzouk SA, Ashraf SS, Tayyari KA (2007) Prototype amperometric biosensor for sialic acid determination. *Anal Chem.* 79:1668-74

McEver RP (1997) Selectin-carbohydrate interactions during inflammation and metastasis. *Glycoconj.* J 14:585-91

Merkoci A (2007) Electrochemical biosensing with nanoparticles. *Febs J.* 274:310-6

Moens S, Vanderleyden J (1997) Glycoproteins in prokaryotes. Arch Microbiol 168:169-75

Monk DJ, Walt DR (2004) Optical fiber-based biosensors. *Anal Bioanal. Chem.* 379:931-45

Morris KN, Jensen KB, Julin CM, Weil M, Gold L (1998) High affinity ligands from in vitro selection: complex targets. *Proc. Natl. Acad. Sci. U S A* 95:2902-7

Mouricout M (1997) Interactions between the enteric pathogen and the host. An assortment of bacterial lectins and a set of glycoconjugate receptors. *Adv. Exp. Med. Biol.* 412:109-23

Muramatsu T (2000) Essential roles of carbohydrate signals in development, immune response and tissue functions, as revealed by gene targeting. *J. Biochem*.127:171-6

Narimatsu H (2006) Human glycogene cloning: focus on beta 3-glycosyltransferase and beta 4-glycosyltransferase families. *Curr.Opin. Struct. Biol.* 16:567-75

Navani NK, Li Y (2006) Nucleic acid aptamers and enzymes as sensors. *Curr. Opin. Chem. Biol.* 10:272-81

Newman JD, Turner AP (2005) Home blood glucose biosensors: a commercial perspective. *Biosens Bioelectron.* 20:2435-53

Ngantung FA, Miller PD, Brushett FR, Tang GL, Wang DIC (2006) RNA interference of sialidase improves glycoprotein sialic acid content consistency. *Biotechnology and Bioengineering* 95:106-119

Palma AS, Feizi T, Zhang Y, Stoll MS, Lawson AM, Diaz-Rodriguez E, Campanero-Rhodes MA, Costa J, Gordon S, Brown GD, Chai W (2006) Ligands for the beta-Glucan Receptor, Dectin-1, Assigned Using "Designer" Microarrays of Oligosaccharide Probes (Neoglycolipids) Generated from Glucan Polysaccharides. *J. Biol. Chem.* 281:5771-5779

Patwa TH, Zhao J, Anderson MA, Simeone DM, Lubman DM (2006) Screening of glycosylation patterns in serum using natural glycoprotein microarrays and multi-lectin fluorescence detection. *Anal. Chem.* 78:6411-21

Pei Z, Anderson H, Aastrup T, Ramström O (2005) Study of real-time lectin-carbohydrate interactions on the surface of a quartz crystal microbalance. *Biosensors and Bioelectronics* 21:60-66

Pejcic B, de Marco R (2006) Impedance spectroscopy: Over 35 years of electrochemical sensor optimization Electrochimica Acta 51:6217-6229

Petrossian K, Banner LR, Oppenheimer SB (2007) Lectin binding and effects in culture on human cancer and non-cancer cell lines: examination of issues of interest in drug design strategies. *Acta Histochem.* 109:491-500

Pilobello KT, Mahal LK (2007) Deciphering the glycocode: the complexity and analytical challenge of glycomics. *Current Opinion in Chemical Biology* 11:300-305

Pusztai A, Ewen SW, Grant G, Peumans WJ, Van Damme EJ, Coates ME, Bardocz S (1995) Lectins and also bacteria modify the glycosylation of gut surface receptors in the rat. *Glycoconj.* J 12:22-35

Qiu Y, Patwa TH, Xu L, Shedden K, Misek DE, Tuck M, Jin G, Ruffin MT, Turgeon DK, Synal S, Bresalier R, Marcon N, Brenner DE, Lubman DM (2008) Plasma glycoprotein profiling for colorectal cancer biomarker identification by lectin glycoarray and lectin blot. *J. Proteome. Res.* 7:1693-703

Raman R, Raguram S, Venkataraman G, Paulson JC, Sasisekharan R (2005) Glycomics: an integrated systems approach to structure-function relationships of glycans. *Nat. Methods.* 2:817-24

Raman R, Venkataraman M, Ramakrishnan S, Lang W, Raguram S, Sasisekharan R (2006) Advancing glycomics: implementation strategies at the consortium for functional glycomics. *Glycobiology* 16:82R-90R

Ramanathan K, Danielsson B (2001) Principles and applications of thermal biosensors. *Biosens Bioelectron.* 16:417-23

Ratner DM, Adams EW, Disney MD, Seeberger PH (2004) Tools for glycomics: mapping interactions of carbohydrates in biological systems. *Chembiochem* 5:1375-83

Reading PC, Tate MD, Pickett DL, Brooks AG (2007) Glycosylation as a target for recognition of influenza viruses by the innate immune system. *Adv. Exp. Med. Biol.* 598:279-92

Reinders J, Sickmann A (2007) Modificomics: posttranslational modifications beyond protein phosphorylation and glycosylation. *Biomol. Eng.* 24:169-77

Rich RL, Myszka DG (2006) Survey of the year 2005 commercial optical biosensor literature. *Journal of Molecular Recognition* 19:478-534

Robinson MJ, Sancho D, Slack EC, LeibundGut-Landmann S, Reis e Sousa C (2006) Myeloid C-type lectins in innate immunity. *Nat Immunol.* 7:1258-65

Rojo J, Diaz V, De la Fuente J, Segura I, Barrientos AG, Riese HH, Bernad A, Penades S (2004) Gold Glyconanoparticles as New Tools in Antiadhesive Therapy. *ChemBioChem.* 5:291-297

Saxon E, Bertozzi CR (2001) Chemical and biological strategies for engineering cell surface glycosylation. Annu Rev Cell Dev Biol 17:1-23

Scanlan CN, Offer J, Zitzmann N, Dwek RA (2007) Exploiting the defensive sugars of HIV-1 for drug and vaccine design. Nature 446:1038-45

Schweppe RE, Haydon CE, Lewis TS, Resing KA, Ahn NG (2003) The characterization of protein post-translational modifications by mass spectrometry. *Acc. Chem. Res.* 36:453-61

Shangguan D, Cao Z, Meng L, Mallikaratchy P, Sefah K, Wang H, Li Y, Tan W (2008) Cell-specific aptamer probes for membrane protein elucidation in cancer cells. *J. Proteome Res.* 7:2133-9

Shangguan D, Cao ZC, Li Y, Tan W (2007) Aptamers evolved from cultured cancer cells reveal molecular differences of cancer cells in patient samples. *Clin. Chem.* 53:1153-5

Sharma R, Schumacher U (1995) The influence of diets and gut microflora on lectin binding patterns of intestinal mucins in rats. *Lab. Invest* 73:558-64

Sharma R, Schumacher U, Ronaasen V, Coates M (1995) Rat intestinal mucosal responses to a microbial flora and different diets. *Gut* 36:209-14

Sharon N (2007) Lectins: carbohydrate-specific reagents and biological recognition molecules. *J. Biol. Chem.* 282:2753-64

Sharon N, Goldstein IJ (1998) Lectins: more than insecticides. *Science* 282:1049

Sharon N, Lis H (2004) History of lectins: from hemagglutinins to biological recognition molecules. *Glycobiology* 14:53R-62R

Shen Z, Huang M, Xiao C, Zhang Y, Zeng X, Wang PG (2007) Nonlabeled Quartz Crystal Microbalance Biosensor for Bacterial Detection Using Carbohydrate and Lectin Recognitions. *Anal. Chem.* 79:2312-2319

Shiau AK, Massari ME, Ozbal CC (2008) Back to basics: label-free technologies for small molecule screening. *Comb. Chem. High Throughput Screen* 11:231-7

Shibuya N, Goldstein IJ, Broekaert WF, Nsimba-Lubaki M, Peeters B, Peumans WJ (1987) The elderberry (Sambucus nigra L.) bark lectin recognizes the Neu5Ac(alpha 2-6)Gal/GalNAc sequence. *J. Biol. Chem.* 262:1596-1601

Shibuya N, Goldstein IJ, Broekaert WF, Nsimba-Lubaki M, Peeters B, Peumans WJ (1987) Fractionation of sialylated oligosaccharides, glycopeptides, and glycoproteins on immobilized elderberry (Sambucus nigra L.) bark lectin. *Archives of Biochemistry and Biophysics* 254:1-8

Shin I, Park S, Lee M-r (2005) Carbohydrate Microarrays: An Advanced Technology for Functional Studies of Glycans. Chemistry - *A European Journal* 11:2894-2901

Slovin SF, Ragupathi G, Musselli C, Olkiewicz K, Verbel D, Kuduk SD, Schwarz JB, Sames D, Danishefsky S, Livingston PO, Scher HI (2003) Fully synthetic carbohydrate-based vaccines in biochemically relapsed prostate cancer: clinical trial results with alpha-N-acetylgalactosamine-O-serine/threonine conjugate vaccine. *J. Clin. Oncol.* 21:4292-8

Spiro RG (2002) Protein glycosylation: nature, distribution, enzymatic formation, and disease implications of glycopeptide bonds. *Glycobiology* 12:43R-56R

Staiano M, Bazzicalupo P, Rossi M, D'Auria S (2005) Glucose biosensors as models for the development of advanced protein-based biosensors. *Mol. Biosyst.* 1:354-62

Suh KY, Khademhosseini A, Jon S, Langer R (2006) Direct confinement of individual viruses within polyethylene glycol (PEG) nanowells. *Nano Lett.* 6:1196-201

Suh KY, Khademhosseini A, Yoo PJ, Langer R (2004) Patterning and separating infected bacteria using host-parasite and virus-antibody interactions. Biomed Microdevices 6:223-9

Tang Z, Shangguan D, Wang K, Shi H, Sefah K, Mallikratchy P, Chen HW, Li Y, Tan W (2007) Selection of aptamers for molecular recognition and characterization of cancer cells. *Anal. Chem.* 79:4900-7

Turnbull JE, Field RA (2007) Emerging glycomics technologies. *Nat. Chem. Biol.* 3:74-7

Umesaki Y, Okada Y, Matsumoto S, Imaoka A, Setoyama H (1995) Segmented filamentous bacteria are indigenous intestinal bacteria that activate intraepithelial lymphocytes and induce MHC class II molecules and fucosyl asialo GM1 glycolipids on the small intestinal epithelial cells in the ex-germ-free mouse. *Microbiol. Immunol.* 39:555-62

Valentiner U, Fabian S, Schumacher U, Leathem AJ (2003) The influence of dietary lectins on the cell proliferation of human breast cancer cell lines in vitro. *Anticancer. Res.* 23:1197-206

van Vliet SJ, den Dunnen J, Gringhuis SI, Geijtenbeek TB, van Kooyk Y (2007) Innate signaling and regulation of Dendritic cell immunity. *Curr. Opin. Immunol.* 19:435-40

Varki A (1993) Biological roles of oligosaccharides: all of the theories are correct. *Glycobiology* 3:97-130

Vigerust DJ, Shepherd VL (2007) Virus glycosylation: role in virulence and immune interactions. *Trends Microbiol.* 15:211-8

Vigerust DJ, Ulett KB, Boyd KL, Madsen J, Hawgood S, McCullers JA (2007) N-linked glycosylation attenuates H3N2 influenza viruses. *J. Virol.* 81:8593-600

Vo-Dinh T, Cullum B (2000) Biosensors and biochips: advances in biological and medical diagnostics. *Fresenius J. Anal .Chem.* 366:540-51

Waggoner PS, Craighead HG (2007) Micro- and nanomechanical sensors for environmental, chemical, and biological detection. *Lab. Chip.* 7:1238-55

Walker WA (2000) Role of nutrients and bacterial colonization in the development of intestinal host defense. *J. Pediatr. Gastroenterol. Nutr.* 30 Suppl 2:S2-7

Walsh CT, Garneau-Tsodikova S, Gatto GJ, Jr. (2005) Protein posttranslational modifications: the chemistry of proteome diversifications. *Angew Chem. Int. Ed. Engl.* 44:7342-72

Walsh G (2006) Biopharmaceutical benchmarks 2006. Nat Biotech 24:769-776

Wang J (2006) Electrochemical biosensors: Towards point-of-care cancer diagnostics. *Biosensors and Bioelectronics* 21:1887-1892

Wang J, Zhou HS (2008) Aptamer-Based Au Nanoparticles-Enhanced Surface Plasmon Resonance Detection of Small Molecules. *Anal. Chem.*

Warren L, Buck CA, Tuszynski GP (1978) Glycopeptide changes and malignant transformation. A possible role for carbohydrate in malignant behavior. *Biochim Biophys Acta* 516:97-127

Weiss AA, Iyer SS (2007) Glycomics Aims To Interpret the Third Molecular Language of Cells. *Microbe* 2:489-497

Werner RG, Kopp K, Schlueter M (2007) Glycosylation of therapeutic proteins in different production systems. *Acta Paediatr. Suppl.* 96:17-22

Willison HJ, Yuki N (2002) Peripheral neuropathies and anti-glycolipid antibodies. *Brain.* 125:2591-625

Yamaoka Y, Ojo O, Fujimoto S, Odenbreit S, Haas R, Gutierrez O, El-Zimaity HM, Reddy R, Arnqvist A, Graham DY (2006) Helicobacter pylori outer membrane proteins and gastroduodenal disease. *Gut* 55:775-81

Yamauchi J, Kawai Y, Yamada M, Uchikawa R, Tegoshi T, Arizono N (2006) Altered expression of goblet cell- and mucin glycosylation-related genes in the intestinal epithelium during infection with the nematode,Nippostrongylus brasiliensis, in rat. *Apmis* 114:270-278

Zhang HQ, Boussaad S, Tao NJ (2003) High-performance differential surface plasmon resonance sensor using quadrant cell photodetector. *Review of Scientific Instruments* 74:150-153

Zheng T, Peelen D, Smith LM (2005) Lectin Arrays for Profiling Cell Surface Carbohydrate Expression. *J. Am. Chem. Soc.* 127:9982-9983

In: Glycobiology Research Trends
Editors: G. Powell and O. McCabe

ISBN: 978-1-60692-841-7

Chapter IV

STRUCTURE AND ACTIVITIES OF NATURAL COMPLEX POLYSACCHARIDES

Nicola Volpi[*] *and Francesca Maccari*
Department of Biologia Animale, Biological Chemistry Section,
University of Modena and Reggio Emilia, Italy

ABSTRACT

Glycosaminoglycans (GAGs), hyaluronic acid or hyaluronan (HA), keratan sulfate (KS), chondroitin sulfates (CSs) and heparin (Hep)/heparan sulfate (HS), are complex ubiquitous natural polysaccharides that exhibit a wide range of biological functions by participating and regulating multiple cellular events and (patho)physiological processes. They are generally present either as free chains (HA) or as side chains of proteoglycans (PGs) (CS/dermatan sulfate (DS), Hep/HS and KS) and are most often found in cell membranes and in the extracellular matrix. The recent emergence of improved analytical tools for the study of these complex sugars has produced a virtual explosion in the field of glycomics. In particular, the well-known therapeutic applications of some of these macromolecules, in particular Hep as an anticoagulant and antithrombotic (macro)molecule and CS in the treatment of osteoarthritis (OA), and the increased understanding of GAG structure-function relationship has led to the discovery of novel drugs for the possible treatment of some serious diseases.

Keywords: Glycosaminoglycans, Heparin, Heparan sulfate, Dermatan sulfate, Chondroitin sulfate, Keratan sulfate

[*] Address for correspondence: Prof. Nicola Volpi, Department of Biologia Animale. University of Modena & Reggio Emilia, Italy. Via Campi 213/d, 41100 Modena, Italy. E-mail: volpi@unimo.it;Phone number: 0039 (0)59 2055543; Fax number: 0039 (0)59 2055548

Abbreviations

CS:	chondroitin sulfate.
DS:	dermatan sulfate.
GAG(s):	glycosaminoglycan(s).
GalNAc:	*N*-acetyl-galactosamine.
GlcA:	glucuronic acid.
GlcNAc:	*N*-acetyl-glucosammine.
HA:	hyaluronic acid.
Hep:	heparin.
HS:	heparan sulfate.
KS:	keratan sulfate.
IdoA:	iduronic acid.
MCD:	macular corneal dystrophies.
OA:	osteoarthritis.
PG(s):	proteoglycan(s).
SF:	synovial fluid.

Introduction

Glycosaminoglycans (GAGs) are very complex, unbranched, polydisperse, natural polysaccharides composed of disaccharide units of glucuronic acid (GlcA) or iduronic acid (IdoA) (keratan sulfate (KS) has galactose instead of uronic acid) linked to a glucosamine (GlcN) or galactosamine (GalN) residue (Figure 1) [1-3]. In general, GAGs are sulfated macromolecules (with the exception of hyaluronic acid (HA)) having different numbers of sulfo groups linked at different positions. They have very heterogeneous structures by considering relative molecular mass, charge density and chemical properties generating various biological and pharmacological activities [1-3]. Based on carbohydrate backbone structure, it is possible to distinguish four classes of GAGs: 1. HA, 2. KS, 3. chondroitin sulfate (CS)/dermatan sulfate (DS), and 4. heparan sulfate (HS)/heparin (Hep). HA is the only GAG containing an unmodified *N*-acetylglucosamine (GlcNAc)-GlcA repeating unit, while the other polysaccharides are generally modified through post-biosynthetic modifications, such as the addition of *O*-sulfo groups, C5-epimerization to form IdoA residues, and de-*N*-acetylation to produce GlcN-sulfo residues. These macro and micro modifications often play a key role in a wide variety of biological and pharmacological processes [1-3].

GAG chains are covalently attached (with the exception of HA) at their reducing end to a core protein to produce macromolecules named proteoglycans (PGs) [4-6] localized at cellular and extracellular levels playing structural and regulatory roles due to their interaction with several proteins. In fact, PGs are not only structural components, but they participate in many cellular events and physiological processes, such as cell proliferation and differentiation, cell-cell and cell-matrix interactions [7-9] and are implicated in regulatory functions of development, angiogenesis, axonal growth, cancer progression, microbial pathogenesis, and anticoagulation [1, 2, 7-9] due to the specific interactions between structural GAGs and numerous proteins. As a consequence, these heteropolysaccharides are macromolecules of great importance in the fields of biochemistry, pathology and pharmacology.

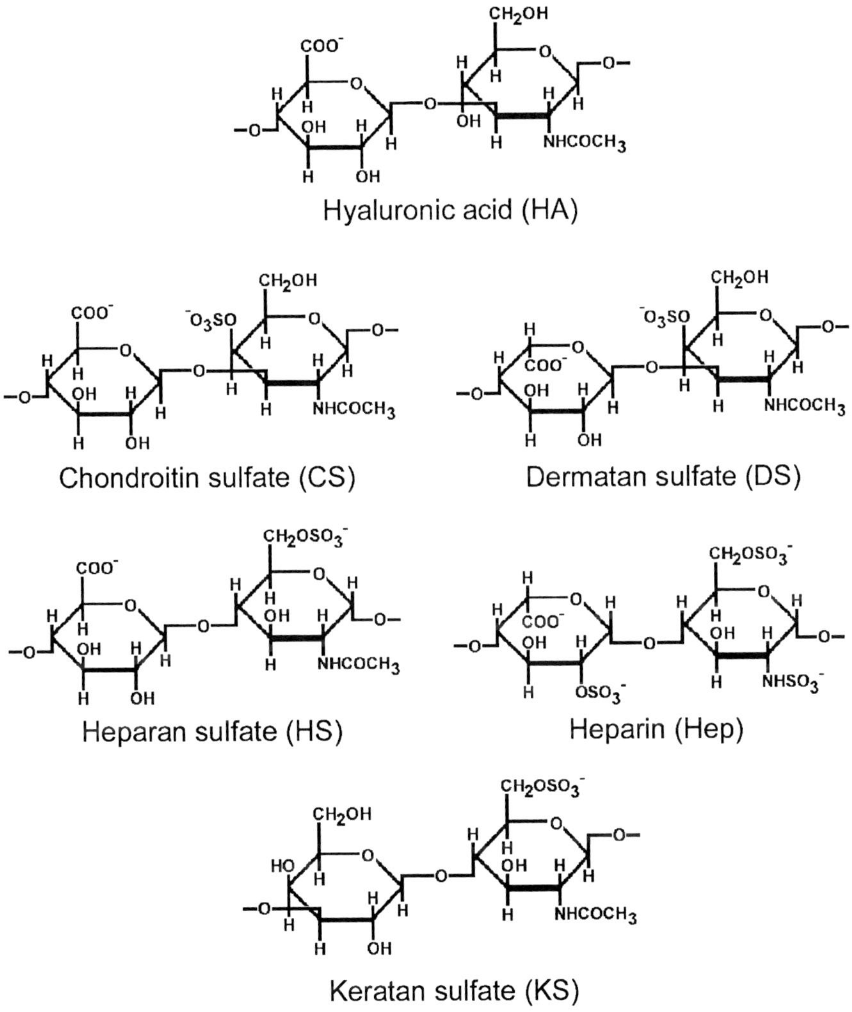

Hyaluronic acid (HA)

Chondroitin sulfate (CS)

Dermatan sulfate (DS)

Heparan sulfate (HS)

Heparin (Hep)

Keratan sulfate (KS)

Figure 1. Structures of disaccharides forming GAGs. Major modifications for each structure are illustrated but minor variations are possible.

HYALURONIC ACID

HA, the only nonsulfated GAG, is composed of repeating unit of [GlcA ($\beta 1\rightarrow 3$) GlcNAc ($\beta 1\rightarrow 4$) $]_n$ (Figure 1). Its is synthetized in the plasma membrane by hyaluronan synthase able to add UDP-GlcA and UDP-GlcNAc units at the reducing end of the chain [10]. To date, HA is the largest natural polysaccharide macromolecule having its molecular mass usually in the order of millions (>2.500 disaccharide units) [10]. At the moment, a large number of specific HA binding sites are known that have evolved in other matrix (macro)molecules and on cell surfaces. Moreover, HA shows other specific properties due to its very peculiar high molecular mass. As a consequence, HA carries out its functions not only in connective tissues but also in all body tissues and fluids. In particular, HA interacts with the hyaladherins, a family of aggregating proteins, able to regulate tissue morphogenesis and remodeling, and angiogenesis [10]. Furthermore, abnormal increased amounts of HA are found in the affected tissues/fluids in many diseases, such as malignant mesothelioma, atherosclerosis and exfoliation syndrome, inflammatory rheumatic diseases and cirrhotic liver diseases [10], due to an its increased synthesis caused by the stimulation of cells by various inflammatory mediators and growth factors. As a consequence, HA determination has been proposed to be useful for the diagnosis of malignant mesothelioma (pleural fluid), rheumatoid arthritis and liver diseases (serum) [11].

HA plays a critical role in stabilizing cartilage matrix due to its capacity to bind to the cartilage PG aggrecan. In fact, large numbers of aggrecan molecules are bound to a relatively short but extended HA chain, with the binding reinforced by small link proteins, producing macro-aggregates having individual mass in the order of several hundred million daltons, able to attract water by osmosis. This mechanism is generally restricted to cartilage, but several other HA-binding PGs (versican, neurocan, brevican and others) have been detected in softer tissues and may have a more not yet well-known general role.

Due to its polymeric nature and high molecular mass previously discussed, HA possesses gel-like properties that provide functional support for tissues [10]. In fact, a decrease in HA molecular size has been reported in osteoarthritis (OA) joints causing a reduction of the rheological properties of synovial fluid (SF) [12] with an increased cartilage attrition and subchondral bone remodeling producing a progression of OA and clinical symptoms. The rationale for the use of purified high molecular mass HA as viscosupplementation drug for the treatment of OA was to restore SF viscoelasticity which would improve joint functionality and consequently decrease symptoms [13].

Besides the positive effect of high molecular mass HA to relief of OA symptoms [13], also lower molecular mass HA fractions were observed to have clinical effectiveness in OA [14] by promoting wound healing and restoring stability to the damaged tissue, preventing infection, and promoting re-growth [14]. HA has recently been implicated in cancer progression and vascular disease [14] as several cell surface proteins bind to the HA surrounding tumor cells and transduce signals that affect the cell's ability to divide and migrate. The successful interruption of these interactions by means of the addition of HA oligomer mimetics offers promise as treatment for primary tumor growth, angiogenesis and metastasis [14]. Finally, as for cancer progression, the HA molecules also interact directly with CD44 and cell surface proteins affecting the growth properties of the surrounding cells and by influencing the interaction of immune cells with inflammatory lesion. As a consequence, the addition of exogenous HA (macro)molecules has also proven useful in the treatment of vascular disease in animal models and appears to be extremely promising as a future clinical therapy [14].

KERATAN SULFATE

KS was extracted and characterized for the first time from the bovine corneal stroma [15] and called KS-I having an alkali-stable bond between GlcNAc and asparagine. Skeletal KS with the alkali-labile bond between GalNAc and serine or threonine was designated as KS-II. This skeletal type has been further subclassified into articular, KS-IIA and KS-IIB [16], for the presence on the former of α (1-3)-fucose and α (2-6)-N-acetylneuraminic acid absent in the latter. The repeating disaccharide unit of KS [Gal (β1→4) GlcNAc $(\beta 1\rightarrow 3)]_n$ contains a galactose residue instead of uronic acid and the glycosidic bonds are reversed in comparison with HA and CS/DS (Figure 1). Sulfate esters are present at the C-6 of one or both of the monosaccharides forming the disaccharide unit, but any other hydroxyl group may carry an esterified sulfate group.

Lumican, keratocan, mimecan and decorin are the major PGs in the corneal stroma with core proteins around 50 kDa [17]. Corneal strength and transparency strongly depend on the development and maintenance of an organized stromal extracellular matrix regulated by a proper ratio of matrix components and appropriate hydration. In corneal stroma, these regularly packed fibrils are further organized into lamellae with adjacent layers perpendicular to one another [17]. The mechanisms that regulate the assembly of the different levels of stromal architecture are not well understood, but PG-collagen and collagen-collagen interactions have been involved. The macular corneal dystrophies (MCD) are a group of autosomal inherited diseases with a severe effect on the structural integrity and function of the corneal stroma [18]. The disease phenotype has been related to defects in KS metabolism by stromal fibroblasts (keratocytes), and at least three biochemical subtypes (types I, IA, and II) of the disease have been identified. Furthermore, the main tissue containing KS is cartilage and thus the determination of KS in serum, urine or synovial fluid has been proposed as a marker for the quantification of cartilage damage in joint diseases.

CHONDROITIN SULFATE/DERMATAN SULFATE

CS and CS B, also known as DS, are constituted by the disaccharide unit [GlcA (β1→3) GalNAc $(\beta 1\rightarrow 4)]_n$ variously sulfated in different positions of the exosamine unit and/or uronic acid. Some, although relatively few, of these positions remain unsulfated. The regular disaccharide sequence of CS A, chondroitin-4-sulfate, is formed by the repeating unit sulfated in position 4 of the GalNAc unit, while CS C, chondroitin-6-sulfate, is composed of a disaccharide unit sulfated in position 6. Disaccharides with different numbers and positions of sulfate groups can be located, in different percentages, inside the polysaccharide chains, such as the disulfated disaccharides in which two sulfate groups are *O*-linked in position 2 of GlcA and 6 of GalNAc (disaccharide D) or in position 4 and 6 of GalNAc (disaccharide E) [19].

In the case of DS, further enzymatic modifications complete the final structure, such as C-5 epimerization of GlcA to IdoA, and *O*-sulfation at C-2 of IdoA. As a consequence, polysaccharide chains of DS are formed of a prevailing disaccharide unit [IdoA (β1→3)

GalNAc(β1$\rightarrow$4)]$_n$ with a minor concentration of disulfated disaccharides, in particular sulfated in position 4 of GalNAc and 2 of the IdoA unit [19] (Figure 1). These heterogeneous structures are responsible for the different and more specialized functions of these GAGs. Furthermore, IdoA imparts conformational flexibility to the DS chain altering the shape and spatial orientation of sulfate residues, endowing the chain with a higher negative charge content than the GlcA [19]. Although the principles of the biosynthetic process have not yet been fully elucidated, it is well known that this process results in the generation, within the polymer chain, of highly modified oligosaccharide domains separated by regions of relatively low-degree structural modifications. Thus, the DS chain has a hybrid co-polymeric structure consisting of low modified (CS) and highly modified (DS) domains [20]. The IdoA-containing units are often sulfated at C-4 of the GalNAc residue, while sulfation at C-6 is frequently associated with GlcA-containing disaccharides [21]. Twenty-three different CS/DS disaccharides have been identified so far [21]. The detailed structure of these GAGs is modified during certain diseases, such as arthritis, atherosclerosis and cancer. The well described structural modifications involve changes in ratios of the two uronic acids and of 4-sulfated disaccharides to 6- and nonsulfated disaccharides. Changes in the size of chains are also described. Many of these alterations can be observed after analysis of CS/DS in biological fluids of patients.

OA and associated musculoskeletal pain and disability are the major causes of infirmity in the United States and other developed nations [22, 23]. Radiographic assessment of OA indicates a prevalence in middle-aged individuals of approximately 80%, and this increases markedly with ageing, with considerable repercussions on the socioeconomic plane. The causes of OA are multifactorial, in particular ageing also associated with other risks, such as mechanical, hormonal, and genetic factors all contributing to different degrees. OA emerges as a clinical syndrome when these determinants result in sufficient joint damage to cause impairment of function and the appearance of symptoms. This clinical syndrome is radiographically demonstrated by joint space narrowing (due to loss of cartilage) and extensive remodeling of subchondral bone with proliferation at the joint margins (osteophytosis). In the late stages of disease, these joints are characterized by extensive cartilage fibrillation, loss of PGs, and bone eburnation at sites of high contact stress. Where cartilage is still intact, invasion of the subchondral vasculature into calcified cartilage occurs, accompanied by advancement of the tidemark [24]. Although some chondrocytes may still proliferate and release PGs territorially, many are non-viable. The subchondral bone beneath the regions lacking in cartilage is generally sclerotic and consists of immature woven bone.

Cartilage breakdown products have been shown to be antigenic [25], and when released into synovial fluid due to excessive catabolism, may provoke synovitis. This synovial inflammation, once established, can alter the metabolism of resident synoviocytes, the major biosynthetic source of HA in synovial fluid. Inflammatory mediators released from local synovial cells and infiltrating leukocytes can promote increased vascular permeability and the accumulation of plasma in synovial fluid, thereby decreasing HA concentration. This dilution of HA and reduction in its molecular weight due to abnormal synthesis by synoviocytes results in a decrease in the viscoelasticity of synovial fluid and thus its ability to lubricate and protect articular cartilage, as above reported. Macrophages of the synovium and the leukocytes that have entered the synovial cavity are also an abundant source of cytokines, procoagulant factors, proteases, oxygen-derived free radicals and nitric oxide. Although much of the excess proteolytic activity released into synovial fluid is neutralized by the endogenous inhibitors present, cytokines and free radicals can freely diffuse into cartilage and downregulate PG and collagen synthesis by chondrocytes. These cytokines can also trigger the production of catabolic proteases, cytokines, and nitric oxide on the part of the cartilage cells, which contribute to further matrix destruction through paracrine pathways [26].

Pharmacological treatment of OA has targeted the symptoms rather than the underlying cause by using analgesics, steroidal, and nonsteroidal antiinflammatory drugs (NSAIDs) [27]. However, the detrimental side effects associated with the use of many of these agents has led to a more conservative approach to their use in recent years.

CS chains are synthesized by cells covalently attached to proteins, which are secreted into the extracellular matrix as PGs, and some of the CS properties are due to its strong charge capable of drawing water into tissues and hydrating them. Articular cartilage is a very specialized tissue with a large expanded extracellular matrix composed of more than 98% of its volume matrix and less than 2% of cells. The properties of this tissue are related to the contribution made by fibrillar collagen and nonfibrillar PGs. The structure of collagen gives it impressive tensile properties and this is utilized in cartilage in a special way to produce a tissue that is not only strong in tension but also resistant to compression by filling the interfibrillar matrix with a very high content of CS-rich PG, primarily aggrecan. The aggrecan at high concentration draws water into the tissue, which swells and expands the matrix placing the collagen network under tension. At this equilibrium, with the tissue swollen with water, it has good compressive resilience. In degenerative joint diseases there is loss of the articular cartilage. A key point in the degenerative process is the loss of PG from the cartilage and the exposure of its collagen network to mechanical disruption. However, results from studies of experimental OA suggest that early in the process there is a hypermetabolic repair response in articular cartilage with increased synthesis of matrix components and increased matrix turnover [28]. The increased synthesis of aggrecan in cartilage is accompanied by interesting changes in the CS, which shows a longer chain length and the chains contain more epitopes recognized by specific antibodies. These CS clusters are rarely found in normal healthy tissue, but their expression increases in cartilage during the early hypermetabolic response in experimental OA. The biological reason for these changes in CS structure in cartilage in early OA is not very clear, but it may be suggested that there are specific biological functions of selected sequences within CS chains that may be important for the processes of tissue repair.

CS also exerts an anti-inflammatory action. Oral administration of CS was shown to significantly decrease the granuloma formation arising from cotton, pellet or sponge implants, the inflammatory response in adjuvant arthritis, and lysosomal enzyme release in carrageenan pleurisy. Furthermore, CS and its fractions inhibit the directional chemotaxis induced by zymosan-activated serum, are able to decrease the phagocytosis and the release of lysozyme induced by zymosan and to protect the plasma membrane from oxygen reactive species. Compared with nonsteroidal anti-inflammatory drugs (indomethacin, ibuprofen), CS appears to be more effective on cellular events of inflammation than on edema formation [29]. CS mainly sulfated in position 4 of the galactosamine unit was shown to up-regulate the antigen-specific Th1 immune response of murine splenocytes sensitized with ovalbumin *in vitro*, and it was able to suppress the antigen-specific IgE responses [30]. In a further recent study [31], a specific sulfation pattern of the CS was demonstrated to be necessary for the Th1-promoted activity. These findings describe a new mechanism for the anti-inflammatory and chondroprotective properties of CS.

Table 1. Summary of results in chondroitin sulphate trials

Author(s)	**Year**	**n° of Patients**	**Follow-up Period**	**Compli-cations**	**Results**
Rovetta [32]	1991	40	25 wk	Well tolerated	Higher therapeutic effect on all symptoms of OA.
Oliviero *et al.* [33]	1991	200	6 mo	3% mild adverse effects	Considerable improvement in both pain and mobility.
Mazieres *et al.* [34]	1992	120	3 mo	Well tolerated	Slowly effective against symptoms of OA and reduction of the need for NSAIDs.
Morreale *et al.* [35]	1996	146	3 mo	Minor	Benefits of CS appeared later but lasted for up to 3 months after end of reatment.
Fleish *et al.* [36]	1997	56	1 yr	Well tolerated	Improvement in mobility, joint effusion and swelling.
Bucsi & Poor [37]	1998	80	6 mo	Minor	43% reduction in joint pain.
Bourgeois *et al.* [38]	1998	127	3 mo	No adverse events	Significant reduction in joint pain.
Uebelhart *et al.* [39]	1998	42	1 yr	None	Decreased joint pain and improved mobility.
Verbruggen *et al.* [40]	1998	119	3 yr	Not documente d	Radiographic demonstration of decrease in number of patients with "new" erosive finger-joint OA.
Uebelhart *et al.* [41]	2004	120	1 yr	Minor adverse events	Decreased pain and improved knee function. Radiological progression of medial femoro-tibial joint space.
Rovetta *et al.* [42]	2004	24	2 yr	Not documente d	Efficacy of oral CS in improving some aspects of erosive OA of the hands.
Clegg *et al.* [43]	2006	1583	24 wk	Mild adverse events	Effective in the Subgroup of patients with moderate-to-severe knee pain.

Mazières *et al.* [44]	2007	307	24 wk	Good tolerance	CS was slightly more effective than placebo on pain.

CS efficacy and tolerance (Table 1) was evaluated in the therapy of tibiofibular arthritis of the knee [32]. Forty patients suffering from this illness at radiological stages 1 and 2 undergoing concomitant therapy with NSAIDs, were randomized into two groups of twenty. The treatment group received the drug under study and the control group received placebo. Treatment was carried out in double blind. The therapy protocol comprised 25 intramuscular injections (one injection twice a week). This cycle was repeated for 6 months, for a total of 50 injections. The patients were visited on days 0, 90, 180, 240, 330 and 360. At each visit the following symptoms were evaluated: spontaneous pain, pain on loading, on passive movement and on pressure; changes in NSAIDs posology were also recorded; lastly any possible side effects were noted. Analysis of results has shown a statistically significant higher therapeutic effect on treatment with CS for all the symptoms taken into consideration. No important side effects were noted, either local or systemic; in two cases only in the group treated with CS and in the same number in the control group slight dyspeptic symptoms were found to occur, but without requiring suspension or reduction in posology. Two patients in the treated group and one in the control group left the study for non-compliance with the type of administration.

Oliviero *et al.* [33] conducted a clinical trial, lasting 6 months, performed on two hundred patients in four different Hospital Departments and one University Center. The results showed a considerable improvement both in pain and in mobility. No relevant side effects were found; only 3% of patients, with oral administration, noticed slight nausea and found it necessary to interrupt treatment.

Mazieres *et al.* treated 120 patients with CS in a placebo-controlled study. Joint pain, as measured on a visual analog scale, the Lequesne index, and patient and physician global assessments all showed significant improvement in the CS-treated group [34]. One hundred and twenty patients with OA of the knees and hips were entered into a randomized, placebo-controlled, double-blind trial designed to evaluate the effectiveness of CS. The three-month treatment phase was followed by a two-month treatment-free phase to allow evaluation of carry-over effects. The main endpoint was use of nonsteroidal antiinflammatory drugs (expressed as mg of diclofenac equivalent). At completion of the three-month treatment phase, patients taking CS (4 capsules/day) were using significantly less NSAIDs; this decrease persisted throughout the two-month treatment-free follow-up phase. The other parameters studied including visual analog scale assessment of pain, the Lequesne pain-function index, and overall patient and physician assessments, all showed a similar significant tendency. Tolerance was outstanding and no patients required premature withdrawal (Table 1). These findings indicate that CS is useful for the treatment of OA, both as an agent slowly effective against symptoms and to reduce the need for NSAIDs.

Morreale and co-workers [35] evaluated 146 patients with knee OA. Seventy four patients received 400 mg CS three times per day for 3 months, followed by 3 months without treatment. These patients were compared with 72 patients who were treated with 50 mg of diclofenac three times a day for 1 month, followed by 5 months off treatment. Those treated with diclofenac had a decrease in joint pain at 10 days, which disappeared shortly after discontinuation of treatment; those treated with CS had a significant response at 30 days, which lasted for 3 months after discontinuation of the drug. The Lequesne index after 3 months of CS treatment was 78% lower than at baseline; after 30 days of diclofenac it was 62.6% lower. Three months after discontinuation of CS, the Lequesne index remained 64.4% lower than the pre-treatment level. Three months after discontinuation of diclofenac the index was only 29.7% lower than at baseline. Thus, the effects of CS treatment persisted longer after discontinuation than those of traditional therapy.

56 patients with knee OA were enrolled in a 1-year double-blind placebo controlled study comparing 800 mg of CS with placebo; forty-seven patients (25 CS-treated, 22 placebo) completed 1 year [36]. Mobility, joint effusion and swelling were significantly better in the CS group (Table 1).

In a study by Bucsi and Poor [37], 40 patients treated with 800 mg of CS were compared with 40 who received placebo. The CS group showed significant improvement in the Lequesne index, visual analog pain scale, 20-meter walk time, and patient and physician efficacy ratings. The CS group used significantly less paracetamol than the placebo group. Patients with OA of the knee were treated with CS (Condrosulf® [IBSA, Lugano, Switzerland]) in a randomized, double-blind, placebo-controlled study, performed in two centres. The efficacy and tolerability of oral CS capsules 2 x 400 mg/day vs placebo was assessed in a 6-month study period. Patients with idiopathic or clinically symptomatic knee OA, with Kellgren and Lawrence radiological scores I-III, were included in this trial. Clinical controls were performed at months 0, 1, 3 and 6. Eighty patients completed the 6-month treatment period. Lequesne's Index and spontaneous joint pain (VAS) decreased constantly in the CS group; on the contrary, slight variations of the scores were reported in the placebo group. The walking time, defined as the minimum time to perform a 20-meter walk, showed a statistically significant constant reduction only in the CS group. ANOVA with repeated measures showed a statistically significant difference in favor of the CS group for these three parameters. During the study, patients belonging to the placebo group reported a higher paracetamol consumption, but this consumption was not statistically different between the two treatment groups. Efficacy judgements were significant in favor of the CS group. Both treatments were very well tolerated. All these results strongly suggest that CS acts as a symptomatic slow-acting drug in knee OA.

Bourgeois *et al.* [38] evaluated 40 patients with knee OA who were treated with CS as a 1200 mg dose oral gel. Forty-three patients were treated with 400 mg capsules three times per day, and 44 received placebo. The Lequesne index and pain score (visual analog scale) decreased significantly in both CS groups, compared with the placebo group, and patient and physician global assessments significantly favored the CS-treated group [38]. This multicenter randomized, double-blind, controlled study was performed to compare the efficacy and tolerability of CS (Condrosulf®) 1200 mg/day oral gel vs CS 3 x 400 mg/day capsules vs placebo, in patients with mono or bilateral knee OA (Kellgren and Lawrence radiographic score grade I to III). A total of 127 patients, 40 of whom were treated with CS 1200 mg/day, 43 with CS 3 x 400 mg/day and 44 with placebo, were included in the statistical analysis of this 3-month treatment study. In the CS groups, Lequesne's Index and spontaneous joint pain (VAS) showed a significant reduction of clinical symptoms ($P < 0.01$ for both parameters), while only a slight reduction was observed in the placebo group (P = ns for Lequesne's Index and P < 0.05 for VAS). The physician's and patient's overall efficacy assessments were significantly in favour of the CS groups (P < 0.01). The treatment carried out with the three formulations was very well tolerated. In conclusion, these results indicate that CS favours the improvement of the subjective symptoms, improving joint mobility (Table 1). An additional consideration is that the efficacy of 1200 mg CS as a single daily dose does not differ from that of 3 x 400 mg daily doses of CS for all the clinical parameters taken into consideration.

A computerized technique was used to measure medial tibiofemoral joint space in patients treated with 800 mg CS per day or with placebo. After 1 year, tibiofemoral joint space width had decreased significantly in placebo-treated patients but had not changed from the baseline value in the CS treatment group [39]. The aim of this study was to assess the clinical, radiological and biological efficacy and tolerability of the chondroitin 4- and 6-sulphate (Condrosulf®), in patients suffering from knee OA. This was a 1-year, randomized, double-blind, controlled pilot study which included 42 patients of both sexes, aged 35-78 years with symptomatic knee OA. Patients were treated orally with 800 mg CS per day or with a placebo administered in identical sachets. The main outcome criteria were the degree of spontaneous joint pain and the overall mobility capacity. Secondary outcome criteria included the actual joint space measurement and the levels of biochemical markers of bone and joint metabolism. This limited study confirmed that CS was well-tolerated and both significantly reduced pain and increased overall mobility capacity. Treatment with CS was also associated in a limited group of patients with a stabilization of the medial femoro-tibial joint width, measured with a digitized automatic image analyzer, whereas joint space narrowing did occur in placebo-treated patients. In addition, the metabolism of bone and joint assessed by various biochemical markers also became stabilized in the CS patients, whereas it was still abnormal in the placebo patients. These results confirm that oral chondroitin 4- and 6-sulphate is an effective and safe symptomatic slow-acting drug for the treatment of knee OA. In addition, CS might be able to stabilize the joint space width and to modulate bone and joint metabolism.

Verbruggen and colleagues [40] evaluated hand radiographs of 119 patients with OA (Table 1). Thirty-four patients who received CS 400 mg three times a day were compared with the results of 85 patients who received placebo. Radiographs were performed annually over a 3-year period. In the CS-treatment group, a significant decrease in the number of patients with new erosive OA was seen. A total of 119 patients were included in a randomized, double-blind, placebo-controlled trial in order to assess the S/DMOAD properties in OA of CS (3 x 400 mg/day, Condrosulf®). Posteranterior roentgenographies of the interphalangeal (IP) joints were carried out at the start of the study and at yearly intervals. This enabled the investigators to document the radiological progression of the anatomical lesions in the pathological finger joints over a 3-year period. It was shown that the progression of OA in the IP finger joints in an individual can be determined by the evolution of his finger joints through previously described anatomical phases: N (not affected), S (classical OA), J (loss of joint space), E (erosive OA) and R (remodeled joint). Structure/disease-modifying anti-OA drug (S/DMOAD) properties were searched for by assaying the number of patients developing OA in previously normal IP joints (N > S), or progressing through the described anatomical phases of the disease (S > J, S > E, J > E, S >

R, J > R, E > R). In the CS group, a significant decrease in the number of patients with new erosive OA finger joints was observed. This result is particularly important since OA of the finger joints becomes a clinical problem (pain, functional loss) when S joints progress to J and especially E phases. During and after these E phases, joints will remodel and show the nodular deformities characteristic of Heberden's and Bouchard's nodes. Treated patients were protected against erosive evolution.

In a very recent randomized, double-blind, multicenter study [41], the efficacy and tolerability of oral CS (Condrosulf®) in knee OA patients in a 3-month duration, twice a-year, intermittent treatment was investigated. A total of 120 patients with symptomatic knee OA were randomized into two groups receiving either 800 mg CS or placebo per day for two periods of 3 months during 1 year. Primary efficacy outcome was Lequesne's algo-functional index (AFI); secondary outcome parameters included VAS, walking time, global judgment, and paracetamol consumption. Radiological progression was assessed by automatic measurement of medial femoro-tibial joint space width on weight-bearing X-rays of both knees. Clinical and biological tolerability was assessed. AFI decreased significantly by 36% in the CS group after 1 year as compared to 23% in the placebo group. Similar results were found for the secondary outcome parameters. Radiological progression at month 12 showed significantly decreased joint space width in the placebo group with no change in the CS group. Tolerability was good with only minor adverse events identically observed in both groups. This study provides evidence that oral CS decreased pain and improved knee function. The 3-month intermittent administration of 800 mg/day of oral CS twice a year does support the prolonged effect known with symptom-modifying agents for OA. The inhibitory effect of CS on the radiological progression of the medial femoro-tibial joint space narrowing could suggest further evidence of its structure-modifying properties in knee OA.

In another recent clinical trial [42], the effect of 800 mg/die of CS per os plus naproxen versus naproxen over 2 years in patients with erosive OA of the hands has been studied. Joint count for erosions, Heberden and Bouchard nodes, Dreiser's algofunctional index and physicians' and patients' global assessment of disease activity were studied. A total of 24 consecutive patients (22 women and 2 men, mean age 53.0 +/- 6) suffering from symptomatic OA with radiographic characteristics of erosive OA were evaluated. The patients were divided into two groups of 12 patients each. The first group took naproxen 500 mg only. The second group was treated with CS 800 mg orally plus naproxen 500 mg. Joint counts, radiological hand examinations and assessment of disease activity were performed at baseline, at 12 months and at 24 months. Heberden, Bouchard and Dreiser scores was recorded. Physician and patient global assessments of disease activity showed no significant difference from baseline scores. The untreated group showed significant worsening in erosion, Heberden and Bouchard nodes, Dreiser index and physician and patient global assessment scores. This study confirms the partial efficacy of oral CS in improving some aspects of erosive OA.

The multicenter, double-blind, placebo- and celecoxib-controlled Glucosamine/CS Arthritis Intervention Trial (GAIT) evaluated the efficacy and safety of glucosamine and CS as a treatment for knee pain from osteoarthritis [43]. The mean age of the patients was 59 years, and 64 percent were women. Overall, glucosamine and CS were not significantly better than placebo in reducing knee pain by 20 percent. As compared with the rate of response to placebo (60.1 percent), the rate of response to glucosamine was 3.9 percentage points higher (P=0.30), the rate of response to CS was 5.3 percentage points higher (P=0.17), and the rate of response to combined treatment was 6.5 percentage points higher (P=0.09). The rate of response in the celecoxib control group was 10.0 percentage points higher than that in the placebo control group (P=0.008). For patients with moderate-to-severe pain at baseline, the rate of response was significantly higher with combined therapy than with placebo (79.2 percent vs. 54.3 percent, P=0.002). Adverse events were mild, infrequent, and evenly distributed among the groups. Glucosamine and CS alone or in combination did not reduce pain effectively in the overall group of patients with osteoarthritis of the knee. Exploratory analyses suggest that the combination of glucosamine and CS may be effective in the subgroup of patients with moderate-to-severe knee pain.

A 24-week, randomised placebo-controlled trial of CS (1 g/day) in patients with symptomatic knee osteoarthritis was performed on a visual analogue scale [44]. Pain on daily activities and Lequesne's Index were the primary efficacy criteria. Secondary outcomes included the rate of responders according to the outcome measures in rheumatoid arthritis clinical trials of the Osteoarthritis Research Society International (OMERACT-OARSI) criteria, quality of life, patient's/physician's global assessments and carry-over effect after treatment. Biochemical markers of bone (CTX-I), cartilage (CTX-II) and synovium (hyaluronic acid) metabolism were also measured. Safety was assessed by recording adverse events (AEs). Statistical analysis was performed on the inter-group differences in the intention-to-treat population. 307 patients were included in the study. 28 (9%) patients discontinued the study because of lack of efficacy or AEs. At the end of treatment, the decrease in pain was -26.2 (24.9) and -19.9 (23.5) mm and improved function was -2.4 (3.4) (-25%) and -1.7 (3.3) (-17%) in the CS and placebo groups, respectively (p = 0.029 and 0.109). The OMERACT-OARSI responder rate was 68% in the CS and 56% in the placebo group (p = 0.03). The investigator's assessments and short form 12 (SF-12) physical component reported improvement more frequently in the CS than in the placebo group (p = 0.044 and 0.021, respectively). No significant difference was observed between treatment groups for changes in biomarkers over 24 weeks. However, there was a significant difference between non-responders and responders according to the OARSI criteria for 24-week changes of CTX-I (p = 0.018) and CTX-II (p = 0.014). Tolerance was considered to be satisfactory. This study failed to show an efficacy of CS on the two primary criteria considered together, although CS was slightly more effective than placebo on pain, OMERACT-OARSI response rate, investigator's assessment and quality of life.

CS is known as a non-toxic substance, since it is a natural component of the connective tissue of both animal and man. Extensive studies with CS were carried out and the results obtained demonstrated that this compound exhibits a good toxicological safety (Table 1). The drug has no genotoxic properties according to a battery of *in vitro* and *in vivo* mutagenicity tests. CS did not affect fertility of rats, reproduction functions and no teratogenic properties at doses causing maternal toxicity. Furthermore, CS has been used in several European countries, such as France (since 1969), Switzerland (since 1982) and Italy (since 1990) and no toxic effect in man has been shown.

HEPARIN/HEPARAN SULFATE

Hep and HS possess a distinctly different repeating disaccharide structure compared with the previous GAGs [GlcA (β1→4) GlcNAc (α1→4)]$_n$ (Figure 1). Hep is sometimes considered to be synonymous with HS, but this is an oversimplification. They have been shown to differ in their degree of sulfation, with Hep being more negatively charged and displaying higher *N*- and *O*-sulfation than HS. They follow different biosynthetic paths in different cells and in different core proteins. The glycosidic linkage between uronic acid and GlcN is (β1→4) instead of (β1→3), and that between GlcN and uronic acid is (α1→4) instead of (β1→4) [8, 9, 45]. The growing GAG polymer chain is *N*-deacetylated and *N*-sulfonylated at the glucosamine residues, yielding regions in the chain particularly available for further structural changes, in particular C-5 epimerization of GlcA and *O*-sulfation mainly at C-2 of IdoA and C-6 of glucosamine [45]. Other more infrequent *O*-sulfations occur at C-2 of GlcA and C-3 of *N*-sulfonylated glucosamine. A few of the glucosamine amino groups may also remain unsubstituted. This process yields hybrid structures with hyper-variable, highly sulfated domains and poorly modified ones. As reported above, Hep has the highest charge density of any known biological macromolecule, while HS is generally less sulfated and with lower IdoA content. Both GAGs are highly polydisperse macromolecules, depending on tissue origin and status. Due to their properties, Hep and HS exhibit diverse biological functions and participate in a large number of interactions with other effective extracellular and cell membrane molecules, such as growth factors, virus proteins, enzymes, adhesion proteins, integrins and thrombin/antithrombin [8, 9, 45]. Therefore, the analysis of Hep/HS, which may identify structural variations due to differential expression of the corresponding PGs and/or of the enzymes responsible for their biosynthesis, may provide evidence for tissue and cell status. Such structural changes correspond mainly to size changes, to the degree of epimerisation and of *O*- and *N*-sulfonylation of glucosamine and they have been observed in cancer and in the urine of patients of many of the known mucopolysaccharidoses due to deficiency in specific degradative enzymes and other diseases.

As previously reported, Hep is a linear polysaccharide consisting of 1→4 linked uronic acid and GlcN repeating units. The uronic acid usually comprises 90% IdoA and 10% GlcA. Hep, with its high content of sulfo and carboxyl groups, is a polyelectrolyte, having the highest negative charge density of any known biological macromolecule. At the disaccharide level, a number of structural variations exist, leading to sequence microheterogeneity within the structure. Furthermore, Hep is polydisperse with a molecular mass range of 5-40, an average molecular mass of approx. 12 kDa, and an average negative charge of about -75.

At the PG level, different numbers of GAG chains (possibly having different saccharide sequences) can be attached to the various serine residues present on the core protein. Hep chains are biosynthesized attached to a unique core protein, serglycin, found primarily in mast cells. On mast cell degranulation, proteases act on the Hep core protein to release peptidoglycan Hep, which is further processed by a β-endoglucuronidase into GAG Hep [46]. The chemical, physical, and biological properties of Hep are primarily ascribed to GAG structure (or sequence), saccharide conformation, chain flexibility, molecular weight, and charge density.

With the discovery of increasing numbers of Hep-binding proteins [9], there was a need to characterize the molecular properties, within the proteins and Hep, responsible for specific recognition (Table 2). The biological activities of Hep (and HS) primarily result from their interaction with hundreds of different proteins. By using modelling, Cardin and Weintraub [47] demonstrated that some Hep binding proteins had defined motifs corresponding to consensus sequences, giving primary evidence for the general structural requirements for GAG-protein interactions. Their results suggested that if the XBBBXXBX (B = basic and X = hydropathic amino acid residue) sequence was contained in an α-helical domain, then the basic amino acids would be displayed on one side of the helix with the hydropathic residues pointing back into the protein core. Hep binding sites, commonly observed on the external surface of proteins, correspond to shallow pockets of positive charge. Thus, the topology of the Hep binding site is also an important factor in binding consensus sequences.

By screening peptide libraries, the conservation of amino acids in Hep binding domains and the importance of spacing between basic amino acids in GAG binding was demonstrated. Peptides enriched in arginine and lysine and polar hydrogen-bonding amino acids were observed to bind Hep with highest affinity [9]. Studies on the role of the pattern and the spacing of the basic amino acids in Hep binding domains showed that Hep interacted more tightly with peptides containing a complementary binding site of high positive charge density while the less sulfated HS interacted more tightly with a complementary site on a peptide that had more widely spaced basic [46].

Of the various activities of pharmaceutical Hep the anticoagulant action (120-180 USP U/mg) is the most widely studied. Anticoagulation occurs when Hep binds to antithrombin III (ATIII), a serine protease inhibitor (serpin). ATIII undergoes a conformational change and becomes activated as an inhibitor of thrombin and other serine proteases in the coagulation cascade. A major breakthrough in the study of Hep-catalyzed anticoagulation resulted from the separation of distinct Hep fractions differing markedly in affinity for ATIII. Low-affinity ATIII binding Hep comprises about two-thirds of porcine intestinal Hep and has a low anticoagulant activity (typically <20 U/mg). In contrast, high ATIII affinity Hep comprises the remaining third of porcine intestinal Hep and has a high anticoagulant activity (typically ~300 U/mg).

NMR studies by several groups confirmed the presence of the 3-*O*-sulfo group within the ATIII binding site, and chemical synthesis of a pentasaccharide containing the 3,6-di-*O*-sulfo group substantiated these findings [48]. The ATIII binding sequences found in certain HSs are partially responsible for the blood compatibility of the vascular endothelium [49].

The ATIII pentasaccharide is sufficient to catalyze the ATIII-mediated inhibition of factor Xa, a critical serine protease in the coagulation cascade. To catalyze the ATIII-mediated inhibition of thrombin, 16-18 saccharide units are required [50]. Thus, the structure-activity relationship of thrombin inhibition has been more difficult to establish because it relied on the synthesis of oligosaccharides substantially larger than ATIII for pharmacological evaluation. Recent studies show that a relatively non-specific but highly charged thrombin binding domain in Hep, localized on the non-reducing side of its ATIII binding site, is required to form a ternary complex. Success in understanding the structure-activity relationship of Hep's inhibition of thrombin has resulted in a new class of potent, synthetic, but still experimental thrombin inhibitors.

Table 2. Characteristics of selected Hep-binding proteins (modified from [47])

Heparin-binding protein	Physiological/Pathological role	Kd	Oligosaccharide size	Sequence features [a]	Function
Proteases/Esterases					
AT III	Coagulation cascade serpin	ca. 20 nM	5-mer	GlcNS6S3S	Enhances
SLPI	Inhibits elastase and cathepsin G	ca. 6 nM	12-mer to 14-mer	IS	Enhances
C1 INH	Inhibits C1 esterase	ca 100 nM	-	HS	Enhances
VCP	Protects host cell from complement	nM	-	-	Unclear
Growth factors					
FGF-1	Cell proliferation, differentiation, Morphogenesis and angiogenesis	nM	4-mer to 6-mer	IdoA2S-GlcNS6S	Activates signal transduction
FGF-2	Same as FGF-1	nM	4-mer to 6-mer	IdoA2S-GlcNS	Same as FGF-1
Chemokines					
PF-4	Inflammation and wound healing	nM	12-mer	HS/LS/HS	Inactivates heparin
IL-8	Pro-inflammatory cytokine	ca 6 uM	18-mer to 20-mer	HS/LS/HS	Promotes
SDF-1α	Pro-inflammatory mediator	ca 20 nM	12-mer to 14-mer	HS	Localizes
Lipid-binding proteins					
Annexin II	Receptor for TPA and plasminogen, CMV and tenascin C	ca. 30 nM	4-mer to 5-mer	HS	unclear
Annexin V	Anticoagulant activity; Influenza and hepatitis B viral entry	ca. 20 nM	8-mer	HS	Assembles
ApoE	Lipid transport; AD risk factor	ca. 100 nM	8-mer	HS	Localizes
Pathogen proteins					
HIV-1 gp120	Viral entry	0.3 μM	10-mer	HS	Inhibits
CypA	Viral entry	-	-	-	Inhibits

Tat	Transactivating factor, primes cells for HIV infection	ca. 70 nM	6-mer	HS	Antagonizes
HSV gB and gC	Viral entry into cells	-	-	-	Inhibits
HSV gD	Viral entry and fusion	-	-	$GlcNH_2 3S$	Inhibits
Dengue virus envelope protein	Viral localization	ca. 15 nM	10-mer	HS	Inhibits
Malaria CS protein	Sporozoite attachment to hepatocytes	ca. 40 nM	10-mer	HS	Inhibits
Adhesion proteins					
Selectins	Adhesion, inflammation and metastasis	μM	> 14-mer	HS with $GlcNH_2$	Blocks
Vitronectin	Cell adhesion and migration	μM	-	-	Removes
Fibronectin	Adhesion	μM	8-mer to 14-mer	HS with GlcNS	Reorganizes
HB-GAM	Neurite outgrowth in development	ca. 10 nM	16-mer to 18-mer	HS	Mediates
AP	In amyloid plaque	μM	4-mer	HS	Assembles

[a] HS: high sulfation; IS: intermediate sulfation; LS: low sulfation.

Table 3. Distribution of heparin in mammalian and other vertebrates (modified from [55])

Tissue	Rabbit	Guinea Pig	Rat	Dog	Cat	Pig	Bovine	Human	Chicken	Frog	Bony Fish	Shark
μg/g dry tissue												
Lung	<1	70	67	217	63	211	300	8	0.5	0.64	0.03	
Liver	<1	<1	<1	141	1	<1	50	<1	0	0.77	1.32	0
Ileum	<1	27	1	400	87	113	1015	32	0.5	0	0	0
Kidney	<1	1	<1	2	6	<1	26	<1	0.1	0.46		0.29
Aorta	<1	<1	9	102	<1	2	150	<1				
Brain	<1	<1	<1	<1	<1	<1	<1	<1	0	0	0	0
Muscle	<1	<1	36	9	<1	5	2	<1	0	11.4	0	0
Spleen	<1	<1	<1	11	<1	<1	19	<1	0			11.9
Skin	<1	<1	175	15	63	2	108	39	0.4	0	0	0
Lymph	<1	11	5	160	74	242	180	41				
Thymus	<1	112	20	20		10	286	35				
Appendix	<1			17	38		20	47				
Branchia											0.03	0

The strong increase in the development of low molecular weight Heps (LMW-Heps) as potential antithrombotic agents came from two observations in the 1970s and 1980s. The first was the finding that LMW-Hep fractions prepared from standard commercial grade Hep progressively lose their ability to prolong the activated partial thromboplastin time (APTT) while retaining their ability to inhibit factor Xa [51]. The second one was the observation that, for an equivalent antithrombotic effect, LMW-Heps produce less bleeding in experimental models than unfractionated Hep [52]. Further studies of platelet function and vascular permeability have provided plausible explanations for the reduced experimental bleeding observed with LMW-Heps. Furthermore, important differences between LMW-Hep and unfractionated Hep have been found in the plasma recovery (bioavailability), in their pharmacokinetics, and in the variability of anticoagulant response to fixed doses [52, 53]. At the moment, several LMW-Heps have been developed commercially and have been shown to be safe and effective for the prevention and treatment of venous thromboembolism.

LMW-Heps are fragments of commercial grade Hep, with a mean molecular masses ranging from 4,000 to 6,500, produced by either chemical or enzymatic depolymerization [52]. Depolymerization is achieved by 1) treatment with nitrous acid, or 2) treatment with the enzyme heparinase, or 3) by hydrolytic cleavage with hydrogen peroxide, or 4) by β-elimination. Depolymerization of natural extract Hep invariably leads to partial loss of the original catalytic activity with the ability to catalyze thrombin inhibition decreasing to a much greater extent than the ability of the fragments to catalyze the inhibition of factor Xa. In fact, the resulting LMW-Heps contain the unique pentasaccharide required for specific binding to ATIII (Figure 2), but in a lower proportion than is contained in the unfractionated Hep.

An updated phylogenetic tree of the distribution of sulfated GAGs in the animal kingdom is shown in Figure 3 [54]. Whereas HSs are ubiquitous components of all tissue-organized metazoan, Hep has shown a very peculiar distribution in mammalian and other vertebrate tissues as well as invertebrates. CS also has a widespread distribution (Figure 3). Since the earlier studies from a variety of mammals, it has been found that lung, intestine, and liver were the organs richest in Hep [55] (Table 3). Except for rabbit tissues, Hep's presence was demonstrated in lung, skin, ileum, lymph nodes, thymus, and appendix of all species studied. The absolute content of Hep varied depending on different tissues. A large variation of the concentration of Hep among species is evident. Thus, bovine and dog tissues contain the highest amounts of Hep. Generally, in non-mammalian vertebrate tissues the amount of Hep is considerably lower [55].

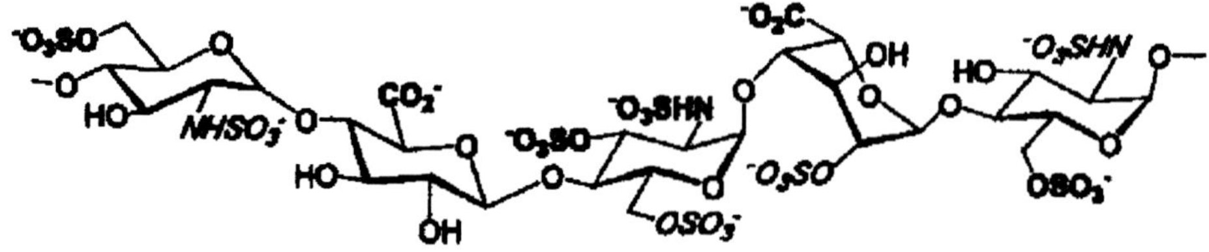

Figure 2. Antithrombin III pentasaccharide binding site. The anionic groups in bold are critical (95% loss in binding energy on removal), and those in italics are important (25-50% loss in binding energy on removal) for interaction with ATIII (modified from [46]).

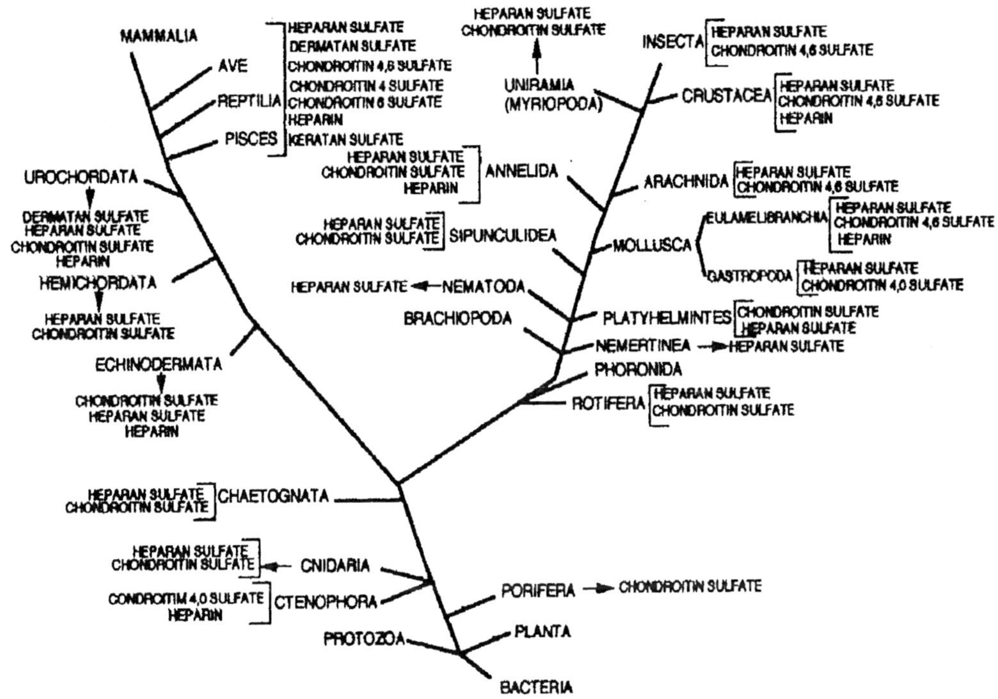

Figure 3. Distribution of sulfated glycosaminoglycans in the animal kingdom (Modified from [54]).

In invertebrates, Hep is found in few taxa, namely molluscs, crustacean, annelida, echinoderma and cnidaria. As observed for vertebrate Hep samples, the anticoagulant activity and molecular mass varied according to the species analyzed. Furthermore, no correlation between molecular mass and anticoagulant activity of the Heps is evident. All these results imply that Heps have a large structural variation depending on their origin [55].

The biological function of the clam Heps and their apparently specific ATIII-binding regions is unclear at the moment. Molluscs do not possess any blood coagulation system similar to that of mammals, yet their Heps are capable of dramatically accelerating the inactivation of mammalian coagulation enzymes by the mammalian protease inhibitor, ATIII. It is possible that the bivalve Hep is designed to interact with an endogenous antithrombin-like protease inhibitor acting on serine protease target enzymes. The existence and function of such an enzyme system remain to be established.

In mammals Hep is released from the mast cells in response to specific inflammatory agents such as IgE antibodies or complement fragments (anaphylatoxins). Since the discovery of mast cells by Paul Ehrlich (1879) and after the demonstration that their metachromatic properties when stained with basic dyes was related to Hep, the question whether this compound was only confined to the mast cells or not has been a matter of controversy. Studies on the concentration of Hep and its content in different fetal and adult bovine tissues [56] have shown that a good correlation between the mast cell number and Hep concentration could be obtained in all tissues analyzed.

Studies on mast-cell deficient mice of the genotypes W/W^v and $S1/S1^d$ established that mast cells originate from hematopoietic stem cells. These experiments also demonstrated that

Hep is present in appreciable amounts in the skin of the breeders and of the normal progeny. On the other hand, no Hep was detected in the skin of the W/W^v genotype, which are deficient in mast cells. No significant differences in the relative amounts of the other sulfated GAGs, namely HS, DS and CS were observed among the breeders analyzed. This suggests that Hep is not replaced by other sulfated GAGs in the animals that lack HEp. These results clearly indicate that Hep is related to the presence of mast cells [57, 58]. Furthermore, in mammals, the Hep-containing mast cells are accumulated in lymphoid organs and in tissues exposed to the external milieu (skin, lungs, intestine) and one suggested role for this polysaccharide in mammalian is to fight external parasites [55].

Hep and other sulfated GAGs as well as histamine were found and quantified in various organs of the mollusc *Anomalocardia brasiliana.* The Hep was present in granules inside the cytoplasm of mast-like cell [59, 60]. A good correlation between Hep and histamine content was found in the labial palp, intestine, ctenide, mantle, and foot tissues.

Some conclusions can be derived from these studies: 1) Hep seems to be present exclusively in mast cells of vertebrates or mast-like cells in the case of molluscs; 2) its primary biological activity is not related with antithrombotic activity since molluscs, which do not possess a coagulation system, contain hep and rabbits that possess this system are devoid of Hep; 3) many indirect evidences suggest that in mammals and molluscs Hep and its mast cells are involved in defense mechanisms independently of the immune system and support the hypothesis that this macromolecule could function as a mechanism for the surveillance of these organisms against certain pathogens.

CONCLUSIONS

The study of complex polysaccharides will certainly extend well into the future with so many questions left still unanswered, such as a) improved preparation and synthesis of GAGs, b) new GAG-based drugs with improved properties, c) new therapeutic applications, d) new GAG mimetics, (e) new biomaterials, and (f) development of an improved understanding of physiology and pathophysiology through glycomics.

The preparation of natural polysaccharides from tissues creates concern particularly after the recent appearance of prion- and virus-based diseases. As a consequence, alternative routes, such as defined, recombinant mammalian cell lines capable of being cultured in large-scale fermentations, or non mammalian tissues, offer an exciting alternative to the present preparations.

New anticoagulants might offer enhanced specificity targeting one or selected groups of coagulation proteases and avoiding undesired interactions with other proteins, thus decreasing the side effects associated with Heps and derivatives use. Furthermore, new therapeutic applications might include the use of GAGs for treatment of infectious diseases, inflammation, and control of cell growth in wound-healing and cancer. These new activities will require the elimination of Hep's anticoagulant activity, the engineering of appropriate pharmacokinetics and pharmacodynamics, and optimally oral bioavailability.

REFERENCES

[1] Sasisekharan, R., Raman, R. & Prabhakar, V. (2006) Glycomics approach to structure-function relationships of glycosaminoglycans. *Annu. Rev. Biomed. Eng.*, *8*, 181-231.

[2] Sugahara, K., Mikami, T., Uyama, T., Mizuguchi, S., Nomura, K. & Kitagawa, H. (2003) Recent advances in the structural biology of chondroitin sulfate and dermatan sulfate. *Curr. Opin. Struct. Biol.*, *13*, 612-20.

[3] Jackson, R.L., Busch, S.J. & Cardin, A.D., (1991) Glycosaminoglycans: molecular properties, protein interactions, and role in physiological processes. *Physiol Rev.*, *71*, 481-539.

[4] Kjellen, L. & Lindahl, U. (1991) Proteoglycans: structures and interactions. *Annu. Rev. Biochem.*, *60*, 443-75.

[5] Ruoslahti, E. (1988) Structure and biology of proteoglycans. *Annu. Rev. Cell Biol.*, 1988, *4*, 229-55.

[6] Heinegard, D. & Oldberg, A. (1989) Structure and biology of cartilage and bone matrix noncollagenous macromolecules. *FASEB J.*, *3*, 2042-51.

[7] Ruoslahti, E. (1989) Proteoglycans in cell regulation. *J. Biol. Chem.*, *264*, 13369-72.

[8] Sasisekharan, R., Shriver, Z., Venkataraman, G. & Narayanasami, U. (2002) Roles of heparan-sulphate glycosaminoglycans in cancer. *Nat. Rev. Cancer*, *2*, 521-8.

[9] Capila, I. & Linhardt, R.J. (2002) Heparin-protein interactions. *Angew Chem. Int. Ed. Engl.*, *41*, 391-412.

[10] Fraser, J.R.E., Laurent, T.C. & Laurent, U.B.G. (1997) Hyaluronan: its nature, distribution, functions and turnover. *J. Intern. Med.*, *242*, 27-33.

[11] Lamari, F., Katsimpris, J., Gartaganis, S. & Karamanos, N.K. (1998) Profiling of the eye aqueous humor in exfoliation syndrome by high-performance liquid chromatographic analysis of hyaluronan and galactosaminoglycans. *J. Chromatogr. B Biomed. Sci. Appl.*, *709*, 173-8.

[12] Ghosh, P. & Guidolin, D. (2002) Potential mechanism of action of intra-articular hyaluronan therapy in osteoarthritis: are the effects molecular weight dependent? *Seminars Arthr. Rheum.*, *32*, 10-37.

[13] Balazs, E.A. & Denlinger, J.L. (1993) Viscosupplementation: a new concept in the treatment of osteoarthritis. *J. Rheumatol.*, *20*, 3-9.

[14] Toole, B.P., Wight, T.N. & Tammi, M.I. (2002) Hyaluronan-cell interactions in cancer and vascular disease. *J. Biol. Chem.*, *277*, 4593-6.

[15] Tai, G.H., Huckerby, T.N. & Nieduszynski, I.A. (1996) Human corneal keratan sulfates. *J. Biol. Chem.*, *271*, 23535-31.

[16] Nieduszynski, I.A., Huckerby, T.N., Dickenson, J.M., Brown, G.M., Tai, G.H., Morris, H.G. & Eady, S. (1990) There are two major types of skeletal keratan sulphates. *Biochem. J.*, *271*, 243-5.

[17] Kao, W.W. & Liu, C.Y. (2002) Roles of lumican and keratocan on corneal transparency. *Glycoconj. J.*, *19*, 275-85.

[18] Nakazawa, K., Hassell, J.R., Hascall, V.C., Lohmander, L.S., Newsome, D.A. & Krachmer, J. (1984) Defective processing of keratan sulfate in macular corneal dystrophy. *J. Biol. Chem.*, *259*, 13751-7.

[19] Volpi N. (Ed.). Chondroitin sulfate: structure, role and pharmacological activity. Amsterdam, Boston, Heidelberg, London, New York, Oxford, Paris, San Diego, San Francisco, Singapore, Sydney, Tokyo, Academic Press, 2006.

[20] Carney, S.L. & Osborne, D.J. (1991) The separation of chondroitin sulfate disaccharides and hyaluronan oligosaccharides by capillary zone electrophoresis. *Anal. Biochem.*, *195*, 132-40.

[21] Karamanos, N.K., Syrokou, A., Vanky, P., Nurminen, M. & Hjerpe, A. (1994) Determination of 24 variously sulfated galactosaminoglycan- and hyaluronan-derived disaccharides by high-performance liquid chromatography. *Anal. Biochem.*, *221*, 189-99.

[22] Croft, P. (1996) The epidemiology of pain: the more you have, the more you get. *Ann. Rheum. Dis.*, *55*, 859-60.

[23] Petersson, I.F. (1996) Occurrence of osteoarthritis of the peripheral joints in European populations. *Ann. Rheum. Dis.*, *55*, 659-61.

[24] Burr, D.B. & Schaffler, M.B. (1997) The involvement of subchondral mineralized tissues in osteoarthrosis: quantitative microscopic evidence. *Microsc. Res. Tech.*, *37*, 343-57.

[25] Glant, T.T., Fulop, C., Cs-Szabo, G., Buzas, E., Ragasa, D. & Mikecz, K. (1995) Mapping of arthritogenic/autoimmune epitopes of cartilage aggrecans in proteoglycan-induced arthritis. *Scand. J. Rheumatol.*, *101*, 43-9.

[26] Evans, C.H., Watkins, S.C. & Stefanovic-Racic, M. (1996) Nitric oxide and cartilage metabolism. *Methods Enzymol.*, *269*, 75-88.

[27] Brandt, K.D. (1993) NSAIDs in the treatment of osteoarthritis. Friends or foes? *Bull. Rheum. Dis.*, *42*, 1-4.

[28] Hardingham, T. & Bayliss, M. (1990) Proteoglycans of articular cartilage: changes in aging and in joint disease. *Semin. Arthritis Rheum.*, *20(Suppl 1)*, 12-33.

[29] Ronca, F., Palmieri, L., Panicucci, P. & Ronca, G. (1998) Anti-inflammatory activity of chondroitin sulfate. *Osteoarthritis Cart.*, *6 Suppl A*, 14-21.

[30] Sakai, S., Akiyama, H., Harikai, N., Toyoda, H., Toida, T., Maitani, T. & Imanari, T. (2002) Effect of chondroitin sulfate on murine splenocytes sensitized with ovalbumin. *Immunol. Lett.*, *84*, 211-6.

[31] Akiyama, H., Sakai, S., Linhardt, R.J., Goda, Y., Toida, T. & Maitani, T. (2004) Chondroitin sulphate structure affects its immunological activities on murine splenocytes sensitized with ovalbumin. *Biochem. J.*, *382*, 269-78.

[32] Rovetta, G. (1991) Galactosaminoglycuronoglycan sulfate (matrix) in therapy of tibiofibular osteoarthritis of the knee. *Drugs Exp. Clin. Res.*, *17*, 53-7.

[33] Oliviero, U., Sorrentino, G.P., De Paola, P., Tranfaglia, E., D'Alessandro, A., Carifi, S., Porfido, F.A., Cerio, R., Grasso, A.M., Policicchio, D. & *et al.* (1991) Effects of the treatment with matrix on elderly people with chronic articular degeneration. *Drugs Exp. Clin. Res.*, *17*, 45-51.

[34] Mazieres, B., Loyau, G., Menkes, C.J., Valat, J.P., Dreiser, R.L., Charlot, J. & Masounabe-Puyanne, A. (1992) Chondroitin sulfate in the treatment of gonarthrosis and coxarthrosis. 5-months result of a multicenter double-blind controlled prospective study using placebo *Rev. Rhum. Mal. Osteoartic.*, *59*, 466-72.

[35] Morreale, P., Manopulo, R., Galati, M., Boccanera, L., Saponati, G. & Bocchi, L. (1996) Comparison of the antiinflammatory efficacy of chondroitin sulfate and diclofenac sodium in patients with knee osteoarthritis. *J. Rheumatol., 23*, 1385-91.

[36] Fleish, A.M., Merlin, C., Imhoff, A. & *et al.* (1997) A one-year randomized, double-blind, placebo-controlled study with oral chondroitin sulfate in patients with knee osteoarthritis. *Osteoarthritis Cart.*, *5*, 70-78.

[37] Bucsi, L. & Poor, G. (1998) Efficacy and tolerability of oral chondroitin sulfate as a symptomatic slow-acting drug for osteoarthritis (SYSADOA) in the treatment of knee osteoarthritis. *Osteoarthritis Cart.*, *6 Suppl A*, 31-39.

[38] Bourgeois, P., Chales, G., Dehais, J., Delcambre, B., Kuntz, J.L. & Rozenberg, S. (1998) Efficacy and tolerability of chondroitin sulfate 1200 mg/day vs chondroitin sulfate 3 x 400 mg/day vs placebo. *Osteoarthritis Cart.*, *6 Suppl A*, 25-33.

[39] Uebelhart, D., Thonar, E.J., Delmas, P.D., Chantraine, A. & Vignon, E. (1998) Effects of oral chondroitin sulfate on the progression of knee osteoarthritis: a pilot study. *Osteoarthritis Cart.*, *6 Suppl A*, 39-46.

[40] Verbruggen, G.; Goemaere, S.; Veys, E.M. Chondroitin sulfate: S/DMOAD (structure/ disease modifying anti-osteoarthritis drug) in the treatment of finger joint OA. *Osteoarthritis Cart.*, 1998, *6 Suppl A*, 37.

[41] Uebelhart, D., Malaise, M., Marcolongo, R., DeVathaire, F., Piperno, M., Mailleux, E., Fioravanti, A., Matoso, L.& Vignon, E. (2004) Intermittent treatment of knee osteoarthritis with oral chondroitin sulfate: a one-year, randomized, double-blind, multicenter study versus placebo. *Osteoarthritis Cart.*, *12*, 269-75.

[42] Rovetta, G., Monteforte, P., Molfetta, G. & Balestra, V. (2004) A two-year study of chondroitin sulfate in erosive osteoarthritis of the hands: behavior of erosions, osteophytes, pain and hand dysfunction. *Drugs Exp. Clin. Res.*, *30* 11-16.

[43] Clegg, D.O., Reda, D.J., Harris, C.L., Klein, M.A., O'Dell, J.R., Hooper, M.M., Bradley, J.D., Bingham, C.O. 3rd, Weisman, M.H., Jackson, C.G., Lane, N.E., Cush, J.J., Moreland, L.W., Schumacher, H.R. Jr, Oddis, C.V., Wolfe, F., Molitor, J.A., Yocum, D.E., Schnitzer, T.J., Furst, D.E., Sawitzke, A.D., Shi, H., Brandt, K.D., Moskowitz, R.W. & Williams, H.J. (2006) Glucosamine, chondroitin sulfate, and the two in combination for painful knee osteoarthritis. N *Engl. J. Med. 354*, 795-808.

[44] Mazières, B., Hucher, M., Zaïm, M. & Garnero, P. (2007) Effect of chondroitin sulphate in symptomatic knee osteoarthritis: a multicentre, randomised, double-blind, placebo-controlled study. *Ann. Rheum. Dis. 66*, 639-45.

[45] Powell, A.K., Yates, E.A., Fernig, D.G. & Turnbull, J.E. (2004) Interactions of heparin/heparan sulfate with proteins: appraisal of structural factors and experimental approaches. *Glycobiology 14*, 17R-30R.

[46] Linhardt, R.J. (2003) Heparin: structure and activity. *J. Med. Chem. 46*, 2551-64.

[47] Cardin, A.D. & Weintraub, H.J. (1989) Molecular modeling of protein-glycosaminoglycan interactions. *Arteriosclerosis 9*, 21-32.

[48] Torri, G., Casu, B., Gatti, G., Petitou, M., Choay, J., Jacquinet, J.C. & Sinay, P. (1985) Mono- and bidimensional 500 MHz 1H-NMR spectra of a synthetic pentasaccharide corresponding to the binding sequence of heparin to antithrombin-III: evidence for

conformational peculiarity of the sulfated iduronate residue. *Biochem. Biophys. Res. Commun. 128*, 134-40.

[49] Marcum, J.A. & Rosenberg, R.D. (1987) Anticoagulantly active heparan sulfate proteoglycan and the vascular endothelium. *Semin. Thromb. Hemost. 13*, 464-74.

[50] Sinay, P. (1999) Sugars slide into heparin activity. *Nature 398*, 377-8.

[51] Andersson, L.O., Barrowcliffe, T.W., Holmer, E., Johnson, E.A. & Sims, G.E. (1976) Anticoagulant properties of heparin fractionated by affinity chromatography on matrix-bound antithrombin iii and by gel filtration. *Thromb. Res. 9*, 575-83.

[52] Hirsh, J. & Levine, M.N. (1992) Low molecular weight heparin. *Blood 79*, 1-17.

[53] Weitz, J.I. (1997) Low-molecular-weight heparins. *New Engl. J. Med. 337*, 688-97.

[54] Medeiros, G.F., Mendes, A., Castro, R.A., Bau, E.C., Nader, H.B. & Dietrich, C.P. (2000) Distribution of sulfated glycosaminoglycans in the animal kingdom: widespread occurrence of heparin-like compounds in invertebrates. *Biochim. Biophys. Acta 1475*, 287-94.

[55] Nader, H.B., Lopes, C.C., Rocha, H.A.O., Santos, E.A. & Dietrich, C.P. (2004) Heparins and heparinoids: occurrence, structure and mechanism of antithrombotic and hemorrhagic activities. *Curr. Pharm. Des. 10*, 951-66.

[56] Nader, H.B., Straus, A.H., Takahashi, H.K. & Dietrich, C.P. (1982) Selective appearance of heparin in mammalian tissues during development. *Biochim. Biophys. Acta 714*, 292-297.

[57] Kitamura, Y. & Go S. (1979) Decreased production of mast cells in S1/S1d anemic mice. *Blood 53*, 492-497.

[58] Marshall, J.S., Kawabori, S., Nielsen, L. & Bienenstock J. (1994) Morphological and functional characteristics of peritoneal mast cells from young rats. *Cell Tissue Res. 276*, 565-570.

[59] Pejler, G., Danielsson, A., Bjork, I., Lindahl, U., Nader, H.B. & Dietrich CP. (1987) Structure and antithrombin-binding properties of heparin isolated from the clams *Anomalocardia brasiliana* and *Tivela mactroides*. *J. Biol. Chem. 262*. 11413-11421.

[60] Dietrich, C.P., de-Paiva, J.F., Moraes, C.T., Takahashi, H.K., Porcionatto, M.A. & Nader HB. (1985) Isolation and characterization of a heparin with high anticoagulant activity from *Anomalocardia brasiliana*. *Biochim. Biophys. Acta 843*, 1-7.

In: Glycobiology Research Trends
Editors: G. Powell and O. McCabe

ISBN: 978-1-60692-841-7

Chapter V

CMP-*N*-ACETYLNEURAMINIC ACID SYNTHETASES AND SIALYLTRANSFERASES FROM BACTERIAL SOURCES

Takeshi Yamamoto*
Glycotechnology Business Unit, Japan Tobacco Inc.
Higashibara 700, Iwata, Shizuoka 438-0802, Japan

ABSTRACT

Sialic acids (Sias) are an important component of carbohydrate chains and are linked to terminal positions of the carbohydrate moiety of glycoconjugates, including glycoproteins and glycolipids. Various studies have focused on clarifying the structure–function relationship of Sias and have revealed that *N*-acetylneuraminic acid (Neu5Ac) is the major Sia component of glycoconjugates, and that the sialylated carbohydrate chains (sialosides) of glycoconjugates play significant roles in many biological processes.

Four main linkage patterns found in the sialosides—Neu5Acα2-6Gal, Neu5Acα2-3Gal, Neu5Acα2-6GalNAc, and Neu5Acα2-8Neu5Ac—are formed by specific sialyltransferases. All known mammalian, bacterial, and viral sialyltransferases use cytidine 5′-monophosphate *N*-acetylneuraminic acid (CMP-Neu5Ac), synthesized by CMP-Neu5Ac synthetase, as the common donor substrate. For this reason, the sialyltransferases and CMP-Neu5Ac synthetases are considered key enzymes in the biosynthesis and *in vitro* enzymatic synthesis of several sialosides.

The sialyltransferases and CMP-Neu5Ac synthetases have been cloned from various sources, including mammalian organs and bacteria. In general, the enzymes with a bacterial origin are more stable and productive in *Escherichia coli* protein expression systems than the mammalian-derived enzymes. In the case of sialyltransferases, the bacterial-derived enzymes show broader acceptor substrate specificity than the mammalian enzymes. These advantages highlight the capacity of bacterial enzymes as efficient tools for the *in vitro* enzymatic synthesis of sialosides.

* Tel: +81-538-32-7389; Fax: +81-538-33-6046; E-mail: takeshi.yamamoto@ims.jti.co.jp

The recent increase in research focusing on sialyltransferases from a diverse range of bacteria has led to the identification of many bacterial sialyltransferases. Several bacterial CMP-Neu5Ac synthetases have also recently been identified. This chapter describes the bacterial CMP-Neu5Ac synthetases and sialyltransferases that show promise as tools for the large-scale production of sialosides.

Introduction

Sialic Acids

Sialic acids (Sias) are a family of monosaccharides comprising over 50 naturally occurring derivatives of neuraminic acid, 5-amino-3,5-dideoxy-D-glycero-D-galacto-non-2-ulopyranonic acid (Neu) [1,2]. Structurally, the Sia derivatives of Neu carry a variety of substitutions at the amino and/or hydroxyl groups. The amino group is often acetylated, glycolylated, or deaminated. The hydroxyl groups can be acetylated at O7, O8, or O9, singly or in combination [3,4], and can also be modified by acetate, lactate, phosphate or sulfate esters. The three major members of the Sia group are *N*-acetylneuraminic acid (Neu5Ac), *N*-glycolylneuraminic acid (Neu5Gc), and 2-keto-3-deoxy-D-glycero-D-galacto-noninic acid (KDN) [1,3]. Although, Sias are widely distributed in higher animals and some microorganisms, only Neu5Ac is ubiquitous, and Neu5Gc is not found in bacteria [3]. Usually, Sias exist in the carbohydrate moiety of glycoconjugates, such as glycoproteins and glycolipids, and are linked to the terminal positions of the carbohydrate chains of the glycoconjugates [5]. Sialylated carbohydrate chains play important roles in many biological processes, including immunological responses, viral infections, cell–cell recognition, and inflammation [6–9]. The structures of typical sialic acids and the related saccharide 2-keto-3-deoxy-D-*manno*-octulosonic acid (KDO) are shown in Figure 1.

Biological Roles of Sialylated Carbohydrate Chains

Guillain-Barré syndrome (GBS) is the prototypic postinfectious autoimmune disease [10]. Epidemiological studies clearly demonstrate that GBS patients develop the syndrome following infection by the Gram-negative bacterium *Campylobacter jejuni* [11]. GBS patients develop neuropathy, which is the most common cause of generalized paralysis [12]. The most common *C. jejuni* serotype associated with GBS is *C. jejuni* O:19 [13]. The serotypes of *C. jejuni* associated with GBS-derived neuropathies express ganglioside-like lipooligosaccharide (LOS) structures on the cell surface. Detailed investigations of LOS structures isolated from GBS patients reveal that the core oligosaccharides mimic gangliosides located in neural tissue. Terminal oligosaccharide moieties identical to those of the gangliosides GM1, GD3, and GT1a have been found in *C. jejuni* O:19 strains [14–17]. The structures of GM1 and LOS are shown in Figure 2. The development of GBS in patients following infection with *C. jejuni* is thought to be related to the molecular mimicry of gangliosides and the cross-reaction of antibodies against LOS, leading to neuropathy. The molecular mimicry basis of GBS is consistent with the finding that most GBS patients develop autoantibodies that react with GM1 subsequent to *C. jejuni* enteritis [18, 19]. Molecular mimicry has been proposed as a pathogenic mechanism underlying autoimmune disease [20]. It has also been demonstrated in the development of Fisher syndrome after infection with *Haemophilus influenzae* in the production of anti-GQ1b autoantibodies mediated by the GQ1b-mimicking LOS on the bacterial surface [21].

Another role of sialylated carbohydrate chains is demonstrated by influenza virus infections. It is well known that the influenza virus binds to cell receptors of host cells via Neu5Ac-linked glycoproteins through viral hemagglutinin recognition of host cell Neu5Ac [22, 23]. The influenza virus also recognizes the carbohydrate chain structure of the host cell [22, 23]. For example, the avian influenza viruses recognize Neu5Acα2-3Galβ1-3/4GlcNAc structures, and the human influenza viruses recognize Neu5Acα2-6Galβ1-3/4GlcNAc structures [24, 25]. The host cell specificity of the influenza viruses is determined by the linkage of Neu5Ac to the galactose residues and by the number of sialic acid residues and core structures [22]. For this reason, the distribution of Neu5Ac on the host cell surface is an important determinant of host tropism. An understanding of the mechanism underlying the control of Neu5Ac expression in human cells and tissues is essential for understanding the process of influenza pathogenesis.

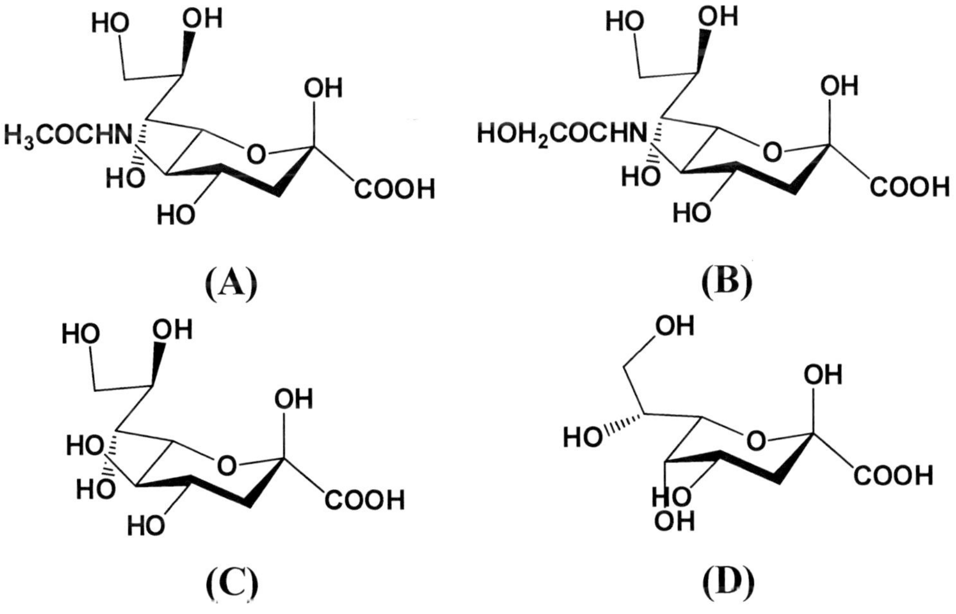

HO OH
OH
H_3COCHN
O
HO
COOH
HO

(A)

HO OH
OH
HOH_2COCHN
O
HO
COOH
HO

(B)

HO OH
OH
HO
O
HO
COOH
HO

(C)

OH
OH
HO
O
COOH
HO
OH

(D)

Figure 1. The structure of sialic acids.
(A) *N*-acetylneuraminic acid (Neu5Ac); (B) *N*-glycolylneuraminic acid (Neu5Gc); (C) 2-keto-3-deoxy-D-glycero-D-galacto-noninic acid (KDN); (D) 2-keto-3-deoxy-manno-octonic acid (KDO).

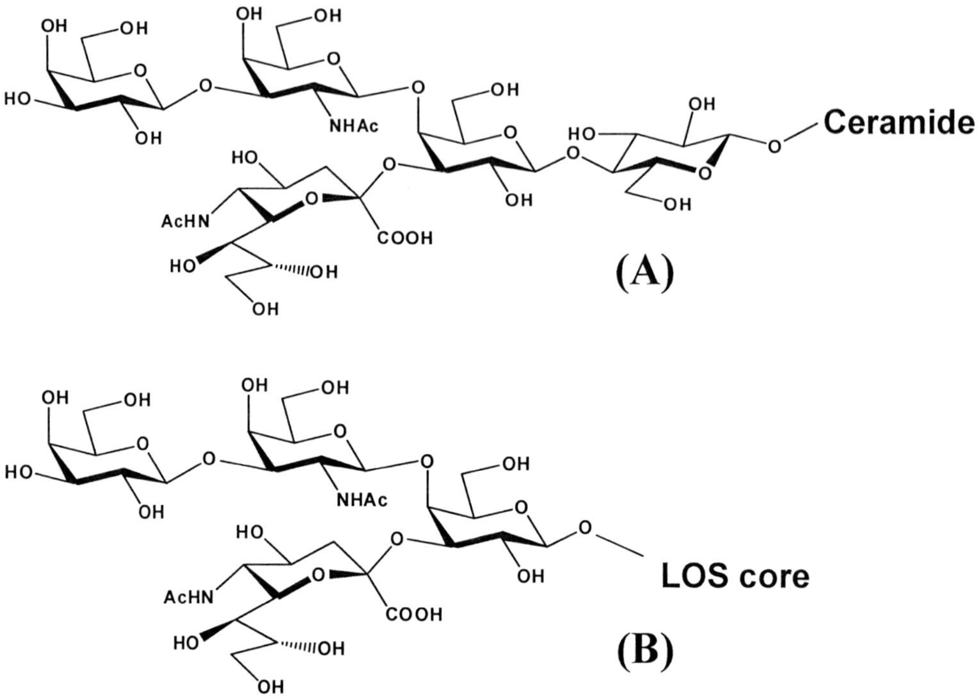

Figure 2. Molecular mimicry of ganglioside, GM1, by the LOS of *C. jejuni*.
GM1 is located in the nerve cell membrane. Many studies demonstrate that the LOS mimics GM1 in the outer part of the cell wall of *C. jejuni*. (A) Structure of ganglioside, GM1. (B) Carbohydrate moieties of LOS obtained from *C. jejuni*.

Importance of Tools for the Investigation of Sialylation

An abundant supply of sialylated oligosaccharides is essential for a detailed investigation of the biological function of sialylation. Chemical and enzymatic glycosylation are the two major routes for the preparation of sialylated oligosaccharides. Although chemical glycosylation has the advantage of high flexibility and wide applicability, the reaction processes are complicated, as the chemical reactions often require multiple protection and de-protection steps [26–28]. By comparison, enzymatic glycosylation using glycosyltransferases is a single-step process with high positional and anomer selectivity and high reaction yield. For example, with sialylation, the transfer of Neu5Ac by sialyltransferases to the appropriate substrate from CMP-Neu5Ac as the donor can be readily achieved in the final step under mild reaction conditions [29–31].

This chapter reviews the pathways for biosynthesis of Neu5Ac and sialylated glycoconjugates, and recent progress in the study of bacterial CMP-Neu5Ac synthetases and sialyltransferases.

Biosynthesis of Sialic Acids and Sialyl-Glycoconjugates

Bacteria

The biosynthesis of sialyl-glycoconjugates involves several steps, including the synthesis of the sugar nucleotide CMP-Neu5Ac from Neu5Ac, a typical sialic acid, by CMP-Neu5Ac synthetase, and the subsequent transfer of Neu5Ac from this donor substrate to appropriate acceptor substrates by sialyltransferase [1].

The biosynthesis of Neu5Ac in bacteria first involves the conversion of *N*-acetylglucosamine 6-phosphate to *N*-acetylmannosamine 6-phosphate by *N*-acylglucosamine 6-phosphate 2-epimerase. *N*-acetylmannosamine 6-phosphate is then dephosphorylated to *N*-acetylmannosamine, which is then conjugated to phospho*enol*pyruvate to form Neu5Ac by Neu5Ac synthase. In an alternative pathway in bacteria, Neu5Ac is generated from *N*-acetylmannosamine and pyruvate by *N*-acetylneuraminate lyase [32]. Neu5Ac is conjugated to cytidine triphosphate (CTP) by CMP-Neu5Ac synthetase to form the activated donor substrate, CMP-Neu5Ac. Neu5Ac is transferred from CMP-Neu5Ac to the appropriate acceptor substrates by sialyltransferase. A schematic diagram of the biosynthesis of Neu5Ac in bacteria is shown in Figure 3.

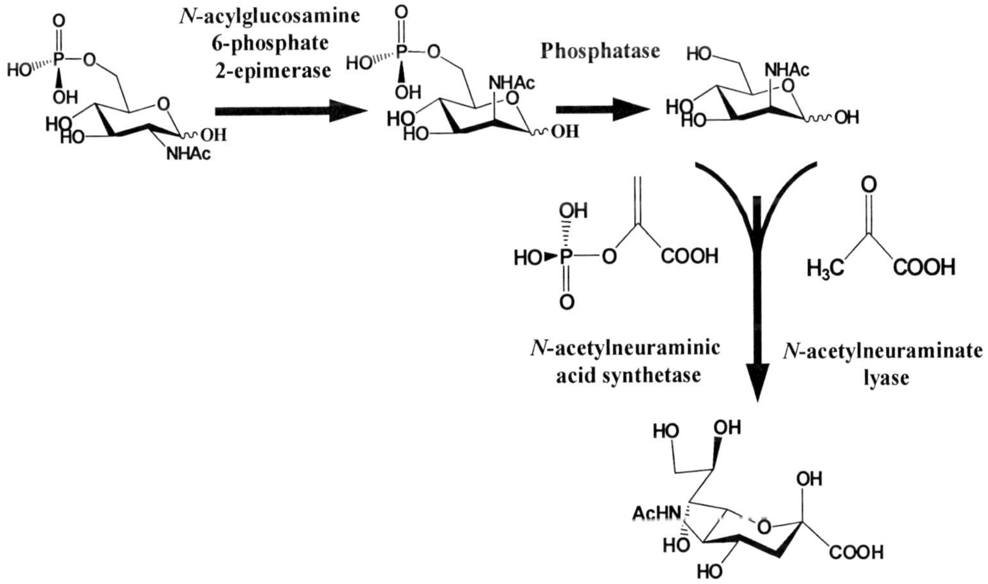

Figure 3. The pathway of Neu5Ac biosynthesis in bacteria. As described in the text, Neu5Ac is synthesized from *N*-acetylglucosamine 6-phosphate in 3 enzymatic reaction steps.

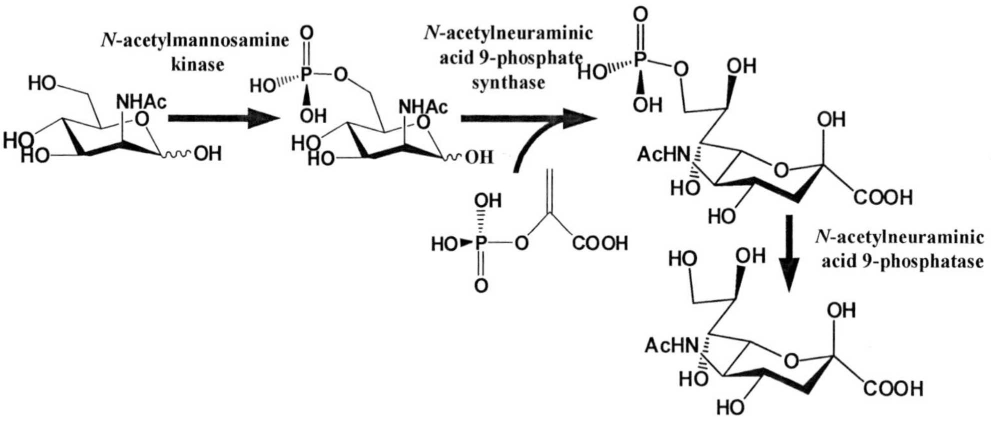

Figure 4. The pathway of Neu5Ac biosynthesis in mammals. As described in the text, Neu5Ac is synthesized from *N*-acetylmannosamine in 3 enzymatic reaction steps.

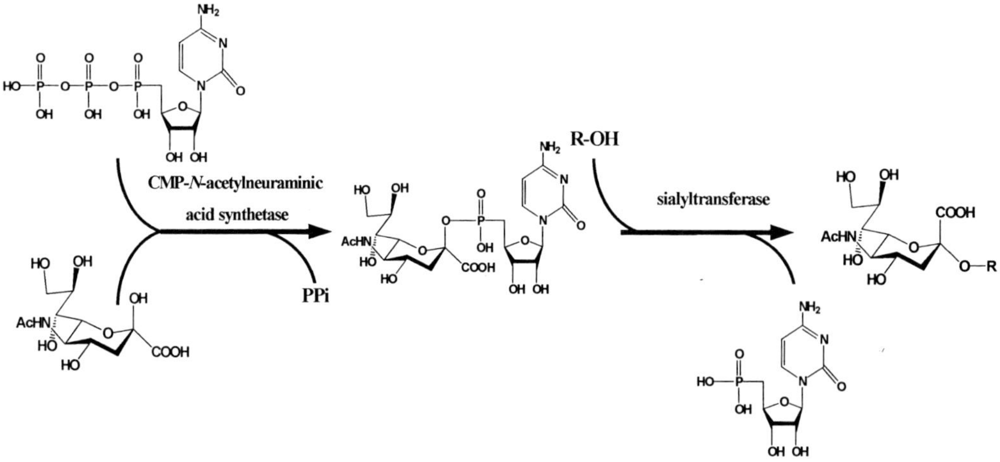

Figure 5. Schematic diagram of the formation of CMP-Neu5Ac and sialylated oligosaccharide. CMP-Neu5Ac is formed by CMP-Neu5Ac synthetase from Neu5Ac and CTP as substrates. Sialylated oligosaccharide is synthesized by sialyltransferase from CMP-Neu5Ac and the acceptor substrate.

Mammals

The pathway for the biosynthesis Neu5Ac in mammals first involves the conversion of *N*-acetylglucosamine to *N*-acetylmannosamine by *N*-acylglucosamine 2-epimerase. *N*-acetylmannosamine is then phosphorylated to *N*-acetylmannosamine 6-phosphate by *N*-acetylmannosamine kinase. The *N*-acetylmannosamine 6-phosphate is conjugated to phospho*enol*pyruvate by *N*-acetylneuraminic acid 9-phosphate synthase to form *N*-acetylneuraminic acid 9-phosphate, which is dephosphorylated to Neu5Ac by *N*-acetylneuraminic acid 9-phosphatase. Neu5Ac is then activated by conjugation to CTP by CMP-Neu5Ac synthetase to form CMP-Neu5Ac. The reactions to generate CMP-Neu5Ac occur in the nucleus. CMP-Neu5Ac is then transported to the lumen of the Golgi apparatus and used as the Neu5Ac donor substrate in sialyltransferase [1, 2]. A schematic diagram of the biosynthesis of Neu5Ac in mammals is shown in Figure 4.

Although bacteria and mammals use different pathways for the biosynthesis of Neu5Ac, the sialylation reactions involving sialyltransferase and CMP-Neu5Ac as the donor substrate are identical.

A schematic diagram of the formation of CMP-Neu5Ac from Neu5Ac and CTP by CMP-Neu5Ac synthetase and of sialylated oligosaccharides from CMP-Neu5Ac and acceptor substrates by sialyltransferase is shown in Figure 5.

CMP-Neu5Ac Synthetase

Mammalian and Bacterial CMP-Neu5Ac Synthetases

CMP-Neu5Ac is a common donor substrate for all known bacterial and mammalian sialyltransferases. CMP-Neu5Ac is produced by CMP-Neu5Ac synthetase (EC 2.7.7.43; also known as *N*-acylneuraminate cytidyltransferase, CMP-sialate pyrophosphorylase, and acylneuraminate cytidyltransferase) from CTP and Neu5Ac as described in the previous section. CMP-Neu5Ac synthetase is considered an important enzyme in the field of glycobiology, as an abundant supply of CMP-Neu5Ac is essential for the synthesis of biologically important sialylated oligosaccharides by means of sialyltransferase [1]. For example, sialylated oligosaccharides are used in the development of clinical medicines and in the study of many biochemical processes [29].

CMP-Neu5Ac synthetases are obtained from several sources [33–35]. Those of bacterial and mammalian origin have many enzymatic and protein properties in common [36]. For example, all the known CMP-Neu5Ac synthetases display maximum enzymatic activity under basic conditions, at around pH 8.5, and require divalent cations for activity. The enzymatic reaction mechanisms for CMP-Neu5Ac synthetase are well characterized [37]. In the presence of a divalent cation, CMP-Neu5Ac synthetase catalyzes the nucleophilic attack of the alpha-phosphate of cytidine triphosphate (CTP) by the anomeric oxygen of beta-Neu5Ac to produce pyrophosphate and CMP-Neu5Ac. Molecular cloning reveals the presence of five conserved regions, known as motifs I to V, among the known CMP-Neu5Ac synthetases [38]. Motifs I to III are also conserved in the CMP-3-deoxy-D-manno-octulosonate (KDO) synthetases (EC 2.7.7.38) [38]. Motif I is involved in nucleotide binding in the CMP-Neu5Ac synthetases and CMP-KDO synthetases [39]. KDO is an eight-carbon monosaccharide that is an essential component of cell wall capsular polysaccharides in Gram-negative bacteria [38]. KDO is activated by CMP to generate CMP-KDO [40], which is the donor substrate of KDO transferase [41, 42]. KDO and Neu5Ac are unique, as the activated forms are nucleoside monophosphate diesters, rather than the more common nucleoside diphosphate diesters. The CMP-NeuAc synthetases and CMP-KDO synthetases are highly specific for their respective sugar substrates: CMP-KDO synthetases do not recognize sialic acid as a substrate, and CMP-Neu5Ac synthetases do not recognize KDO as a substrate [43].

While the CMP-Neu5Ac synthetases have similar enzymatic features, several obvious differences have been demonstrated between mammalian- and bacterial-derived enzymes, including differences in substrate specificity and tertiary structure [44].

Origin of Bacterial CMP-Neu5Ac Synthetase

Genes encoding CMP-Neu5Ac synthetase have been cloned from several types of bacteria, including *Escherichia coli* [45], *Neisseria meningitidis* [46], *Haemophilus ducreyi* [47], *Streptococcus agalactiae* [48], and *Clostridium thermocellum* [44]. The CMP-Neu5Ac synthetase cloned from *C. thermocellum*, an anaerobic, Gram-positive, nonpathogenic, thermophilic bacterium, is thermostable. It is highly active at 50 °C and retains full activity following incubation for 24 h at 37 and 50 °C [44]. For this reason, CMP-Neu5Ac synthetase derived from *C. thermocellum* is considered a powerful tool for the preparation of CMP-Neu5Ac.

The crystallographic structure of bacterial CMP-Neu5Ac synthetase, obtained from *N. meningitidis*, has been determined in the presence and absence of the substrate analog cytidine diphosphate CDP [49]. The crystal structure of a mammalian equivalent, murine CMP-Neu5Ac synthetase, has also been resolved [50]. For reference, the crystal structure of CMP-KDO synthetase, obtained from *E. coli*, has been determined in the presence of substrates and substrate analogs [39].

Bacterial CMP-Neu5Ac Synthetases

CMP-Neu5Ac Synthetases from E. Coli

CMP-Neu5Ac synthetase has been purified to apparent homogeneity from the cytoplasmic fraction of *E. coli* O18:K1 [34], and the gene encoding this enzyme has been identified [34]. Several CMP-Neu5Ac synthetase genes have subsequently been cloned from various strains of *E. coli*.

The characteristics of *E. coli* synthetase have been revealed by use of the native enzyme and several recombinant enzymes. *E. coli* synthetase requires divalent cations for enzyme activity [34]. Of the divalent cations tested, Mg^{2+} and Mn^{2+} are essential for enzyme activity, but Ca^{2+}, Cd^{2+}, and Zn^{2+} have no effect. Maximum *E. coli* synthetase activity is observed with 20–40 mM Mg^{2+} at a pH of 9.0–10.0 [34]. Although the crystallographic structure of *E. coli* CMP-Neu5Ac synthetase has not yet been determined, site-directed mutagenesis and chemical modification studies have revealed the functional role of specific amino acids. For example, Arg12 may be located within the active site of *E. coli* synthetase and play an important role in substrate binding [51]. Furthermore, the mutation of Cys129 to Ser enhanced the sensitivity of the enzyme to heat and chemical denaturation [52]. Recent evidence reveals that *E. coli* CMP-Neu5Ac synthetase is bifunctional, with both CMP-Neu5Ac synthetase and platelet-activating factor acetylhydrolase activity [53].

Systems for the high expression of native and recombinant *E. coli* CMP-Neu5Ac synthetases have been established to generate functional levels of this enzyme. For example, over 25 000 units of wild-type CMP-Neu5Ac synthetase from *E. coli* K-235 has been produced by using optimized culture medium [54].

CMP-Neu5Ac Synthetases from N. Meningitidis

N. meningitidis CMP-Neu5Ac synthetase has been cloned and expressed in *E. coli* [55, 56]. Recombinant *N. meningitidis* CMP-Neu5Ac synthetase has been expressed at very high levels of over 70 000 units of enzyme per liter of culture medium, and has been purified with a high yield [57]. Purified *N. meningitidis* CMP-Neu5Ac synthetase has been used in the gram-scale synthesis of CMP-Neu5Ac [57].

N. meningitidis and *E. coli* CMP-Neu5Ac synthetase have specific enzymatic properties in common. For example, both require divalent cations for activity and have an optimal pH of 9.0. Several differences have also been demonstrated. Firstly, the substrate specificity of *N. meningitidis* CMP-Neu5Ac synthetase is broader than that of *E. coli* synthetase. *E. coli* CMP-Neu5Ac synthetase does not recognize *N*-glycolylneuraminic acid as a substrate, whereas *N. meningitidis* CMP-Neu5Ac synthetase uses *N*-glycolylneuraminic acid and *N*-acetylneuraminic acid as substrates [58]. Secondly, *N. meningitidis* CMP-Neu5Ac synthetase synthesizes several sugar nucleotides, including CMP-Neu5Ac and derivatives with a substitution at C5, C8, and C9 of Neu5Ac [56]. By comparison, *E. coli* synthetase accepts only a few C9-substituted derivatives of Neu5Ac as substrate [59].

CMP-Neu5Ac synthetase from *N. meningitidis* has been crystallized as a dimer, and structural information has revealed the amino acid residues comprising the active site, the mononucleotide-binding pocket, the substrate-binding site, and the dimerization domain of this class of enzyme [49].

SIALYLTRANSFERASES

Mammalian and Bacterial Sialyltransferases

Many mammalian sialyltransferases have been reported [60]. They are classified into four families according to the carbohydrate linkages they synthesize: the beta-galactoside α2,3-sialyltransferases (ST3Gal I–VI), the beta-galactoside α2,6-sialyltransferases (ST6Gal I, II), the GalNAc α2,6-sialyltransferases (ST6GalNAc I–VI), and the α2,8-sialyltransferases (ST8Sia I–VI) [61]. Mammalian and bacterial sialyltransferases catalyze the transfer of Sia from CMP-Sia to the acceptor substrates. It has been recognized for some time that mammalian sialyltransferases contain conserved sequence motifs, including sialyl motifs L, S, and VS [62, 63]. Recently, a fourth motif, motif III, has been identified in human sialyltransferases [64]. Some functions of these motifs have been clarified. For example, mutagenesis studies have demonstrated that motif L is involved in donor substrate binding [65] and that motif S is involved in donor and acceptor substrate binding [62]. As the crystal structures of the mammalian sialyltransferases have not been determined, the reaction mechanism and the interactions between acceptor substrate, donor substrate, and enzyme remain unclear.

Table.1. Bacterial sialyltransferases: GT family, origins and enzyme type

GT family number	Bacteria name	Enzyme type
GT family 38	*Escherichia coli K1*	α2,8/2,9-sialyltransferase
		α2,8-sialyltransferase
	Neisseria meningitidis	α2,8-sialyltransferase
GT family 42	*Campylobacter jejuni*	α2,3/2,8-sialyltransferase
		α2,3-sialyltransferase
	Haemophilus influenzae	α2,3/2,8-sialyltransferase
GT family 52	*Neisseria gonorrhoeae*	α2,3-sialyltransferase
		α2,3/2,6-sialyltransferase
	Neisseria meningitidis	α2,3-sialyltransferase
GT family 80	*Photobacterium damselae*	α2,6-sialyltransferase
	Photobacterium leiognathi	α2,6-sialyltransferase
	Photobacterium phosphoreum	α2,3-sialyltransferase
	Pasteurella multocida	α2,3/2,6-sialyltransferase
	Vibrio sp.	α2,3-sialyltransferase

The known bacterial sialyltransferases do not contain the sialyl motifs found in mammalian sialyltransferases, despite the specificity of the bacterial sialyltransferases for CMP-Neu5Ac, a donor substrate recognized by mammalian sialyltransferases [60]. Two short motifs, referred to as the D/E-D/E-G and HP motifs, have been recently identified in the bacterial sialyltransferases and are shown to be functionally important for enzyme catalysis and donor substrate binding [66]. These two motifs are structurally distinct from the motifs found in mammalian sialyltransferases.

Like the mammalian sialyltransferases, viral α2,3-sialyltransferase, obtained from myxoma-virus–infected RK13 cells, contains the motifs L and S in its catalytic domain and is closely related to mammalian α2,3-sialyltransferase (ST3Gal IV) on the basis of amino acid sequence similarity [67]. The substrate specificity of the viral and mammalian sialyltransferases is very different, despite the common presence of the sialyl motifs. Viral α2,3-sialyltransferase transfers Neu5Ac to the fucosylated substrates Lewis^X and Lewis^a [67].

Origin and Classification of the Bacterial Sialyltransferases

Genes encoding sialyltransferases have been cloned from various types of bacteria, including *Neisseria gonorrhoeae* [68], *N. meningitidis* [69], *C. jejuni* [70], *E. coli* [71], *Photobacterium damselae* [72], *Photobacterium phosphoreum* [73], *Photobacterium leiognathi* [74], *Photobacterium* sp. [75], *Vibrio* sp. [76], *Pasteurella multocida* [77], *H. influenzae* [78], and *Streptococcus agalactiae* [79]. All bacterial sialyltransferases are classified into one of four families in the CAZy (carbohydrate-active enzymes) database [80]: (1) glycosyltransferase (GT) family 38 (polysialyltransferase from *E. coli* and *N. meningitidis*); (2) GT family 42 (lipooligosaccharide α2,3-sialyltransferase and α2,3-/α2,8-sialyltransferase from *C. jejuni* and *H. influenzae*); (3) GT family 52 (α2,3-sialyltransferase from *H. influenzae*, *N. gonorrhoeae*, and *N. meningitidis*); and (4) GT family 80 (α2,6-sialyltransferase and α2,3-/α2,6-sialyltransferase from *Ph. damselae* and *Ph. multocida*). The relationship between classification of the enzymes in the CAZy database and enzyme origin is summarized in Table 1.

Three-dimensional Structure of Bacterial Sialyltransferases

Despite the relatively few reports on bacterial sialyltransferases compared with those on mammalian sialyltransferases, the crystal structures of five bacterial sialyltransferases have been described. The first was that of the bifunctional enzyme α2,3-/α2,8-sialyltransferase (CstII) from *C. jejuni* OH4384, complexed with a substrate analog [81]. The second was that of the multifunctional enzyme α2,3-sialyltransferase (Δ24PmST1) from *P. multocida* strain P-1059, in the presence and absence of CMP. This enzyme shows α2,3-sialyltransferase activity, α2,6-sialyltransferase activity, sialidase activity, and trans-sialidase activity. In addition to a description of the crystal structure of this enzyme in complexes with both acceptor and donor substrate analogs, the substrate binding sites and catalytic mechanism have also been reported [82, 83]. The third structure described was that of monofunctional α2,3-sialyltransferase (CstI) from *C. jejuni* in apo- and substrate-analog–bound forms [84]. The fourth was that of monofunctional α2,6-sialyltransferase (Δ16pspST6) from *Photobacterium* sp. as a CMP donor product and lactose-bound complex [85]. The fifth was that of α2,3-sialyltransferase from *P. multocida* in a complex with CMP-3FNeuAc and lactose [86].

The overall structures of multifunctional α2,3-sialyltransferase (Δ24PmST1) and monofunctional α2,6-sialyltransferase (Δ16pspST6) are similar [85]. These enzymes are both classified in GT family 80 in the CAZy database. There is similarity in the overall architecture of bifunctional α2,3-/α2,8-sialyltransferase (CstII) and monofunctional α2,3-sialyltransferase (CstI), which are both classified in GT family 42. However, there is no structural similarity between the GT family 42 and GT family 80.

Reaction of GT Family 80 Sialyltransferases

In general, the mechanism of reaction of the inverting glycosyltransferases resembles that of the inverting glycosylhydrolases, as both possess an acidic amino acid that activates the acceptor hydroxyl group by deprotonation [87]. As mentioned previously, sialyltransferases use CMP-Neu5Ac (cytidine 5'-monophospho-β-D-*N*-acetylneuraminic acid) as a donor substrate and transfer Neu5Ac to an acceptor substrate through α-linkage. Therefore, sialyltransferases are classified into inverting glycosyltransferases. Superimposition of the substrate binding sites of Δ16pspST6 and Δ24PmST1 reveals the catalytic role of specific amino acid residues in the sialyltransferases [85]. Firstly, Asp232 of Δ16pspST6, which corresponds to Asp141 of Δ24PmST1 in the primary amino acid sequence, may play a key role as the catalytic base in the transfer reaction [85]. Secondly, His405 of Δ16pspST6, which corresponds to His311 of Δ24PmST1, may act as a catalytic residue by protonating phosphate oxygen as a catalytic acid [85].

In the crystal structure, CMP is located in a deep cleft between domains 2 and 3 of Δ16pspST6 and Δ24PmST1, and nine amino acid residues (Gly361, Lys403, His405, Pro406, Ser430, Phe431, Ser449, Ser450, and Leu451 of Δ16pspST6) interact with it. Moreover, all except Ser430 and Leu451 are conserved among the sialyltransferases in GT family 80 [88].

Lactose is recognized by five amino acid residues (Trp270, Asp141, His112, Asn85, and Met144) in Δ24PmST1. These residues make contact with only the galactose moiety of lactose. Three of the five in Δ24PmST1—Trp270, Asp141, and His112—are conserved in Δ16pspST6 as Trp365, Asp232, and His204, respectively, and occupy structurally identical positions [85]. The main difference in the galactose binding site between Δ24PmST1 and Δ16pspST6 is the presence of Asn85 in Δ24PmST1 and His123 in Δ16pspST6. His123 in Δ16pspST6 lies within the hydrogen-bonding distance of O3 and O5 of the galactose binding site and is thought to determine the orientation of lactose [85]. Pro34, the corresponding residue in Δ24PmST1, cannot form a hydrogen bond with the galactose binding site. His123 of Δ16pspST6 is conserved among other α2,6-sialyltransferases belonging to GT family 80 [88]. However, in these α2,3-sialyltransferases, including those cloned from *P. phosphoreum*, *Photobacterium* sp., and *Vibrio* sp., the amino acids corresponding to the His123 of Δ16pspST6 are substituted for arginine. For this reason, the conserved amino acid residues of the sialyltransferases belonging to GT family 80 are thought to be involved in the linkage of sialic acid [85, 88].

Sialyltransferases Belong to GT Family 80

As described in the previous section, bacterial sialyltransferases are classified into four groups as GT families 38, 42, 52, and 80 in the CAZy database. Many bacterial sialyltransferases in GT family 80 show unique enzymatic characters. The primary structure of sialyltransferases belonging to the GT family 80 and conserved motifs are shown in Figure 6.

α2,6-Sialyltransferase Obtained from P. Damselae

The monofunctional α2,6-sialyltransferase produced by *P. damselae* JT0160 has unique acceptor substrate specificity compared with the equivalent mammalian enzymes. The *P. damselae* α2,6-sialyltransferase recognizes lactose and *N*-acetyllactosaminide as acceptor substrates with almost equal K_m values. By comparison, rat liver α2,6-sialyltransferase has a K_m value approximately 33 times higher for lactose than for *N*-acetyllactosaminide [89]. These results indicate that *P. damselae* α2,6-sialyltransferase does not recognize the 2-acetamido group in the *N*-acetylglucosaminyl residue. Although fucosylated acceptors are not the native substrates of mammalian sialyltransferases, the *P. damselae* α2,6-sialyltransferase transfers Neu5Ac to the galactose residue of carbohydrate chains at position 6 in 2′-fucosyllactose and 3′-sialyllactose [90]. Furthermore, the *P. damselae* α2,6-sialyltransferase also transfers Neu5Ac to both asialo-*N*-linked and asialo-*O*-linked glycoproteins [91]. *P. damselae* α2,6-sialyltransferase recognizes both CMP-Neu5Ac and CMP-KDN as donor substrates [92]. It also recognizes many CMP-sialic acid derivatives with the non-natural modification of an azido or acetylene group at positions C5, C7, C8, and/or C9 [93]. This feature highlights this enzyme as a valuable tool for the *in vitro* enzymatic synthesis of sialosides.

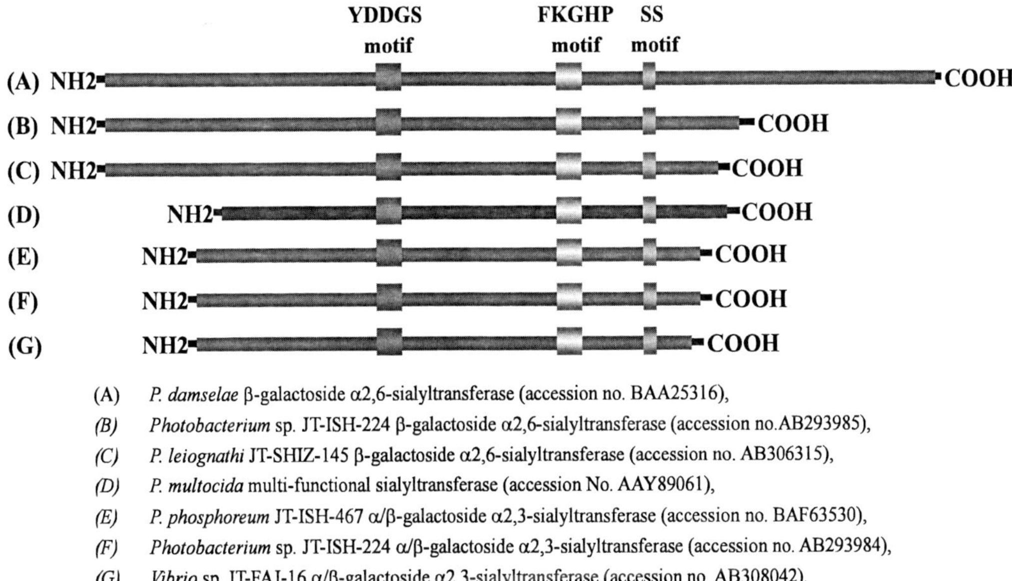

Figure 6. Primary structure of sialyltranaferases belonging to the GT family 80. As described in the text, there are three conserved motifs in these enzymes.

α2,6-Sialyltransferase Obtained from P. Leiognathi

The maximal enzyme activity of most sialyltransferases of mammalian and bacterial origin is achieved under acidic conditions, at around pH 6.0. By comparison, the maximal activity of monofunctional α2,6-sialyltransferase produced by *P. leiognathi* is achieved at pH 8 [74]. CMP-Neu5Ac (the common donor substrate) and the sialylglycoconjugates (the sialyltransferase reaction products) are both more stable under basic conditions than under acidic conditions. The high functionality of *P. leiognathi*-derived α2,6-sialyltransferase under basic conditions provides a unique advantage of this enzyme for the efficient production of sialosides.

An increase in the concentration of sodium chloride also increases the activity of *P. leiognathi* α2,6-sialyltransferase. Approximately 500 mM NaCl in the reaction mixture is required for maximum activity of this enzyme. The conditions in the periplasm are thought to be similar to those of the environment in which the bacterium grows. The optimal conditions for the α2,6-sialyltransferase from *P. leiognathi* are very similar to those of average seawater (pH 8.0, 500 mM NaCl). The enzyme of *P. leiognathi* JT-SHIZ-145, a Gram-negative marine bacterium, has a candidate signal peptide in the NH_2-terminal region of its deduced sequence. Thus, this enzyme is thought to be translocated across the cytoplasmic membrane to the periplasm [74].

α/β-Galactoside α2,3-sialyltransferase from Vibrio sp.

The α/β-galactoside α2,3-sialyltransferase from *Vibrio* sp. has optimal enzyme activity at pH 5.5 and 20 °C. One of the most striking features of this enzyme is its similar specificity for both methyl-α-D-galactopyranoside and methyl-β-D-galactopyranoside [76]. Similar specificity has been reported for α2,3-sialyltransferase from *Neisseria meningitidis* immunotype L1. Since the α2,3-sialyltransferase transfers Neu5Ac to both the α- and β-anomers of sugars, it can be described as an α/β-galactoside α2,3-sialyltransferase.

Among the monosaccharides and disaccharides, the preferred acceptor substrate of α/β-galactoside α2,3-sialyltransferase from *Vibrio* sp. is methyl-α-D-galactopyranoside. There is no significant difference in enzyme activity for lactose, *N*-acetyllactosamine, and methyl-β-D-galactopyranosyl-β-1,3-*N*-acetylglucosaminide. This finding suggests a lack of sensitivity to the second sugar at the non-reducing terminus and to the linkage between the terminal two sugars. Moreover, α/β-galactoside α2,3-sialyltransferase from *Vibrio* sp. shows twice the relative activity for methyl-α/β-D-galactopyranosides than for methyl-α/β-D-glucopyranosides. Together, these findings reveal that the affinity of this enzyme for the acceptor substrate is affected by the orientation of the hydroxyl group at the C4 position of the sugar. It has also been shown that the relative activity of α/β-galactoside α2,3-sialyltransferase from *Vibrio* sp. for methyl-α-D-galactopyranoside is twice that for *N*-acetylgalactosamine and methyl-β-D-mannopyranoside. Subsequently, the enzyme may be affected by the acetoamido group at the C2 position and by the orientation of the hydroxyl group at that position. This enzyme shows a wide range of acceptor substrate specificities and transfers Neu5Ac to both *N*-linked and *O*-linked asialo-glycoprotein [76].

Multi-functional α2,3-sialyltransferase from P. Multocida

This enzyme is a putative sialyltransferase encoded by the *Pm0188* gene from *P. multocida* genomic strain Pm70 and was identified by a BLAST search using the amino acid sequence of a *P. damselae* α2,6-sialyltransferase as a probe [77]. Recombinant *P. multocida* sialyltransferase shows four functions: (1) α2,3-sialyltransferase activity that efficiently transfers a Neu5Ac from CMP-Neu5Ac to galactosides at a wide pH range (pH 6.0–10.0), with optimal activity at pH 7.5–9.0; (2) α2,6-sialyltransferase activity that efficiently transfers a Neu5Ac from CMP-Neu5Ac to galactosides at pH 4.5–7.0; (3) *trans*-sialidase activity that transfers the Neu5Ac residue from α2,3-linked sialyl galactosides to another galactoside, with optimal activity at pH 5.5–6.5; and (4) neuraminidase activity that specifically cleaves α2,3-sialyl linkages, with optimal activity at pH 5.0–5.5 [77].

P. multocida sialyltransferase shows flexible donor substrate specificity when functioning as a sialyltransferase, and accepts CMP-Neu5Ac (the common donor substrate of both mammalian and bacterial sialyltransferases), CMP-KDN, and CMP-Neu5Gc. By comparison, CMP-Neu5Gc and CMP-KDN are poor donor substrates for *N. meningitidis* α2,3-sialyltransferase, and *N. gonorrhoeae* α2,3-sialyltransferase is not able to recognize CMP-KDN as a donor substrate.

CONCLUSION

Known systems for the glycosylation of proteins have historically been restricted to those in eukaryotes. Recent studies, however, have revealed that glycoproteins are expressed also in prokaryotes [94, 95], although the bacterial *N*-linked sugar chains differ structurally from their eukaryotic counterparts [96, 97]. Many bacterial enzymes involved in glycan synthesis have been reported, and in general, they are more stable and productive than their mammalian equivalents. For this reason, the efficient production of glycans and the modification of glycoconjugates through the use of bacterial enzymes have become more widely applicable. Several methods for the large-scale preparation of oligosaccharides with bacterial enzymes have been reported [98, 99]. For example, the large-scale synthesis of important oligosaccharides, including the carbohydrate moieties of gangliosides GM1, GM2, and H-antigen, have been synthesized by using metabolically engineered *E. coli* strains that overexpress the heterologous bacterial glycosyltransferase gene [100–103]. Developments should allow more types of important oligosaccharides to be prepared in large quantities with bacterial enzymes.

As described in the Introduction, sialylated glycoconjugates play a significant role in many biological processes. In particular, the importance of complex-type oligosaccharides for protein function has been demonstrated [104]. Glycoprotein expression systems using yeast or CHO cells have been developed for the generation of glycoproteins for therapeutic use. The main disadvantage of these methods for the preparation of therapeutic glycoproteins is the lack of an intact oligosaccharide moiety. This disadvantage has been overcome with the development of a chemical glycoprotein preparation method with the capacity to generate glycopeptides having intact oligosaccharides [105]. This method has been used recently for the synthesis of glycoproteins with intact human complex-type sialyloligosaccharides [105].

In summary, bacterial enzymes provide a powerful tool for the production of large quantities of glycopeptides with intact oligosaccharides and for the development of therapeutic proteins.

REFERENCES

[1] Angata, T., & Varki, A. (2002) Chemical diversity in the sialic acids and related α-keto acids: an evolutionary perspective. *Chem. Rev. 102*, 439–469.

[2] Vimr, E. R., Kalivoda, K. A., Denzo, E. L. & Steenbergen, S. M. (2004) Diversity of microbial sialic acid metabolism. *Microbiol. Mol. Biol. R., 68*, 132-153.

[3] Schauer, R. (2004) Sialic acid: fascinating sugars in higher animals and man. *Zoology, 107*, 49-64.

[4] Corfield, A.P. & Schauer, R. Occurrence of sialic acids. In: Schauer, R., editor. *Sialic Acids: Chemistry, metabolism and function.* New York: Springer-Verlag Vienna; 1982; 5–50.

[5] Schauer, R. (1982) Chemistry, metabolism, and biological functions of sialic acids in *Advances in Carbohydrate Chemistry and Biochemistry,* Vol. 40, pp. 131-234, Academic Press, New York.

[6] Paulson, J. C. (1989) Glycoproteins: What are the sugar chains for? *Trends. Biochem. Sci., 14*, 272-276.
[7] Hakomori, S. (1990) Bifunctional role of glycosphingolipids. *J. Biol. Chem., 265*, 18713-18716.
[8] Lasky, L.A. (1992) Selectins: Interpreters of cell-specific carbohydrate information during inflammation. *Science, 258*, 964-969.
[9] Feizi, T. (1985) Demonstration by monoclonal antibodies that carbohydrate structures of glycoproteins and glycolipids are onco-developmental antigens. *Nature, 314*, 53-57.
[10] Yuki, N. (2001) Infectious origins of, and molecular mimicry in, Guillain-Barré and Fisher syndromes. *Lancet Infect. Dis., 1*, 29-37.
[11] Yuki, N., Suzuki, K., Koga, M., Nishimoto, Y., Odaka, M., Hirata, K., Taguchi, K., Miyatake, T., Furukawa, K., Kobata, T. & Yamada, M. (2004) Carbohydrate mimicry between human ganglioside GM1 and *Campylobacter jejuni* lipo-oligosaccharide causes Guillain-Barré Syndrome. *Proc. Natl. Acad. Sci. USA, 101,* 11404-11409.
[12] Ropper, A. H. (1992) The Guillain-Barré-syndrome. *N. Engl. J. Med., 326*, 1130-1136.
[13] Kuroki, S., Saida, T., Nukina, M., Haruta, T., Yoshida, M., Kobayashi, Y. & Nakanishi, H. (1993) *Campylobacter jejuni* strains from patients with Guillain-Barré syndrome belong mostly to penner serogroup 19 and contain beta-*N*-acetylglucosamine residues. *Ann. Neurol., 33*, 243-247.
[14] Aspinall, G. O., McDonald, A. G., Raju, T. S., Pang, H., Mills, S. D., Kurjanezyk, L. A. & Penner, J. L. (1992) Serological diversity and chemical structures of *Campylobacter jejuni* low-molecular-weight lipopolysaccharides. *J. Bacteriol., 174*, 1324-1332.
[15] Aspinall, G. O., McDonald, A. G., Pang, H., Kurjanezyk, L. A. & Penner, J. L. (1994) Lipopolysaccharides of *Campylobacter jejuni* serotype O:19: structures of core oligosaccharide regions from the serostrain and two bacterial isolates from patients with the Guillain-Barré syndrome. *Biochemistry, 33*, 241-249.
[16] Aspinall, G. O., McDonald, A. G. & Pang, H. (1994) Lipopolysaccharides of *Campylobacter jejuni* serotype O:19: structures of O antigen chains from the serostrain and two bacterial isolates from patients with the Guillain-Barré syndrome. *Biochemistry, 33*, 250-255.
[17] Gilbert, M., Brisson, J. R., Karwaski, M. F., Michniewicz, J., Cunningham, A. M., Wu, Y., Young, N. M. & Wakarchuk, W. W. (2000) Biosynthesis of ganglioside mimics in *Campylobacter jejuni* OH4384. *J. Biol. Chem., 275*, 3896-3906.
[18] Ho, T. W., Willison, H. J., Nachamkin, I., Li, C. Y., Veitch, J., Ung, H., Wang, G. R., Liu, R. C., Cornblath, D. R., Asbury, A. K., Griffin, J. W. & McKhann, G. M. (1999) AntiGD1a antibody is associated with axonal but not demyelinating forms of Guillain-Barré syndrome. *Ann. Neurol., 45*, 168-173.
[19] Ogawara, K., Kuwabara, S., Mori, M., Hattori, T., Koga, M. & Yuki, N. (2000) Axonal Guillain-Barré syndrome: relation to anti-ganglioside antibodies and *Campylobacter jejuni* infection in Japan. *Ann. Neurol., 48*, 624-631.
[20] Yuki, N. (2005) Carbohydrate mimicry: a new paradigm of autoimmune diseases. *Curr.Opin.Immunol., 17*, 577-582.

[21] Koga, M., Gilbert, M., Li, J., Koike, S., Takahashi, M., Furukawa, K., Hirata, K. & Yuki, N. (2005) Antecedent infections in Fisher syndrome: a common pathogenesis of molecular mimicry. *Neurology, 64,* 1605-1611.

[22] Suzuki, Y. (2005) Sialobiology of influenza: molecular mechanism of host range variation of influenza virus. *Biol. Pharm. Bull., 28,* 399-408.

[23] Weis, W., Brown, J. H., Cusack, S., Paulson, J. C., Skehel, J. J. & Wiley, D. C. (1988) Structure of influenza virus haemagglutinin complexed with its receptor, sialic acid. *Nature, 333,* 426-431.

[24] Connor, R. J., Kawaoka, Y., Webster, R. G. & Paulson, J. C. (1994) Receptor specificity in human, avian, and equine H2 and H3 influenza virus isolates. *Virology, 205,* 17–23.

[25] Matrosovich, M., Tuzikov, A., Bovin, N., Gambaryan, A., Klimov, A., Castrucci, M. R., Donatelli, I. & Kawaoka, Y. (2000) Early alterations of the receptor-binding properties of H1, H2, and H3 avian influenza virus hemagglutinins after their introduction into mammals. *J. Virol., 74,* 8502–8512.

[26] Kanie, O. & Hindsgaul, O. (1992) Synthesis of oligosaccharides, glycolipids and glycopeptides. *Curr. Opin. Struct. Biol., 2,* 674-681.

[27] Ito, Y., Nunomura, S., Shibayama, S. & Ogawa, T. (1992) Studies directed toward the synthesis of polysialogangliosides: The regio- and stereocontrolled synthesis of rationally designed fragments of the tetrasialoganglioside GQlb. *J. Org. Chem., 57,* 1821–1831.

[28] Wang, Z., Zhang, X.-F., Ito, Y., Nakahara, Y. & Ogawa, T. (1996) A new strategy for stereoselective synthesis of sialic acid-containing glycopeptide fragment. *Bioorg. Med. Chem., 4,* 1901–1908.

[29] Izumi, M., and Wong, C. -H. (2001) Microbial sialyltransferases for carbohydrate synthesis. *Trends Glycosci. Glycotechnol., 13,* 345–360.

[30] Koeller, K. M. & Wong, C.–H. (2001) Enzymes for chemical synthesis. *Nature, 409,* 232–240.

[31] Sears, P. & Wong, C.-H. (2001) Toward automated synthesis of oligosaccharides and glycoproteins. *Science, 291,* 2344–2350.

[32] Vann, W. F., Tavarez, J. J., Crowley, J., Vimr, E. & Silver, R. P. (1997) Purification and characterization of *Escherichia coli* K1 *neuB* gene product *N*-acetylneuraminic acid synthetase. *Glycobiology, 7,* 697-701.

[33] Rodriguez-Aparicio, L. B., Luengo, J. M., Gonzalez-Clemente, C. & Reglero, A. (1992) Purification and characterization of nuclear cytidine 5′-monophosphate *N*-acetylneuraminic acid synthetase from rat liver. *J. Biol. Chem., 267,* 9257-9263.

[34] Vann, W. F., Silver, R. P., Abeijon, C., Chang, K., Aaronson, W., Sutton, A., Finn, C. W., Lindner, W. & Kotsatos, M. (1987) Purification, properties, and genetic location of *Escherichia coli* cytidine 5′-monophosphate *N*-acetylneuraminic acid synthetase. *J. Biol. Chem., 262,* 17556-17562.

[35] Munster, A. K., Eckhardt, M., Potvin, B., Muhlenhoff, M., Stanley, P. & Gerardy-Schahn, R. (1998) Mammalian cytidine 5′-monophosphate *N*-acetylneuraminic acid synthetase: A nuclear protein with evolutionarily conserved structural motifs. *Proc. Natl. Acad. Sci. USA, 95,* 9140-9145.

[36] Kean, E. L. (1991) Sialic acid activation. *Glycobiology, 1*, 441-447.

[37] Bravo, I. G., Barralo, S., Ferrero, M. A., Rodriguez-Aparicio, L. B., Martinez-Blanco, H. & Regleo, A. (2001) Kinetic properties of the acylneuraminate cytidyltransferase from *Pasteurella haemolytica* A2. *Biochem. J., 358*, 585-598.

[38] Munster-Kuhnel, A. K., Tiralongo, J., Krapp, S., Weinhold, B., Ritz-Sedlacek, V., Jacob, U. & Gerardy-Schahn, R. (2004) Structure and function of vertebrate CMP-sialic acid synthetases. *Glycobiology, 14*, 43R-51R.

[39] Jelakovic, S. & Schlz, G. E. (2001) The structure of CMP: 2-keto-3-deoxy-manno-octonic acid synthetase and of its complexes with substrates and substrate analogs. *J. Mol. Biol., 312*, 143-155.

[40] Raetz, C. R. H. & Whitfield, C. (2002) Lipopolysaccharide endotoxins. *Annu. Rev. Biochem., 71*, 635-700.

[41] Clementz, T. & Raetz, C. R. H. (1991) A gene coding for 3-deoxy-manno-octonic acid transferase in *Escherichia coli*. *J. Biol. Chem., 266*, 9687-9696.

[42] White, K. A., Kaltashov, I. A., Cotter, R. J. & Raetz, R. H. (1997) A mono functional 3-deoxy-D-manno-octulosonic acid (Kdo) transferase and a Kdo kinase in extracts of *Haemophilus influenzae*. *J. Biol. Chem., 272*, 16555-16563.

[43] Ambrose, M. G., Freese, S. J., Reinhold, M. S., Warner, T. G. & Vann, W. F. (1992) 13C NMR investigation of the anomeric specificity of CMP-*N*-acetylneuraminic acid synthetase from *Escherichia coli*. *Biochemistry, 31*, 775-780.

[44] Mizunar, R. M. & Pohl, N. (2007) Cloning and characterization of a heat-stable CMP-*N*-acetylneuraminic acid synthetase from *Clostridium thermocellum*. *Appl. Microbiol.Biotechnol., 76*, 827-834.

[45] Zapata, G., Vann, W. F., Aaronson, W., Lewis, M. S. & Moos, M. (1989) Sequence of the cloned *Escherichia coli* K1 CMP-N-acetylneuraminic acid synthetase gene. *J. Biol. Chem., 264*, 14769-14774.

[46] Edwards, U. & Frosch, M. (1992) Sequence and functional analysis of the cloned *Neisseria meningitidis* CMP-NeuAc synthetase. *FEMS Microbiol.Lett., 96*, 161-166.

[47] Tullius, M. V., Munson, R. S. Jr., Wang, J. & Gibson, W. (1996) Purification, cloning, and expression of a cytidine 5′-monophosphate *N*-acetylneuraminic acid synthetase from *Haemophilus ducreyi*. *J. Biol. Chem., 271*, 15373-15380.

[48] Yu, H., Ryan, W., Yu, H. & Chen, X. (2006) Characterization of a bifunctional cytidine 5′-monophosphate *N*-acetylneuraminic acid synthetase cloned from *Streptococcus agalactiae*. *Biotechnology letters, 28*, 107-113.

[49] Mosimann, S. C., Gilbert, M., Dombroswki, D., To, R., Wakarchuk, W. W. & Strynadka, N. C. J. (2001) Structure of a sialic acid-activating synthetase, CMP-acylneuraminate synthetase in the presence and absence of CDP. *J. Biol. Chem., 276*, 8190-8196.

[50] Krapp, S., Munster-Kuhnel, A. K., Kaiser, J. T., Huber, R., Tiralongo, J., Gerardy-Schahn, R. & Jacob, U. (2003) The crystal structure of murine CMP-5-*N*-acetylneuraminic acid synthetase. *J. Mol. Biol., 334*, 625-637.

[51] Stoughton, D. M., Zapata, G., Picone, R. & Vann, W. F. (1999) Identification of Arg-12 in the active site of *Escherichia coli* K1 CMP-sialic acid synthetase. *Biochem. J., 343*, 397-402.

[52] Zapata, G., Roller, P. P., Crowley, J. & Vann, W. F. (1993) The role of cysteine residues 129 and 329 in *Escherichia coli* K1 CMP-NeuAc synthetase. *Biochem. J., 295*, 485-491.

[53] Liu, G., Jin, C. & Jin, C. (2004) CMP-*N*-acetylneuraminic acid synthetase from *Escherichia coli* K1 is a bifunctional enzyme. *J. Biol. Chem., 279*, 17738-17749.

[54] Kittelmann, M., Klein, T., Kragl, U., Wandrey, C. & Ghisalba, C. (1995) CMP-*N*-acetyl neuraminic-acid synthetase from *Escherichia coli*; fermentative production and application for the preparative synthesis of CMP-neuraminic acid. *Appl. Microbiol. Biotechnol., 44*, 59-67.

[55] Gilbert, M., Watson, D. C. & Wakarchuk, W. W. (1997) Purification and characterization of the recombinant CMP-sialic acid synthetase from *Neisseria meningitidis*. *Biotechnol.Lett., 19*, 417-420.

[56] Yu, H., Yu, H., Kaepel, R. & Chen, X. (2004) Chemoenzymatic synthesis of CMP-sialic acid derivatives by a one-pot two-enzyme system: comparison of substrate flexibility of three microbial CMP-sialic acid synthetases. *Bioorg. Med. Chem., 12*, 6427-6435.

[57] Karwaski, M. F., Wakarchuk, W. W. & Gilbert, M. (2002) High-level expression of recombinant Neisseria CMP-sialic acid synthetase in *Escherichia coli*. *Protein expression and Purification, 25*, 237-240.

[58] Gilbert, M., Bayer, R., Cunningham, A. M., DeFrees, S., Gao, Y., Watson, D. C., Young, N. M. & Wakarchuk, W. W. (1998) The synthesis of sialylated oligosaccharides using CMP-Neu5Ac synthetase/sialyltransferase fusion. *Nat. Biotechnol., 16*, 769-772.

[59] Shames, S. L., Simon, S. L., Christopher, C. W., Schmid, W., Whitesides, G. M. & Yang, L. L. (1991) CMP-*N*-acetylneuraminic acid synthetase of *Escherichia coli*: high level expression, purification and use in the enzymatic synthesis of CMP-*N*-acetylneuraminic acid and CMP-neuraminic acid derivatives. *Glycobiology, 1*, 187-191.

[60] Yamamoto, T., Takakura, Y. & Tsukamoto, H. (2006) Bacterial sialyltransferase. *Trends. Glycosci. Glycotechnol., 18*, 253–265.

[61] Taniguchi, N., Honke, K. & Fukuda, M. Handbook of Glycosyltransferases and related genes. 1st Edition. Tokyo: Springer-Verlag Tokyo; 2002.

[62] Datta, A. K., Sinha, A. & Paulson, J. C. (1998) Mutation of the sialyltransferase S-sialylmotif alters the kinetics of the donor and acceptor substrates. *J. Biol. Chem., 273*, 9608 9618.

[63] Sasaki, K. (1996) Molecular cloning and characterization of sialyltransferase. *Trends. Glycosci. Glycotechnol., 8*, 195–215.

[64] Jeanneau, C., Chazalet, V., Augé, C., Soumpasis, D. M., Harduin-Lepers, A., Delannoy, P., Imberty, A. & Breton, C. (2004) Structure-function analysis of the human sialyltransferase ST3Gal I: role of n-glycosylation and a novel conserved sialylmotif. *J. Biol. Chem., 279*, 13461-13468.

[65] Datta, A. K. & Paulson, J. C. (1995) The sialyltransferase “Sialylmotif” participates in binding the donor substrate CMP-NeuAc. *J. Biol. Chem., 270*, 1497–1500.

[66] Freiberger, F., Claus, H., Gunzel, A., Oltmann-Norden, I., Vionnet, J., Muhlenhoff, M., Vogel, U., Vann, W.F., Gerardy-Schahn, R. & Stummeyer, K. (2007) Biochemical characterization of a *Neisseria meningitidis* polysialyltransferase reveals novel functional motifs in bacterial sialyltransferases. *Mol. Microbiol., 65*, 1258–1275.

[67] Sujino, K., Jackson, R. J., Chan, N. W. C., Tsuji, S. & Palcic, M. M. (2000) A novel viral a2,3-sialyltransferase (v-ST3Gal I): transfer of sialic acid to fucosylated acceptors. *Glycobiology, 10*, 313-320.

[68] Gilbert, M., Watson, D.C., Cunningham, A. M., Jennings, M. P., Young, N. M., Wakarchuk, W. W. (1996) Cloning of the lipooligosaccharide α-2,3-sialyltransferase from the bacterial pathogens *Neisseria meningitidis* and *Neisseria gonorrhoeae*. *J. Biol. Chem., 271*, 28271–28276.

[69] Edwards, U., Muller, A., Hammerschmidt, S., Gerardy-Schahn, R. & Frosch, M. (1994) Molecular analysis of the biosynthesis pathway of the (α-2,8)polysialic acid capsule by *Neisseria meningitidis* serogroup B. *Mol. Microbiol., 14*, 141–149.

[70] Gilbert, M., Brisson, J. R., Karwaski, M.F., Michniewicz, J., Cunningham, A. M., Wu, Y., Young, N. M. & Wakarchuk, W. W. (2000) Biosynthesis of ganglioside mimics in *Campylobacter jejuni* OH4384: identification of the glycosyltransferase genes, enzymatic synthesis of model compounds, and characterization of nanomole amounts by 600-MHz ^{1}H and ^{13}C NMR analysis. *J. Biol. Chem., 275*, 3896–3906.

[71] Shen, G.J., Datta, A.K., Izumi, M., Koeller, K.M. & Wong, C.-H. (1999) Expression of α2,8/2,9-Polysialyltransferase from *Escherichia coli* K92. characterization of the enzyme and its reaction products. *J. Biol. Chem., 274*, 35139–35146.

[72] Yamamoto, T., Nakashizuka, M. & Terada, I.(1998) Cloning and expression of a marine bacterial beta-galactoside alpha 2,6-sialyltransferase gene from *Photobacterium damsela* JT0160. *J. Biochem. (Tokyo), 123*, 94–100.

[73] Tsukamoto, H., Takakura, Y. & Yamamoto, T. (2007) Purification, cloning and expression of an α-/β-galactoside α2,3-sialyltransferase from a luminous marine bacterium, *Photobacterium phosphoreum*. *J. Biol. Chem., 282*, 29794–29802.

[74] Yamamoto, T., Hamada, Y., Ichikawa, M., Kajiwara, H., Mine, T., Tsukamoto, H. & Takakura, Y. (2007) A β-galactoside α2,6-sialyltransferase produced by a marine bacterium, *Photobacterium leiognathi* JTSHIZ-145, is active at pH 8. *Glycobiology, 17*, 1167–1174.

[75] Okino, N., Kakuta, Y., Kajiwara, H., Ichikawa, M., Takakura, Y., Ito, M. & Yamamoto, T. (2007) Purification, crystallization and preliminary crystallographic characterization of the alpha 2,6-sialyltransferase from Vibrionaceae *Photobacterium* sp. JT-ISH-224. *Acta Crystallogr. Sect. F. Struct. Biol. Cryst. Commun., 1*, 662–664.

[76] Takakura, Y., Tsukamoto, H. & Yamamoto, T. (2007) Molecular cloning, expression and properties of an α/β-galactoside α2,3-sialyltransferase from *Vibrio* sp. JT-FAJ-16. *J. Biochem. (Tokyo), 142*, 403–412.

[77] Yu, H., Chokhawala, H., Karpel, R., Yu, H., Wu, B., Zhang, J., Zhang, Y., Jia, Q. & Chen, X. (2005) A multifunctional *Pasteurella multocida* sialyltransferase: a powerful tool for the synthesis of sialoside libraries. *J. Am. Chem. Soc., 127*, 17618–17619.

[78] Hood, D.W., Cox, A.D., Gilbert, M., Makepeace, K., Walsh, S., Deadman, M. E., Cody, A., Martin, A., Mansson, M., Schweda, E. K., Brisson, J. R., Richards, J.C.,

Moxon, E.R. & Wakarchuk, W. W. (2001) Identification of a lipopolysaccharide α-2,3-sialyltransferase from *Haemophilus influenzae*. *Mol. Microbiol., 39*, 341–350.

[79] Watanabe, M., Miyake, K., Yamamoto, S., Kataoka, Y., Koizumi, S., Endo, T., Ozaki, A. & Iijima, S. (2002) Identification of sialyltransferases of *Streptococcus agalactiae*. *J. Biosci. Bioeng., 93*, 610-613.

[80] Coutinho, P. M., Deleury, E., Davies, G. J. & Henrissat, B. (2003) An evolving hierarchical family classification for glycosyltransferases. *J. Mol. Biol., 328*, 307–317.

[81] Chiu, C. P., Watts, A.G., Lairson, L. L., Gilbert, M., Lim, D., Wakarchuk, W. W., Withers, S. G. & Strynadka, N. C. (2004) Structural analysis of the sialyltransferase CstII from *Campylobacter jejuni* in complex with a substrate analog. *Nat. Struct. Mol. Biol., 11*, 163–170.

[82] Ni, L., Sun, M., Yu, H., Chokhawala, H., Chen, X. & Fisher, A. J. (2006) Cytidine 5′-monophosphate (CMP)-induced structural changes in a multifunctional sialyltransferase from *Pasteurella multocida*. *Biochemistry, 45*, 2139–2148.

[83] Ni, L., Chokhawala, H., Cao, H., Henning, R., Ng, L., Huang, S., Yu, H., Chen, X. & Fisher, A. J. (2007) Crystal structures of *Pasteurella multocida* sialyltransferase complexes with acceptor and donor analogues reveal substrate binding sites and catalytic mechanism. *Biochemistry, 46*, 6288–6298.

[84] Chiu, C. P., Lairson, L. L., Gilbert, M., Wakarchuk, W. W., Withers, S. G. & Strynadka, N. C. (2007) Structural analysis of the alpha-2,3-sialyltransferase Cst-I from *Campylobacter jejuni* in apo and substrate-analogue bound forms. *Biochemistry, 46*, 7196–7204.

[85] Kakuta, Y., Okino, N., Kajiwara, H., Ichikawa, M., Takakura, Y., Ito, M. & Yamamoto, T. (2008) Crystal structure of *Vibrionaceae Photobacterium* sp. JT-ISH-224 α2,6-sialyltransferase in a ternary complex with donor product CMP and acceptor substrate lactose: catalytic mechanism and substrate recognition. *Glycobiology, 18*, 66-73.

[86] Kim, D. U., Yoo, J. H., Lee, Y. J., Kim, K. S. & Cho, H. S. (2007) Structural analysis of sialyltransferase PM0188 from *Pasteurella multocida* complexed with donor analogue and acceptor sugar. *BMB Report, 41*, 48-54.

[87] Lairson, L. L. & Withers, S. G. (2004) Mechanistic analogies amongst carbohydrate modifying enzymes. *Chem. Commun.*, 2243–2248.

[88] Yamamoto, T., Ichikawa, M. & Takakura, Y. (2008) Conserved amino acid sequences in the bacterial sialyltransferases belonging to Glycosyltransferase family 80. *Biochem Biophys Res Commun., 365*, 340-343.

[89] Yamamoto, T., Nakashizuka, M., Kodama, H., Kajihara, Y. & Terada, I. (1996) Purification and characterization of a marine bacterial beta-galactoside alpha 2,6-sialyltransferase from *Photobacterium damsela* JT0160. *J. Biochem.(Tokyo), 120*, 104–110.

[90] Kajihara, Y., Yamamoto, T., Nagae, H., Nakashizuka, M., Sakakibara, T. & Terada, I. (1996) A novel α-2,6-sialyltransferase: Transfer of sialic acid to fucosyl and sialyl trisaccharides. *J. Org. Chem., 61*, 8632–8635.

[91] Yamamoto, T., Nagae, H., Kajihara, Y. & Terada, I. (1998) Mass production of bacterial alpha2,6-sialyltransferase and enzymatic syntheses of sialyloligosaccharides. *Biosci. Biotechnol. Biochem., 62*, 210–214.

[92] Kajihara, Y., Akai, S., Nakagawa, T., Sato, R., Ebata, T., Kodama, H. & Sato, K. (1999) Enzymatic synthesis of Kdn oligosaccharides by a bacterial α-(2→6)-sialyltransferase. *Carbohydr. Res., 315*, 137–141.

[93] Yu, H., Huang, S., Chokhawala, H., Sun, M., Zheng, H. & Chen, X. (2006) Highly efficient chemoenzymatic synthesis of naturally occurring and non-natural alpha-2,6-linked sialosides: a P. damsela alpha-2,6-sialyltransferase with extremely flexible donor-substrate specificity. *Angew Chem Int Ed Engl., 45*, 3938-3944.

[94] Erickson, P. R. & Herzberg, M. C. (1993) Evidence for the covalent linkage of carbohydrate polymers to a glycoprotein from *Streptococcus sanguis. J. Biol. Chem., 268*, 23780–23783.

[95] Young, N. M., Brisson, J. R., Kelly, J., Watson, D. C., Tessier, L., Lanthier, P. H., Jarrell, H. C., Cadotte, N., St Michael, F., Aberg, E. & Szymanski, C. M. (2002) Structure of the *N*-linked glycan present on multiple glycoproteins in the gram-negative bacterium, *Campylobacter jejuni. J. Biol. Chem., 277*, 42530–42539.

[96] Wacker, M., Linton, D., Hitchen, P. G., Nita-Lazar, M., Haslam, S. T., North, S. J., Panico, M., Morris, H. R., Dell, A., Wren, B. W. & Aebi, M. (2002) N-linked glycosylation in *Campylobacter jejuni* and its functional transfer into *E. coli. Science, 298*, 1790–1793.

[97] Power, P. M., Roddam, L. F., Dieckelmann, M., Srikhanta, Y. N., Tan, Y. C., Berrington, A. W. & Jennings, M. P. (2000) Genetic characterization of pilin glycosylation in *Neisseria meningitidis. Microbiology, 146*, 967–979.

[98] Koizumi, S. Endo, T. Tabata, K. Nagano, H. Ohnishi, J. & Ozaki, A. (2000) Large-scale production of GDP-fucose and Lewis X by bacterial coupling. *J. Ind. Microbiol. Biotechnol., 25*, 213-217.

[99] Koizumi, S. Endo, T. Tabata, K. & Ozaki, A. (1998) Large-scale production of UDP-galactose and globotriose by coupling metabolically engineered bacteria. *Nat. Biotechnol., 16*, 847-850.

[100] Dumon, C., Bosso, C., Utille, J. P., Heyraud, A. & Samain, E. (2006) Production of Lewis X tetrasaccharides by metabolically engineered *Escherichia coli. Chembiochemistry, 7*, 59-65.

[101] Antoine, T., Priem, B., Heyraud, A., Greffe, L., Gilbert, M., Wakarchuk, W. W., Lam, J. S. & Samain, E. (2003) Large-scale in vivo synthesis of the carbohydrate moieties of gangliosides GM1 and GM2 by metabolically engineered *Escherichia coli. Chembiochemistry, 4*, 406-412.

[102] Drouillard, S., Driguez, H. & Samain, E. (2006) Large-scale synthesis of H-antigen oligosaccharides by expressing *Helicobacter pylori* alpha1,2-fucosyltransferase in metabolically engineered *Escherichia coli* cells. *Angew. Chem. Int. Ed. Engl., 45*, 1778-1780.

[103] Fierfort, N. & Samain, E. (2008) Genetic engineering of *Escherichia coli* for the economical production of sialylated oligosaccharides. *J. Biotechnol., 134*, 261-265.

[104] Varki, A. (1993) Biological roles of oligosaccharides: all of the theories are correct. *Glycobiology, 3,* 97–130.

[105] Yamamoto, N., Tanabe, Y., Okamoto, R., Dawson, P. E. & Kajihara, Y. (2008) Chemical synthesis of a glycoprotein having an intact human complex-type sialyloligosaccharide under the Boc and Fmoc synthetic strategies. *J. Am. Chem. Soc., 130,* 501-510.

In: Glycobiology Research Trends
Editors: G. Powell and O. McCabe

ISBN: 978-1-60692-841-7

Chapter VI

Heparan Sulfate D-glucosaminyl 3-*O*-sulfotransferases

Tomio Yabe

Department of Applied Life Science, Gifu University, 1-1 Yanagido, Gifu, Gifu 501-1193, Japan

Nobuaki Maeda

Department of Developmental Neuroscience, Tokyo Metropolitan Institute for Neuroscience, 2-6 Musashidai, Fuchu, Tokyo 183-8526, Japan

Abstract

Heparan sulfate (HS) is a polysaccharide belonging to the family of glycosaminoglycans that is ubiquitously expressed on the cell surface and as an extracellular substance. The sulfated domains of HS give specific binding affinities for ligands and regulate the biological functions by binding with various proteins, including growth factors, morphogens, and extracellular matrix molecules. These regulatory interactions of HS are considered to be structure-dependent and are determined by HS sulfotransferases. HS has a disaccharide unit comprising *N*-acetylglucosamine (GlcNAc) and glucuronic acid (GlcA). In the biosynthesis of HS, after the elongation of the skeletal chains involved in the sequential addition of alternating GlcNAc and GlcA residues, further modification by HS sulfotransferases produces the fine structures of HS. One of the HS sulfotransferases is 3-*O*-sulfotransferase (3-OST), which catalyzes the transfer of sulfate groups to the 3-*O*-position of D-glucosamine residues of HS from 3'-phosphoadenosine 5'-phosphosulfate. This is a rare, but essential HS chain modification required to produce the specific binding affinities between active HS and the target proteins. Recent studies on HS biosynthesis have indicated that 3-OST plays a pivotal role in bioactivity.

I. INTRODUCTION

Recent research has increasingly indicated that the heparan sulfate (HS) ubiquitously distributed on the cell surface and in the extracellular matrix is involved in a large number of biological processes, such as blood coagulation, viral infection, tumor metastasis, and various developmental processes [1]. This has stimulated interest in elucidating the detailed structure of HS polysaccharide chains. Biosynthetic processing of HS chains comprising alternating 1–4-linked glucosamine and hexuronic acid residues produces enormous structural diversity from the relatively simple structure through differences in the extent and position of *N*-and *O*-sulfation as well as uronic acid epimerization (for review, see [2]). Structurally different HS variants exhibit different affinities for a variety of proteins such as growth factors, enzymes, and extracellular matrix components; thus they play a specific role in regulating these biological processes [3].

The structural distinctions of functional domains in HS are due to enzymatic modifications in the Golgi apparatus of nascent polymers composed of the repeating disaccharide units of *N*-acetylglucosamine (GlcNAc) and glucuronic acid (GlcA), which are synthesized by HS copolymerase, EXTs. The polysaccharide chains are partly *N*-deacetylated and *N*-sulfated by glucosaminyl *N*-deacetylase/*N*-sulfotransferase. Sometimes, the *N*-deacetylation/*N*-sulfation reaction is partial, giving rise to *N*-unsubstituted glucosamine units. Subsequently, some GlcA units are C5 epimerized to iduronic acid (IdoA) by glucuronyl C5-epimerase, after which *O*-sulfation of GlcA/IdoA by HS 2-*O*-sulfotransferase (2-OST), and 6- and 3-*O*-sulfation of glucosamine by 6-OSTs and 3-OSTs, respectively (for review, see [4]; Figure 1). As most HS sulfotransferases are composed of several isoforms, the functional diversity of HS is increased. This suggests that each isoform of the sulfotransferases is involved in the production of specific functional domains of HS having distinct biological activities.

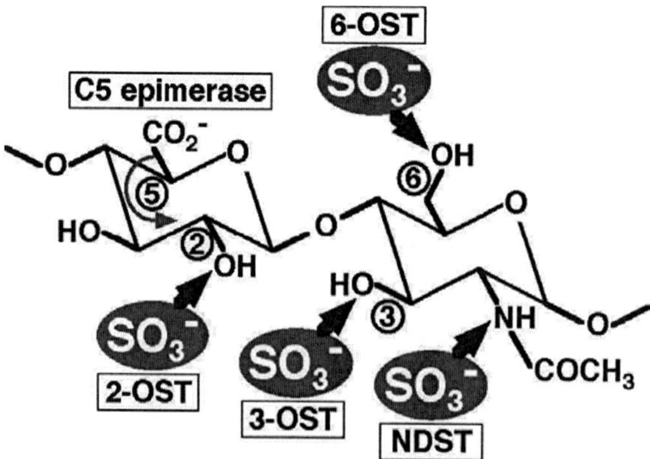

Figure 1. Schematic structure of the HS disaccharide unit. HS is synthesized in the Golgi body through a sequential modification by many enzymes after the polymerization of repeating disaccharide units consisting of GlcNAc (right) and GlcA (left). The nascent polysaccharide chains are partly sulfated by NDST. Then, some GlcA units are C5 epimerized to IdoA by glucuronyl C5-epimerase, and this is

followed by *O*-sulfation at various positions. The *O*-sulfation includes 2-*O*-sulfation of GlcA/IdoA by 2-OST, and 6-and 3-*O*-sulfation of GlcN units by 6-OSTs and 3-OSTs, respectively. Figure reprinted from [67].

II. Characterization of Heparan Sulfate D-glucosaminyl 3-*O*-sulfotransferases

Heparan sulfate (HS) D-glucosaminyl 3-*O*-sulfotransferase (3-OST; EC 2.8.2.23) catalyzes the transfer of sulfate from 3'-phosphoadenosine 5'-phosphosulfate (PAPS) to the 3-*O*-position of D-glucosamine (GlcN) residues in heparin and HS chains [5]. Heparin is usually stored within the secretory granules of mast cells and differs from HS only in the degree of sulfation and fraction of highly flexible iduronic acid. The biosynthesis method of the antithrombin (AT) binding region in the heparin/HS proposed by Kusche *et al.* [6] uses oligosaccharide sulfate acceptors that incorporate the critical GlcN 3-*O*-sulfate group at the conclusion of the biosynthetic process to make the final modification to the saccharide sequence already endowed with all other structural components required for AT binding. Later, the sequence of hexuronic acid units within the 3-*O*-sulfate acceptor structure was found to be essential for 3-*O*-sulfation and generation of a functional AT binding region [7]. Also, the 3-*O*-sulfated $GlcNSO_3$ units that conclude the polymer modification and formation of AT binding regions in the biosynthesis of heparin/HS were carried out by detergent-solubilized mouse mastocytoma microsomal proteins as a microsomal 3-*O*-sulfotransferase [8].

1. Cloning of HS D-glucosaminyl 3-*O*-sulfotransferase Isoform 1

3-OST was originally identified as a purified component from the culture medium of mouse LTA cell clone 33^+ (L-33^+ cells) [9, 10]. Secreted protein from L-33^+ cells in serum-free medium were found to exhibit HS conversion for anticoagulant activity when isolated by subjecting large volumes of conditioned medium of suspension culture to heparin-AF Toyopearl affinity chromatography, Mono Q-FPLC, TSK SW3000-HPLC, and 3', 5'-ADP agarose affinity chromatography. The purified component was identified as HS D-glucosaminyl 3-*O*-sulfotransferase based on its activity to transfer sulfate from [^{35}S]PAPS to the 3-*O*-position of GlcN residues [10]. Then, mouse cDNA encoding 3-OST was isolated by proteolyzing the purified enzyme with Lys-C, sequencing the resultant peptides, and employing degenerate polymerase chain reaction (PCR) primers corresponding to the sequences of the peptides as well as the amino terminus to amplify a fragment of LTA cDNA to use as a probe in order to obtain full-length enzyme cDNAs [11]. In addition, human 3-OST cDNA was isolated by using a probe based on a search using the mouse sequence of the expressed sequence tag data bank [11]. The encoded enzymes were predicted to be intraluminal Golgi residents, presumably interacting via their C-terminal regions with an integral membrane protein. Furthermore, the affinity and kinetics of the interaction between 3-OST and HS were examined using surface plasmon resonance with the result that 3-OST bound with micromolar affinity to HS and this interaction was apparently independent of the presence of the coenzyme, PAPS [12]. Otherwise, 3-OST enzyme is not only identified in mammals, but also in other species, including nematode, fruit fly, and zebrafish [13–15].

2. Other 3-OST Isoforms

Presently, 3-OSTs comprise the largest multigene family of HS biosynthetic enzymes, with a total of seven different 3-OST isoforms, designated as 3-OST-1, -2, -3A, -3B, -4, -5, and -6, having been identified [16–20]. Shworak *et al.* identified expressed sequence tag clones homologous to the sulfotransferase domain of 3-OST-1 and subsequently isolated human cDNAs encoding 3-OST-2, -3A, -3B, and an incomplete clone of 3-OST-4 [16] (later, full-length 3-OST-4 cDNA was reported by Lawrence *et al.* [20]). Expression of full-length enzymes confirmed the 3-*O*-sulfation of specific glucosaminyl residues within heparan sulfate [21]. Also, 3-OST-3 provides sites for the binding of herpes simplex virus type 1 (HSV-1) envelope glycoprotein D (gD) and for the initiation of HSV-1 entry [22]. For structural analysis of the key enzyme generating an entry receptor for HSV-1, Moon *et al.* concluded that residues Gln^{255} and Lys^{368} were essential for the sulfotransferase activity and these residues participated in the substrate recognition of 3-OST-3 based on the obtained crystal structure of 3-OST-3 at 1.95 Å in a ternary complex with 3'-phosphoadenosine 5'-phosphate and a tetrasaccharide substrate [23]. Further biophysical investigation of 3-OST-3 revealed that since 3-OST-3A is able to bind non-substrate glycosaminoglycan ligands with high affinity, discrimination among ligands is triggered by protein oligomerisation [24].

Three years after the identification of 3-OST-2, -3, and -4, an additional 3-OST isoform cDNA, designated as 3-OST-5, was isolated from a human placenta cDNA library and expressed in COS-7 cells based on probing the non-redundant database with the deduced amino acid sequence of human 3-OST-1 [18]. The binding of 3-OST-5-modified HS to both gD and AT suggested that 3-OST-5 generates both an anticoagulant HS and an entry receptor for HSV-1. In addition, Chen and Liu characterized the structure of AT-binding HS generated by 3-OST-5 [25]. Three years after the identification of 3-OST-5, Xu *et al.* reported the isolation and characterization of another 3-OST isoform, designated 3-OST-6 [19]. The cDNA was identified by a homology search using the amino acid sequence of 3-OST-3. Investigation of entry of HSV-1 into cells using Chinese-hamster ovary cell line stably expressing 3-OST-6 revealed that 3-OST-6 cDNA encoded a protein that had 3-OST activity and synthesized a receptor for entry and cell-cell fusion of HSV-1, but not for other subtypes of the alphaherpesvirus family [19].

These 3-OST isoforms share greater than 60% identity in the sulfotransferase domain and exhibit distinct tissue expression patterns [16, 19, 26]. In addition, these isoforms recognize the unique saccharide sequences around the modification site as substrate [16,21]. Tissue distribution of human 3-OST isoforms shows that these members are expressed in a wide variety of human cell types and tissues [16,18,19,21,26]. Moreover, the extensive number of 3-OST genes with diverse expression patterns of multiple transcripts suggests that the 3-OST isoforms regulate important biologic properties of HS.

III. FUNCTION OF 3-*O*-SULFATED HEPARAN SULFATE

3-*O*-Sulfation of D-glucosamine residues in heparan sulfate (HS) is the rarest modification for HS biosynthesis and is a key regulator for generating HS sequences which are expected to define specific protein interaction patterns. Thus, the activities of several effectors associated HS are influenced by selective binding to 3-*O*-sulfated HS motifs. The best characterized interaction between HS and effector protein is that involving the antithrombin (AT)-binding site, which accelerates anticoagulant activity [10]. Additionally, 3-*O*-sulfated HS has been found to bind to the envelope glycoprotein D (gD) of herpes simplex virus type 1 (HSV-1), to fibroblast growth factor 7 (FGF-7), and to a receptor for fibroblast growth factors [10,22,27,28]. Furthermore, some researchers have reported that function is accelerated by 3-*O*-sulfated HS and is evident in the predominant restriction of daytime pineal glands, regulation of Notch signaling, and cyclophilin B binding to responsive cells [14, 29, 30]. Each biosynthesis of discrete 3-*O*-sulfated motifs is controlled by distinct forms of HS 3-*O*-sulfotransferase (3-OST).

1. Binding to Antithrombin

The blood anticoagulant activity of heparin and HS depends on their ability to bind to AT, which is the most important plasma protein inhibiting the proteinases in hemostasis, with high affinity [31, 32]. Heparin differs from HS only in its degree of sulfation and fraction of highly flexible iduronic acid. HS is highly heterogeneous and contains clusters of high sulfation and iduronic acid-rich regions, with interactions with AT [4]. Therefore, only a small fraction of HS chains possess the high affinity site to AT, although some 30% of heparin chains bind AT with high affinity [33]. Both chains contain a specific sequence constituting the AT binding domain, which is characterized to contain a rare 3-*O*-sulfated $GlcNSO_3$ residue.

The occurrence of 3-*O*-sulfated D-glucosamine residues in heparin/HS was, in fact, suggested by Danishefsky *et al.* in 1969 on the basis of methylation analysis [34]. In 1976, three research groups independently found that most of the anticoagulant activity of heparin is attributable to species with high affinity for AT [35–37]. This was followed by the discovery that the rare 3-*O*-sulfated D-glucosamine residue is an essential component of AT binding [5]. Furthermore, the pentasaccharide sequence (Figure 2) of the structure of the active domain, containing 3-*O*-sulfated $GlcNSO_3$, was elucidated [38]. Finally, the total synthesis of pentasaccharide was achieved as was the synthesis of heparin corresponding to the actual minimal sequence required for heparin binding to AT [39].

The search for the binding site precursor of heparin/HS and AT presents considerable difficulties for structural analysis because of the high diversity in structure. However, many researchers have been trying to construct a molecular model of the interaction between AT and a potent heparin/HS analogue. Grootenhuis *et al.* reported initial results of model building and molecular dynamics studies that provide insight into the interaction between pentasaccharide and AT at the molecular level [40]. Furthermore, Kuberan *et al.* reported the biosynthesis of bioactive HS oligosaccharides, including AT-binding pentasaccharide, using an engineered set of cloned enzymes that mimics the Golgi apparatus *in vitro* [41].

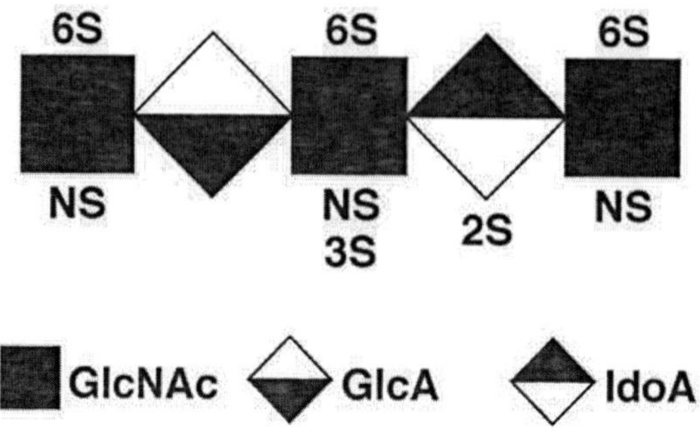

Figure 2. Pentasaccharidic antithrombin-binding sequence. The characters of 6S, NS, 2S, and 3S indicates 6-*O*-, *N*-, 2-*O*-, and 3-*O*-sulfation, respectively.

2. Binding to Envelope Glycoprotein D of HSV-1

HSV-1 is a member of the herpesvirus family and infection in humans is prevalent. A specific 3-*O*-sulfated HS was suggested to be involved in assisting HSV-1 entry [22], and infection was reported to be accomplished in a two-step process, attachment to cells and entry into cells [42]. The 3-*O*-sulfated HS involved in assisting HSV-1 entry is generated by 3-OST-3, but not by 3-OST-1 [22]. As 3-OST-3-modified HS is rarely found in HS from natural sources, HSV-1 was suggested to recognize a unique saccharide structure specifically [43]. Biochemical study revealed that HS heptasulfated octasaccharide (Figure 3) binds to viral gD and assists viral entry [44]. In several reports on the role of 3-*O*-sulfated HS as the receptor for HSV-1 entry, Tiwari *et al.* suggested that 3-*O*-sulfated HS could play a crucial role in virus entry and cell fusion [45]. They identified the novel role of 3-*O*-sulfated HS in mediating infection of corneal fibroblasts [46] and discovered that the association of gD with cell surface-bound receptor is not essential for HSV-1 entry and spread [47]. Based on these advances in knowledge, an attempt to inhibit HSV-1 infection by blocking viral entry with a specific oligosaccharide was also reported [48].

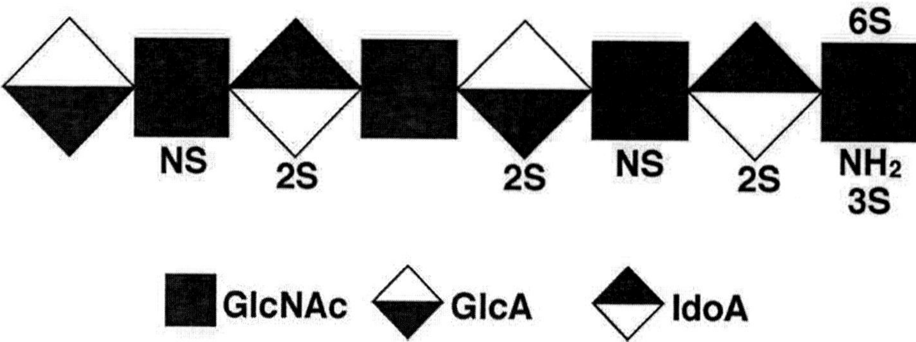

Figure 3. Octasaccharidic HSV-1 gD-binding sequence. The characters of 6S, NS, 2S, and 3S indicates 6-*O*-, *N*-, 2-*O*-, and 3-*O*-sulfation, respectively.

Recently, 3-OST-2, -4, -5, and -6 were also reported to be key enzymes in generating an entry receptor of HSV-1 [18–20, 49]. In addition, the potential role of 3-OST-2 in the spread of HSV-1 infection in the brain [50] and the role of 3-OST-4 in HSV-1 pathogenesis [51] were reported. Based on these characterizations of 3-OST enzymes, Lawrence *et al.* suggested two major sequence specificities within the large 3-OST multigene family [20]. First, 3-OST-2, -3A, -3B, -4, and -6 preferentially generate gD-binding HS. These isoforms show highly homologous sulfotransferase domains that are consistent with the domain defining enzymatic sequence specificity [52]. Second, the most structurally distinct enzyme, 3-OST-1, preferentially generates AT-binding HS [11]. In other case, 3-OST-5 exhibits both specificities [18] and must be considered a member of both classes.

3. Other Functions

Leder found that human urine contains a novel 3-*O*-sulfatase, which is specific for the 3-*O*-sulfate ester of sulfaminoglucopyranoside 3-sulfate. The specificity of this enzyme suggests that a 3-*O*-sulfated D-glucosamine moiety might play a role in the physiological activity of heparin or HS [53]. In addition, Edge *et al.* reported that fragmentation of heparan sulfate chains from bovine glomerular basement membrane by hydrazine/nitrous acid treatment followed by NaB^3H_4-reduction yielded novel disaccharide units containing 3-*O*-sulfated D-glucosamine [54].

The fibroblast growth factor (FGF) signal transduction system is ubiquitous, and the secreted signaling molecules, FGF, play crucial roles in developmental process by intracellular signaling events (for review, see [55]). Heparin or HS interacts independently with both activating FGFs and the ectodomain of their receptors (FGFRs) and forming FGF/HS/FGFR signaling complexes [4, 56–58]. FGF-2 associates more specifically with a pentasaccharide containing $GlcNSO_3$ and a single iduronic acid-2-*O*-sulfate [59, 60]. The 6-*O*-sulfate of $GlcNSO_3$ residues may contribute to the interaction with FGFs [61, 62]. In contrast to FGF-1 and FGF-2, protection of FGF-7 was enhanced by heparin oligosaccharides of increased length and a 3-*O*-sulfate was the most effective [27, 28].

Furthermore, Kuberan *et al.* reported that light induces changes of HS fine structure in pineal gland and that the occurrence of the rare 3-*O*-sulfation catalyzed by 3-OST-2 is predominantly restricted to daytime [29], demonstrating for the first time the involvement of an HS fine structure in circadian rhythm. Also, Kamimura *et al.* reported that *Drosophila melanogaster* HS 3-*O*-sulfotransferase-b (Hs3st-B), which is similar to 3-OST-3, plays a role in Notch signaling by affecting stability or intracellular trafficking of Notch protein [14]. Interestingly, they showed that suppression of Hs3st-B function using transgenic RNAi caused various morphological defects by compromising N signaling. This finding demonstrated, for the first time, that 3-*O*-sulfated HS controls a particular developmental pathway required for morphogenesis. Most recently, Vanpuille *et al.* used these results to strongly support the hypothesis that 3-*O*-sulfation of *N*-unsubstituted glucosamine residues by 3-OST-3 could be a key modification providing specialized HS structures for cyclophilin B, which was initially identified as a cyclosporin A-binding protein that binds to HS on the responsive cells [30]. A role for secreted cyclophilins in the development and/or progression of physiological inflammatory responses and pathological disorders has been suggested by many reports [63–66]. Therefore, the interaction of cyclophilins with cell surface HS could be a key step for inducing cellular signaling events.

IV. 3-*O*-Sulfotransferases in Brain Development

1. Distribution of *3-OST* mRNAs Detected by *In situ* Hybridization in the Developing Brain

To better understand the functions of heparan sulfate (HS) sulfotransferases in the development of the nervous system, we have recently examined the expression of several enzymes by *in situ* hybridization and real-time reverse transcription-polymerase chain reaction (RT-PCR) in embryo, juvenile, and adult mouse brain [67]. We conclude that the structure of HS is altered spatiotemporally in the regulation of various biological activities in the developing brain including the proliferation of neuronal progenitors, extension of axons, and formation of dendrites [67]. In this section, we discuss the possible functional roles of HS 3-*O*-sulfotransferases (3-OSTs) among the investigated enzymes in the signaling of several HS-binding proteins related to the development of the nervous system.

The expression of four genes of 3-OST isoforms (*3-OST-1*, *-2*, *-3B*, and *-4*) were investigated, along with another seven genes including *6-OST* and *NDST* isoforms. The signals of *3-OST-3*B mRNA were found to be not significant at any developmental stage in the brain, consistent with finding that the expression of this gene is rarely detected in the brain by Northern blot analysis [16].

Our *in situ* hybridizatin experiments indicated that multiple isoforms of each 3-OST were typically expressed simultaneously in the same regions. *3-OST-1*, *3-OST-2*, and *3-OST*-4 mRNAs were expressed in the cortical plate (CP) and ventricular zone (VZ) of E16 cerebral cortex (Figure 4) [67]. *3-OST-1*, *3-OST-2*, and *3-OST*-4 transcripts were expressed in the external granular layer (EGL) and Purkinje cell layer (PCL) of P1 cerebellum (Figure 5) [67]. This might suggest that brain cells synthesize complex HS chains with many different functional domains generated by multiple HS sulfotransferase isoforms including 3-OSTs. However, it should be noted that neurons show highly polarized structure with functionally distinct regions, such as dendrites, axons, cell bodies, spines, and axon terminals. The development of these distinct regions is considered to be regulated by different signaling pathways, which might require structurally different HS chains. Each of the multiple 3-OST isoforms expressed in one neuron might differentially contribute to the biosynthesis of HS chains required for the development of specific regions of that cell. Future fine structural analyses are necessary to evaluate whether distinct regions in one neuron express structurally different HS chains.

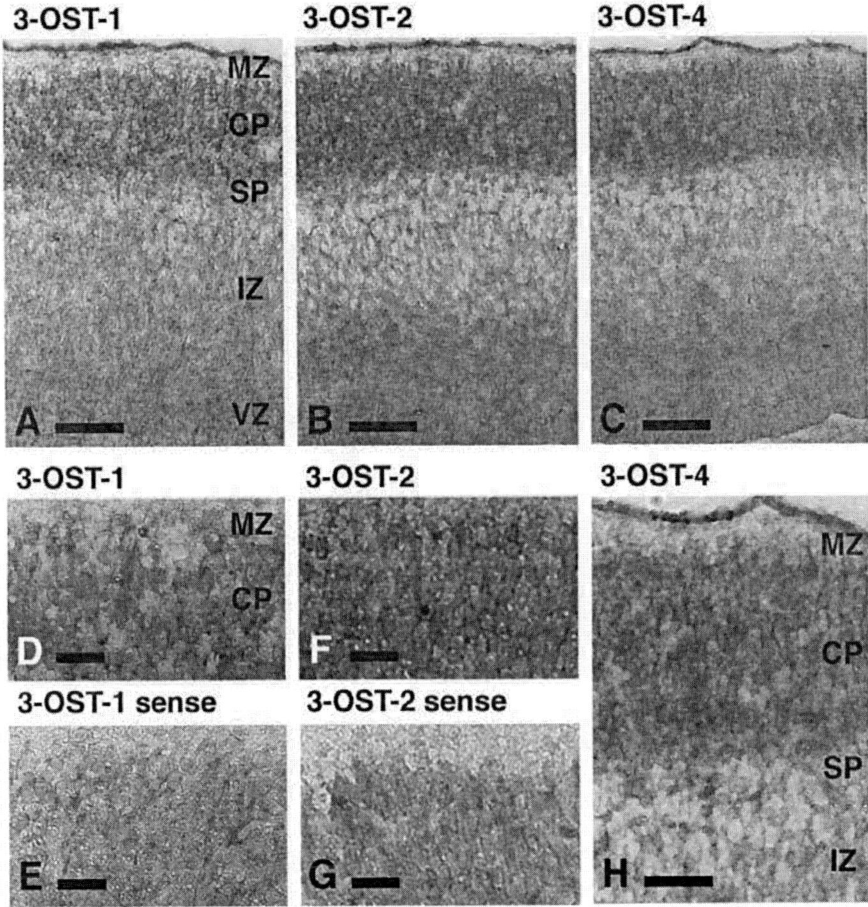

Figure 4. *In situ* hybridization analysis of the expression of *3-OST*-1 (A, D, and E), *3-OST*-2 (B, F, and G), and *3-OST*-4 (C and H) mRNAs in the E16 cerebral cortex. (E and G) Hybridization with a sense probe showed no signals. (H) Higher magnification of C is represented. CP, cortical plate; IZ, intermediate zone; MZ, marginal zone; SP, subplate; VZ, ventricular zone. Scale bars show 100 μm (A–C), 50 μm (H), and 25 μm (D–G). Figure modified from [67].

Examination of the Purkinje cell layer showed that expression of 3-OST genes at P1 was downregulated at P7. Subsequently, expression of 3-OSTs in the Purkinje cells increased [67]. At P7, weak expression of only *3-OST*-2 was observed on Purkinje cells, but at P14, the expression of *3-OST*-1 and *3-OST*-4 was also detected on this type of cells in addition to that of *3-OST-2*. In the adult brain, Purkinje cells were found to strongly express these three 3-OSTs. The significance of such expression is unclear, but mature Purkinje cells might require a more complex HS structure for physiological function. Salinas *et al.* reported that Wnt-3 expression in Purkinje cells increases postnatally when parallel fibers of granule cells make contacts with Purkinje cells. They suggested that Wnt-3 plays a role in the growth and maturation of Purkinje cell dendrites [68]. Because HS plays crucial roles in Wnt signaling [69], HS modified by these 3-OSTs in Purkinje cells might control the Wnt signaling.

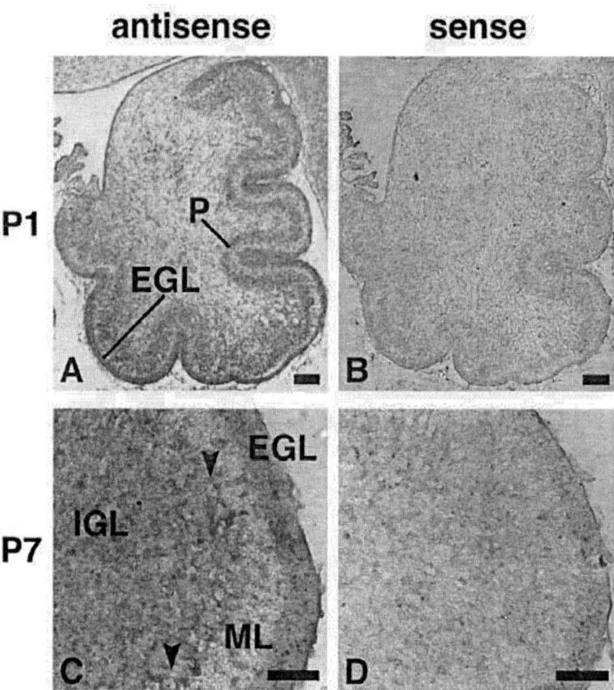

Figure 5. *In situ* hybridization analysis of the expression of *3-OST*-2 mRNAs in P1 (A and B) and P7 (C and D) cerebellum. (A and B) Signals were observed in the external granular layer (EGL) and Purkinje cell layer (P). (C and D) Significant expression of *3-OST*-2 was observed in the EGL and internal granular layer (IGL). In addition, the transcript of *3-OST*-2 was detected in some of the cells in the molecular layer (ML) and a few Purkinje cells (arrowheads). (B and D) Hybridization with a sense probe showed no signals. Scale bars show 100 μm (A and B) and 50 μm (C and D). Figure reprinted from [67].

2. 3-OSTs and Development of Brain

Strong signals of *3-OST-1*, *3-OST-2*, and *3-OST*-4 expression were also observed in CP neurons at E16, when these neurons extend axons, making major tracts and later differentiating into deep layer neurons [67]. Nes-*EXT1*-null mice lacked a corpus callosum, hippocampal commissure, and anterior commissure [70], and these defects resemble those of netrin-1-deficient mice [71], suggesting that HS is required for netrin-1 signaling. Netrins are a family of heparin-binding secreted proteins and thus appear to be HS-binding proteins. Nes-*EXT1*-null mice also display defects in the guidance of retinal axons at the optic chiasm, a phenotype similar to that of *Slit1*/*Slit*2 double-null mice, and genetic interaction between *EXT*1 and *Slit*2 was demonstrated [70,72]. Slit2 protein was shown to bind to HS; this is necessary for the repulsive activities of this protein [73, 74]. Slit proteins were also involved in the development of several major forebrain tracts, such as the corticofugal, callosal, and thalamocortical tracts [75]. *3OST-1*, *3-OST-2*, and *3-OST-4*, which are expressed in the early CP neurons [67], might be involved in axonal guidance in collaboration with netrin and slit proteins. Further study is necessary to investigate whether 3-*O*-sulfated HSs are present on the axons and to analyze the structural requirement of HS for the signaling of netrin and slit proteins.

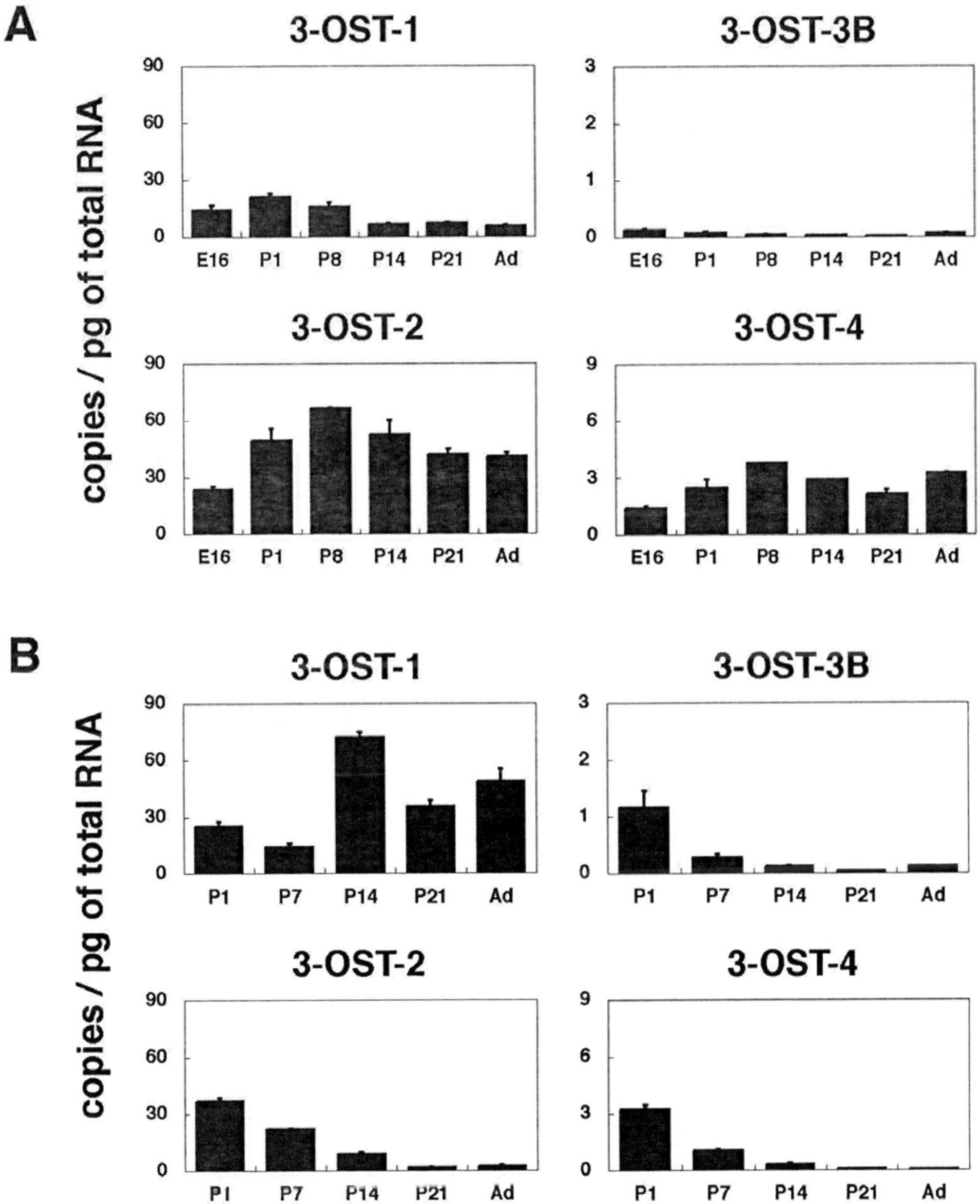

Figure 6. Quantitative analysis of 3-OST transcripts in the cerebrum (A) and cerebellum (B) by RT-PCR. The expression level of each 3-OST gene was normalized to that of the glyceraldehyde-3-phosphate dehydrogenase (GAPDH) transcript. Data were obtained from triplicate experiments and were given as the mean ± standard deviation (SD). Figure modified from [67].

In the postnatal period, high levels of *3-OST*-1 and *3-OST*-2 mRNAs were observed in the brain (Figure 6). Notably, the expression level of *3-OST*-2 peaked at P8, suggesting the importance of this enzyme in the late development of cortical neurons (Figure 6A). At P7, relatively strong signals for this gene were observed on the neurons in the superficial layers, but at P14, the signal became rather uniform among almost all neurons [67]. This postnatal stage is a period of extensive dendritic growth and branching, terminal branching of afferent axons, and synaptogenesis. Recently, Whitford *et al.* indicated that slit proteins are involved in the dendritic patterning of cortical neurons [76]. In addition to axonal guidance, these 3-OSTs might play a role in dendritic growth by controlling slit activity.

Furthermore, cortical neurons show highly diversified morphologies with variable extension patterns of dendrites and axons. As previously discussed, this morphogenesis is under the control of several heparin-binding proteins such as slit and netrin. Variable HS structures generated by different combinations of HS sulfotransferases, including 3-OSTs, might diversify the behavior of neurites in response to slit and netrin proteins, leading to the diversified patterns of neurite extension.

3. Future Perspectives

In this section, we have demonstrated that the expression of 3-OST mRNAs is regionally and temporally regulated in the developing brain [67]. The observed complex patterns of cellular and regional expression of 3-OST genes are consistent with the current paradigm that different sulfation patterns of HS are associated with distinct functions. This section suggests that the structure of HS changes dynamically during the development of the brain through transcriptional control of the 3-OST genes regulating the activities of many HS-binding proteins. However, it should be noted that posttranscriptional regulation also plays a critical role in the determination of the protein levels of 3-OSTs. Thus, immunohistochemical studies using antibodies against each 3-OST and specific HS structural domains are the next important project. Furthermore, our results provide a valuable basis for the design of functional experiments, such as the knockdown of specific enzymes.

V. Conclusion

To gain insights into the function and biological activities of 3-OSTs, further research is necessary to investigate the regulation of HS biosynthesis through the expression of HS sulfotransferases involving 3-OSTs. In addition, identifying molecules that are the targets for specific HS structures constructed by 3-OST enzymes would be critical for understanding the role of 3-OSTs.

References

[1] Bernfield, M., Gotte, M., Park, P. W., Reizes, O., Fitzgerald, M. L., Lincecum, J. & Zako, M. (1999). Functions of cell surface heparan sulfate proteoglycans. *Annu. Rev. Biochem.*, *68*, 729–777.

[2] Bishop, J. R., Schuksz, M. & Esko, J. D. (2007). Heparan sulphate proteoglycans fine-tune mammalian physiology. *Nature*, *446*, 1030–1037.

[3] Kinnunen, T., Huang, Z., Townsend, J., Gatdula, M. M., Brown, J. R., Esko, J. D. & Turnbull, J. E. (2005). Heparan 2-O-sulfotransferase, hst-2, is essential for normal cell migration in *Caenorhabditis elegans*. *Proc. Natl. Acad. Sci. U. S. A.*, *102*, 1507–1512.

[4] Esko, J. D. & Selleck, S. B. (2002). Order out of chaos: assembly of ligand binding sites in heparan sulfate. *Annu. Rev. Biochem.*, *71*, 435–471.

[5] Lindahl, U., Backstrom, G., Thunberg, L. & Leder, I. G. (1980). Evidence for a 3-O-sulfated D-glucosamine residue in the antithrombin-binding sequence of heparin. *Proc. Natl. Acad. Sci. U. S. A.*, *77*, 6551–6555.

[6] Kusche, M., Backstrom, G., Riesenfeld, J., Petitou, M., Choay, J. & Lindahl, U. (1988). Biosynthesis of heparin. O-sulfation of the antithrombin-binding region. *J. Biol. Chem., 263, 15474–15484.*

[7] Kusche, M., Oscarsson, L. G., Reynertson, R., Roden, L. & Lindahl, U. (1991). Biosynthesis of heparin. Enzymatic sulfation of pentasaccharides. *J. Biol. Chem.*, *266*, 7400–7409.

[8] Razi, N. & Lindahl, U. (1995). Biosynthesis of heparin/heparan sulfate. The D-glucosaminyl 3-O-sulfotransferase reaction: target and inhibitor saccharides. *J. Biol. Chem.*, *270*, 11267–11275.

[9] Shworak, N. W., Fritze, L. M., Liu, J., Butler, L. D. & Rosenberg, R. D. (1996). Cell-free synthesis of anticoagulant heparan sulfate reveals a limiting converting activity that modifies an excess precursor pool. *J. Biol. Chem.*, *271*, 27063–27071.

[10] Liu, J., Shworak, N. W., Fritze, L. M., Edelberg, J. M. & Rosenberg, R. D. (1996). Purification of heparan sulfate D-glucosaminyl 3-O-sulfotransferase. *J. Biol. Chem.*, *271*, 27072–27082.

[11] Shworak, N. W., Liu, J., Fritze, L. M., Schwartz, J. J., Zhang, L., Logeart, D. & Rosenberg, R. D. (1997). Molecular cloning and expression of mouse and human cDNAs encoding heparan sulfate D-glucosaminyl 3-O-sulfotransferase. *J. Biol. Chem.*, *272*, 28008–28019.

[12] Munoz, E., Xu, D., Kemp, M., Zhang, F., Liu, J. & Linhardt, R. J. (2006). Affinity, kinetic, and structural study of the interaction of 3-O-sulfotransferase isoform 1 with heparan sulfate. *Biochemistry*, *45*, 5122–5128.

[13] Cadwallader, A. B. & Yost, H. J. (2006). Combinatorial expression patterns of heparan sulfate sulfotransferases in zebrafish: I. the 3-O-sulfotransferase family. *Dev. Dyn.*, *235*, 3423–3431.

[14] Kamimura, K., Rhodes, J. M., Ueda, R., McNeely, M., Shukla, D., Kimata, K., Spear, P. G., Shworak, N. W. & Nakato, H. (2004). Regulation of Notch signaling by Drosophila heparan sulfate 3-O sulfotransferase. *J. Cell Biol.*, *166*, 1069–1079.

[15] Turnbull, J., Drummond, K., Huang, Z., Kinnunen, T., Ford-Perriss, M., Murphy, M. & Guimond, S. (2003). Heparan sulphate sulphotransferase expression in mice and *Caenorhabditis elegans*. *Biochem. Soc. Trans.*, *31*, 343–348.

[16] Shworak, N. W., Liu, J., Petros, L. M., Zhang, L., Kobayashi, M., Copeland, N. G., Jenkins, N. A. & Rosenberg, R. D. (1999). Multiple isoforms of heparan sulfate D-glucosaminyl 3-O-sulfotransferase. Isolation, characterization, and expression of

human cDNAs and identification of distinct genomic loci. *J. Biol. Chem.*, *274*, 5170–5184.

[17] Daniels, R. J., Peden, J. F., Lloyd, C., Horsley, S. W., Clark, K., Tufarelli, C., Kearney, L., Buckle, V. J., Doggett, N. A., Flint, J. & Higgs, D. R. (2001). Sequence, structure and pathology of the fully annotated terminal 2 Mb of the short arm of human chromosome 16. *Hum. Mol. Genet.*, *10*, 339–352.

[18] Xia, G., Chen, J., Tiwari, V., Ju, W., Li, J.-P., Malmstrom, A., Shukla, D. & Liu, J. (2002). Heparan sulfate 3-O-sulfotransferase isoform 5 generates both an antithrombin-binding site and an entry receptor for herpes simplex virus, type 1. *J. Biol. Chem.*, *277*, 37912–37919.

[19] Xu, D., Tiwari, V., Xia, G., Clement, C., Shukla, D. & Liu, J. (2005). Characterization of heparan sulphate 3-O-sulphotransferase isoform 6 and its role in assisting the entry of herpes simplex virus type 1. *Biochem. J.*, *385*, 451–459.

[20] Lawrence, R., Yabe, T., Hajmohammadi, S., Rhodes, J., McNeely, M., Liu, J., Lamperti, E. D., Toselli, P. A., Lech, M., Spear, P. G., Rosenberg, R. D. & Shworak, N. W. (2007). The principal neuronal gD-type 3-O-sulfotransferases and their products in central and peripheral nervous system tissues. *Matrix Biol.*, *26*, 442–455.

[21] Liu, J., Shworak, N. W., Sinay, P., Schwartz, J. J., Zhang, L., Fritze, L. M. & Rosenberg, R. D. (1999). Expression of heparan sulfate D-glucosaminyl 3-O-sulfotransferase isoforms reveals novel substrate specificities. *J. Biol. Chem.*, *274*, 5185–5192.

[22] Shukla, D., Liu, J., Blaiklock, P., Shworak, N. W., Bai, X., Esko, J. D., Cohen, G. H., Eisenberg, R. J., Rosenberg, R. D. & Spear, P. G. (1999). A novel role for 3-O-sulfated heparan sulfate in herpes simplex virus 1 entry. *Cell*, *99*, 13–22.

[23] Moon, A. F., Edavettal, S. C., Krahn, J. M., Munoz, E. M., Negishi, M., Linhardt, R. J., Liu, J. & Pedersen, L. C. (2004). Structural analysis of the sulfotransferase (3-O-sulfotransferase isoform 3) involved in the biosynthesis of an entry receptor for herpes simplex virus 1. *J. Biol. Chem.*, *279*, 45185–45193.

[24] Wille, I., Rek, A., Krenn, E. & Kungl, A. J. (2007). Biophysical investigation of human heparan sulfate D-glucosaminyl 3-O-sulfotransferase-3A: a mutual effect of enzyme oligomerisation and glycosaminoglycan ligand binding. *Biochim. Biophys. Acta.*, *1774*, 1470–1476.

[25] Chen, J. & Liu, J. (2005). Characterization of the structure of antithrombin-binding heparan sulfate generated by heparan sulfate 3-O-sulfotransferase 5. *Biochim. Biophys. Acta.*, *1725*, 190–200.

[26] Mochizuki, H., Yoshida, K., Gotoh, M., Sugioka, S., Kikuchi, N., Kwon, Y.-D., Tawada, A., Maeyama, K., Inaba, N., Hiruma, T., Kimata, K. & Narimatsu, H. (2003). Characterization of a heparan sulfate 3-O-sulfotransferase-5, an enzyme synthesizing a tetrasulfated disaccharide. *J. Biol. Chem.*, *278*, 26780–26787.

[27] McKeehan, W. L., Wu, X. & Kan, M. (1999). Requirement for anticoagulant heparan sulfate in the fibroblast growth factor receptor complex. *J. Biol. Chem.*, *274*, 21511–21514.

[28] Ye, S., Luo, Y., Lu, W., Jones, R. B., Linhardt, R. J., Capila, I., Toida, T., Kan, M., Pelletier, H. & McKeehan, W. L. (2001). Structural basis for interaction of FGF-1,

FGF-2, and FGF-7 with different heparan sulfate motifs. *Biochemistry*, *40*, 14429–14439.

[29] Kuberan, B., Lech, M., Borjigin, J. & Rosenberg, R. D. (2004). Light-induced 3-O-sulfotransferase expression alters pineal heparan sulfate fine structure. A surprising link to circadian rhythm. *J. Biol. Chem.*, *279*, 5053–5054.

[30] Vanpouille, C., Deligny, A., Delehedde, M., Denys, A., Melchior, A., Lienard, X., Lyon, M., Mazurier, J., Fernig, D. G. & Allain, F. (2007). The heparin/heparan sulfate sequence that interacts with cyclophilin B contains a 3-O-sulfated N-unsubstituted glucosamine residue. *J. Biol. Chem.*, *282*, 24416–24429.

[31] Bourin, M. C. & Lindahl, U. (1993). Glycosaminoglycans and the regulation of blood coagulation. *Biochem. J.*, *289*, 313–330.

[32] Bjork, I. & Olson, S. T. (1997). Antithrombin. A bloody important serpin. *Adv. Exp. Med. Biol.*, *425*, 17–33.

[33] Casu, B. & Lindahl, U. (2001). Structure and biological interactions of heparin and heparan sulfate. *Adv. Carbohydr. Chem. Biochem.*, *57*, 159–206.

[34] Danishefsky, I., Steiner, H., Bella, A. J. & Friedlander, A. (1969). Investigations on the chemistry of heparin. VI. Position of the sulfate ester groups. *J. Biol. Chem.*, *244*, 1741–1745.

[35] Andersson, L. O., Barrowcliffe, T. W., Holmer, E., Johnson, E. A. & Sims, G. E. (1976). Anticoagulant properties of heparin fractionated by affinity chromatography on matrix-bound antithrombin III and by gel filtration. *Thromb. Res.*, *9*, 575–583.

[36] Hook, M., Bjork, I., Hopwood, J. & Lindahl, U. (1976). Anticoagulant activity of heparin: separation of high-activity and low-activity heparin species by affinity chromatography on immobilized antithrombin. *FEBS Lett.*, *66*, 90–93.

[37] Lam, L. H., Silbert, J. E. & Rosenberg, R. D. (1976). The separation of active and inactive forms of heparin. *Biochem. Biophys. Res. Commun.*, *69*, 570–577.

[38] Thunberg, L., Backstrom, G. & Lindahl, U. (1982). Further characterization of the antithrombin-binding sequence in heparin. *Carbohydr. Res.*, *100*, 393–410.

[39] Choay, J., Petitou, M., Lormeau, J. C., Sinay, P., Casu, B. & Gatti, G. (1983). Structure-activity relationship in heparin: a synthetic pentasaccharide with high affinity for antithrombin III and eliciting high anti-factor Xa activity. *Biochem. Biophys. Res. Commun.*, *116*, 492–499.

[40] Grootenhuis, P. D. J. & Boeckel, C. A. A.v (1991). Constructing a molecular model of the interaction between antithrombin III and a potent heparin analogue. *J. Am. Chem. Soc.*, *113*, 2743–2747.

[41] Kuberan, B., Lech, M. Z., Beeler, D. L., Wu, Z. L. & Rosenberg, R. D. (2003). Enzymatic synthesis of antithrombin III-binding heparan sulfate pentasaccharide. *Nat. Biotechnol.*, *21*, 1343–1346.

[42] Spear, P. G., Eisenberg, R. J. & Cohen, G. H. (2000). Three classes of cell surface receptors for alphaherpesvirus entry. *Virology*, *275*, 1–8.

[43] Liu, J., Shriver, Z., Blaiklock, P., Yoshida, K., Sasisekharan, R. & Rosenberg, R. D. (1999). Heparan sulfate D-glucosaminyl 3-O-sulfotransferase-3A sulfates N-unsubstituted glucosamine residues. *J. Biol. Chem.*, *274*, 38155–38162.

[44] Liu, J., Shriver, Z., Pope, R. M., Thorp, S. C., Duncan, M. B., Copeland, R. J., Raska, C. S., Yoshida, K., Eisenberg, R. J., Cohen, G., Linhardt, R. J. & Sasisekharan, R. (2002). Characterization of a heparan sulfate octasaccharide that binds to herpes simplex virus type 1 glycoprotein D. *J. Biol. Chem.*, *277*, 33456– 33467.

[45] Tiwari, V., Clement, C., Duncan, M. B., Chen, J., Liu, J. & Shukla, D. (2004). A role for 3-O-sulfated heparan sulfate in cell fusion induced by herpes simplex virus type 1. *J. Gen. Virol.*, *85*, 805–809.

[46] Tiwari, V., Clement, C., Xu, D., Valyi-Nagy, T., Yue, B. Y. J. T., Liu, J. & Shukla, D. (2006). Role for 3-O-sulfated heparan sulfate as the receptor for herpes simplex virus type 1 entry into primary human corneal fibroblasts. *J. Virol.*, *80*, 8970–8980.

[47] Tiwari, V., O'donnell, C., Copeland, R. J., Scarlett, T., Liu, J. & Shukla, D. (2007). Soluble 3-O-sulfated heparan sulfate can trigger herpes simplex virus type 1 entry into resistant Chinese hamster ovary (CHO-K1) cells. *J. Gen. Virol.*, *88*, 1075– 1079.

[48] Copeland, R., Balasubramaniam, A., Tiwari, V., Zhang, F., Bridges, A., Linhardt, R. J., Shukla, D. & Liu, J. (2008). Using a 3-O-sulfated heparin octasaccharide to inhibit the entry of herpes simplex virus type 1. *Biochemistry*, *47*, 5774– 5783.

[49] Shukla, D. & Spear, P. G. (2001). Herpesviruses and heparan sulfate: an intimate relationship in aid of viral entry. *J. Clin. Invest.*, *108*, 503–510.

[50] O'Donnell, C. D., Tiwari, V., Oh, M.-J. & Shukla, D. (2006). A role for heparan sulfate 3-O-sulfotransferase isoform 2 in herpes simplex virus type 1 entry and spread. *Virology*, *346*, 452–459.

[51] Tiwari, V., O'Donnell, C. D., Oh, M.-J., Valyi-Nagy, T. & Shukla, D. (2005). A role for 3-O-sulfotransferase isoform-4 in assisting HSV-1 entry and spread. *Biochem. Biophys. Res. Commun.*, *338*, 930–937.

[52] Yabe, T., Shukla, D., Spear, P. G., Rosenberg, R. D., Seeberger, P. H. & Shworak, N. W. (2001). Portable sulphotransferase domain determines sequence specificity of heparan sulphate 3-O-sulphotransferases. *Biochem. J.*, *359*, 235– 241.

[53] Leder, I. G. (1980). A novel 3-O sulfatase from human urine acting on methyl-2-deoxy-2-sulfamino-alphs-D-glucopyranoside 3-sulfate. *Biochem. Biophys. Res. Commun.*, *94*, 1183–1189.

[54] Edge, A. S. & Spiro, R. G. (1990). Characterization of novel sequences containing 3-O-sulfated glucosamine in glomerular basement membrane heparan sulfate and localization of sulfated disaccharides to a peripheral domain. *J. Biol. Chem.*, *265*, 15874–15881.

[55] Itoh, N. (2007). The Fgf families in humans, mice, and zebrafish: their evolutional processes and roles in development, metabolism, and disease. *Biol. Pharm. Bull.*, *30*, 1819–1825.

[56] Yayon, A., Klagsbrun, M., Esko, J. D., Leder, P. & Ornitz, D. M. (1991). Cell surface, heparin-like molecules are required for binding of basic fibroblast growth factor to its high affinity receptor. *Cell*, *64*, 841–848.

[57] Rapraeger, A. C., Krufka, A. & Olwin, B. B. (1991). Requirement of heparan sulfate for bFGF-mediated fibroblast growth and myoblast differentiation. *Science*, *252*, 1705–1708.

[58] Spivak-Kroizman, T., Lemmon, M. A., Dikic, I., Ladbury, J. E., Pinchasi, D., Huang, J., Jaye, M., Crumley, G., Schlessinger, J. & Lax, I. (1994). Heparin-induced oligomerization of FGF molecules is responsible for FGF receptor dimerization, activation, and cell proliferation. *Cell, 79*, 1015–1024.

[59] Maccarana, M., Casu, B. & Lindahl, U. (1993). Minimal sequence in heparin/heparan sulfate required for binding of basic fibroblast growth factor. *J. Biol. Chem., 268*, 23898–23905.

[60] Turnbull, J. E., Fernig, D. G., Ke, Y., Wilkinson, M. C. & Gallagher, J. T. (1992). Identification of the basic fibroblast growth factor binding sequence in fibroblast heparan sulfate. *J. Biol. Chem., 267*, 10337–10341.

[61] DiGabriele, A. D., Lax, I., Chen, D. I., Svahn, C. M., Jaye, M., Schlessinger, J. & Hendrickson, W. A. (1998). Structure of a heparin-linked biologically active dimer of fibroblast growth factor. *Nature, 393*, 812–817.

[62] Guimond, S., Maccarana, M., Olwin, B. B., Lindahl, U. & Rapraeger, A. C. (1993). Activating and inhibitory heparin sequences for FGF-2 (basic FGF). Distinct requirements for FGF-1, FGF-2, and FGF-4. *J. Biol. Chem., 268*, 23906– 23914.

[63] Arora, K., Gwinn, W. M., Bower, M. A., Watson, A., Okwumabua, I., MacDonald, H. R., Bukrinsky, M. I. & Constant, S. L. (2005). Extracellular cyclophilins contribute to the regulation of inflammatory responses. *J. Immunol., 175*, 517– 522.

[64] De Ceuninck, F., Allain, F., Caliez, A., Spik, G. & Vanhoutte, P. M. (2003). High binding capacity of cyclophilin B to chondrocyte heparan sulfate proteoglycans and its release from the cell surface by matrix metalloproteinases: possible role as a proinflammatory mediator in arthritis. *Arthritis Rheum., 48*, 2197–2206.

[65] Gwinn, W. M., Damsker, J. M., Falahati, R., Okwumabua, I., Kelly-Welch, A., Keegan, A. D., Vanpouille, C., Lee, J. J., Dent, L. A., Leitenberg, D., Bukrinsky, M. I. & Constant, S. L. (2006). Novel approach to inhibit asthma-mediated lung inflammation using anti-CD147 intervention. *J. Immunol., 177*, 4870–4879.

[66] Tegeder, I., Schumacher, A., John, S., Geiger, H., Geisslinger, G., Bang, H. & Brune, K. (1997). Elevated serum cyclophilin levels in patients with severe sepsis. *J. Clin. Immunol., 17*, 380–386.

[67] Yabe, T., Hata, T., He, J. & Maeda, N. (2005). Developmental and regional expression of heparan sulfate sulfotransferase genes in the mouse brain. *Glycobiology, 15*, 982–993.

[68] Salinas, P. C., Fletcher, C., Copeland, N. G., Jenkins, N. A. & Nusse, R. (1994). Maintenance of Wnt-3 expression in Purkinje cells of the mouse cerebellum depends on interactions with granule cells. *Development, 120*, 1277–1286.

[69] Lin, X. (2004). Functions of heparan sulfate proteoglycans in cell signaling during development. *Development, 131*, 6009–6021.

[70] Inatani, M., Irie, F., Plump, A. S., Tessier-Lavigne, M. & Yamaguchi, Y. (2003). Mammalian brain morphogenesis and midline axon guidance require heparan sulfate. *Science, 302*, 1044–1046.

[71] Serafini, T., Colamarino, S. A., Leonardo, E. D., Wang, H., Beddington, R., Skarnes, W. C. & Tessier-Lavigne, M. (1996). Netrin-1 is required for commissural axon guidance in the developing vertebrate nervous system. *Cell, 87*, 1001–1014.

[72] Plump, A. S., Erskine, L., Sabatier, C., Brose, K., Epstein, C. J., Goodman, C. S., Mason, C. A. & Tessier-Lavigne, M. (2002). Slit1 and Slit2 cooperate to prevent premature midline crossing of retinal axons in the mouse visual system. *Neuron, 33*, 219–232.

[73] Hu, H. (2001). Cell-surface heparan sulfate is involved in the repulsive guidance activities of Slit2 protein. *Nat. Neurosci.*, *4*, 695–701.

[74] Ronca, F., Andersen, J. S., Paech, V. & Margolis, R. U. (2001). Characterization of Slit protein interactions with glypican-1. *J. Biol. Chem.*, *276*, 29141–29147.

[75] Bagri, A., Marin, O., Plump, A. S., Mak, J., Pleasure, S. J., Rubenstein, J. L. R. & Tessier-Lavigne, M. (2002). Slit proteins prevent midline crossing and determine the dorsoventral position of major axonal pathways in the mammalian forebrain. *Neuron*, *33*, 233–248.

[76] Whitford, K. L., Marillat, V., Stein, E., Goodman, C. S., Tessier-Lavigne, M., Chedotal, A. & Ghosh, A. (2002). Regulation of cortical dendrite development by Slit-Robo interactions. *Neuron*, *33*, 47–61.

In: Glycobiology Research Trends
Editors: G. Powell and O. McCabe

ISBN: 978-1-60692-841-7

Chapter VII

Click Chemistry in Carbohydrate Based Drug Development and Glycobiology: An Update

Brendan L. Wilkinson,[1] Laurent F. Bornaghi,[1] Todd A. Houston[2] and Sally-Ann Poulsen[1]

[1] Eskitis Institute for Cell and Molecular Therapies,
Griffith University, 170 Kessels Road, Nathan, Queensland 4111, Australia
[2] Institute for Glycomics, Gold Coast Campus,
Griffith University, Queensland 4222, Australia

Abstract

The 1,3-dipolar cycloaddition reaction (1,3-DCR) of a 1,3-dipole to a dipolarophile for the synthesis of heterocycles is a ubiquitous transformation in synthetic organic chemistry. Recently, the Sharpless and Meldal groups have reported the dramatic rate enhancement (up to 10^7 times) and improved regioselectivity of the Huisgen 1,3-DCR of an organic azide to a terminal acetylene to afford, specifically a 1,4-disubstituted-1,2,3-triazole in the presence of a Cu^I catalyst. This Cu^I catalysed 1,3-DCR has successfully fulfilled the requirements of "click chemistry" as prescribed by Sharpless and within the past few years has become a premier component of the click chemistry paradigm. Click chemistry has proven to be of remarkable utility and broad scope, not only in organic synthesis, but in chemical biology and drug discovery. Click chemistry is highly modular and simplifies difficult syntheses. The biocompatibility of the reaction, tolerance towards a broad range of pH and relative inertness of acetylenes and azides within highly functionalised biological milieus has allowed click chemistry to become a viable bioconjugation strategy both for labelling biomolecules and for *in situ* lead discovery applications. More recently, click chemistry has emerged as a powerful conjugation strategy for the preparation of structurally diverse neoglycoconjugates of biomedical interest. This chapter aims to highlight recent developments appearing within the literature concerning the use of click chemistry in carbohydrate based drug discovery and glycobiology. Topics range from small molecule probes and drug leads, multivalent neoglycoconjugates acting as lectin inhibitors and potential vaccines, to bioconjugation strategies for labelling of engineered cell surface glycans. The chapter aims to be comprehensive with commentary on future perspective.

1. Introduction

1.1. Background and Significance

Protein and lipid glycosylation is an omnipresent and life governing process. Oligosaccharides and polysaccharides, either free or anchored to proteins and lipids to form glycoconjugates, exert a multitude of biological effects ranging from nascent protein folding and stabilisation, energy storage, structural support and protection, and as molecular recognition motifs for cell-cell communication events underlying physiological processes including cellular differentiation and development, hormone trafficking and immune surveillance [1]. Overwhelming evidence is now available to implicate both glycan binding and defective metabolic pathways in pathological processes including inflammation and microbial virulence, neurodegenerative conditions and tumour metastasis [2]. The clinical intervention of a defective metabolic pathway, the inhibition of a glycan recognition or pathogen binding event and the elucidation of aberrant glycosylation patterns expressed on certain cancer cell surfaces, are each primary avenues for rational carbohydrate based drug design and vaccine development [3].

The acquisition of a well defined, single glycoform is a prerequisite for correlating structure and function and ultimately developing carbohydrate based therapeutics. Glycan biosynthesis occurs within the secretory pathway of the ER-golgi complex and is not template driven nor under direct transcriptional control [1a]. Naturally occurring glycoproteins are therefore expressed as collections of glycoforms which contain the same underlying peptide sequence but display variable patterns of glycosylation [4]. Due to this considerable microheterogeneity, the isolation of a single glycoform for structure-activity purposes from a biological extract and/or from genetic engineering methods alone is often exceedingly difficult. Whilst genetic engineering and "knock-out/knock-in" methods have contributed enormously to understanding the implications of protein glycosylation, arguably the most reliable means of procuring homogeneous glycoforms is via their *de novo* chemical synthesis [5].

Chemical methods have emerged as powerful tools for elucidating cellular mechanisms of health and pathology associated with glycosylation [4-6]. Improvements in our understanding of glycan metabolism and structure-function relationships using molecular tools allow for the development novel drug leads for the treatment of carbohydrate related disease processes. Two possible pathways can be envisaged, either through interrogation of cell surface glycoprotein/lipid binding events with artificial oligosaccharides, glycoconjugates and their analogues; or the perturbation of glycan metabolic pathways with small molecule enzyme inhibitors, leading to downstream changes in glycosylation patterns [5].

1.2. Glycomimetics as Probes and Drug Leads

Although the knowledge base concerning the structure and function of glycans has increased dramatically in recent years, our ability to routinely apply this knowledge in a drug development context is still far from complete [3b]. The sheer numbers of potential carbohydrate related drug targets and the inherent difficulties associated with the synthesis of natural glycans have fuelled efforts to prepare carbohydrate mimics from natural or synthetic origins as novel drug leads or as biochemical tools [7]. Natural and synthetic carbohydrates are often considered "privileged" combinatorial scaffolds because of the unrivalled structural diversity arising from a high chiral density, variable branching patterns and polyfunctionality [8]. The synthetic modification and inherent reactivity of these powerful building blocks can in turn lead to a tremendous explosion in structural and stereochemical diversity.

Many of the carbohydrate based drugs currently in clinical use are indeed mimics of endogenous carbohydrates and act as inhibitors of carbohydrate-protein binding events or modulators of glycan metabolism (Table 1) [7d,9]. Carbohydrates linked to many natural products are often a prerequisite for biological activity and can thus heavily influence the pharmacokinetics, drug targeting and mechanism of action. Indeed, the literature is replete with examples of carbohydrate containing natural products possessing antimicrobial and anticancer properties, such as the aminoglycosides, saponins and anthracyclines [10]. Despite the plethora of disease states that involve carbohydrate recognition phenomena, comparatively few carbohydrate based drugs are in clinical use. The failure of carbohydrate based therapeutics during the 1980's and 1990's can however be attributed to a poor understanding of their mechanism and pharmacokinetic profiles, rather than inherent flaws in carbohydrate based drugs [11]. When a detailed knowledge is applied in a drug design setting, excellent results have been achieved [12]. Heparin, for example, a polysaccharide displaying potent anticoagulant properties is amongst the worlds most widely used drug [3b,13]. The aminoglycoside and glycopeptide antibiotics remain the frontline defence against a wide range of drug resistant bacterial infections [3c]. Owing to their unique physicochemical and biochemical properties, glycoconjugates have also been employed as vehicles and prodrugs for cell selective drug targeting [3c,14].

Despite the widespread physiological and pathophysiological effects of carbohydrates, there are relatively few clinically used carbohydrate based drugs. Carbohydrates possess less than ideal therapeutic profiles concerning solubility, *in vivo* stability, target binding affinity and cell permeability (although notable exceptions do exist) [9] and as such have received little attention compared to other small molecule drug classes. The synthetic modification of natural polyfunctional carbohydrate scaffolds can however lead to analogues with improved bioactivity and pharmacokinetic profiles. Unnatural functionalities can be introduced into the carbohydrate ring to exploit a specific structural element within the target receptor that is otherwise neglected in the native ligand-receptor binding event [7,8]. The substitution or elimination of ring hydroxyl groups often leads to improved solubility and *in vivo* clearance. In addition, the bioavailability of carbohydrate based therapeutics can be improved by the substitution of a natural, labile glycosidic linkage with a robust, artificial linkage that is impervious to digestive degradation and enzymatic catabolism. Importantly, the installation of artificial linkages can potentially simplify complex multistep syntheses by exploiting chemistry which is facile, high yielding, orthogonal and stereoselective, whilst retaining as much as possible the geometry and electronics of the native linkage [16].

Table 1. Clinically used carbohydrate based drugs and glycomimetic drugs [3,6b,8]

Compound	Target	Indication
Acarbose, Voglibiose, Miglitol (eg. Glyset®)	Intestinal α-glucoamylase, α-glucosidase	Non-insulin dependent Diabetes mellitus (NIDDM)
Heparin analogues (eg. Lovenox®)	Antithrombin III (factor Xa)	Thrombosis
2-Deoxy-2,3-deanhydro-*N*-acetylneuraminic acid derivatives (eg. Relenza®, Tamiflu®)	Influenza sialidase	Antiinfluenza
N-butyl deoxynojirimycin (Zavesca®)	Glucosylceramide synthase, glycan processing α-glucosidase	Lysosomal storage diseases Type 1 Gauchers disease
Vancomycin (Vancocyn®, Lyphocin®)	Bacterial transpeptidase	Methicillin-resistant *Staphylococcus aureus* (MRSA)
Topirimate (Topamax®)	Brain carbonic anhydrase (CA) isozymes*	Antiseizure, migraine

* Postulated anticonvulsant mechanism is inhibition of brain CA isozymes [15].

1.3. Click Chemistry: Background and Perspective

The 1,3-dipolar cycloaddition reaction (1,3-DCR) of a 1,3-dipole to a dipolarophile (i.e. an acetylene or alkene) for the synthesis of five membered heterocycles are ubiquitous transformations in synthetic organic chemistry [17]. Recently, the Sharpless [18a] and Meldal [18b] groups have reported the dramatic rate enhancement (up to 10^7 times) and improved regioselectivity of the Huisgen 1,3-DCR of an organic azide to a terminal acetylene to afford, specifically, the 1,4-disubstituted-1,2,3-triazole in the presence of a Cu^I catalyst (Scheme 1). The Cu^I catalysed 1,3-DCR has successfully fulfilled the requirements of "click chemistry" as prescribed by Sharpless, and within the past few years has become a premier component of the click chemistry paradigm [19].

The classical non-catalysed process proceeds by a concerted mechanism under thermal conditions to afford a mixture of 1,4- and 1,5-disubstituted 1,2,3-triazole regioisomers. The relative proportion of regioisomers and the rate can be predicted from electronic and steric effects [20]. Increased 1,4-regioselectivity is increasingly observed with electron deficient acetylene dipolarophiles (increased dipolarophilicity) and electron rich azides. The Cu^I catalysed ("click") process has been postulated to occur by a stepwise mechanism on the basis of recent thermal and kinetic studies [21]. The substantial rate increase of the Cu^I catalysed process in aqueous solvents is rationalised in terms of a stepwise process which lowers the activation barrier relative to that of the non-catalysed process by as much as 11.8 kcal mol^{-1} [21]. The proposed catalytic cycle involves several postulated and transient Cu^I-

acetylide complexes, starting with the complexation of the alkyne to the Cu^I metal centre to form a Cu^I-alkyne π-complex (A) (Scheme 2) [21]. The enhanced reaction rate in water relative to organic solvents can be rationalised in terms of the endothermic ligand disassociation in organic media, for example, acetonitrile (endothermic by 0.6 kcal mol^{-1}) relative to water (exothermic by 11.8 kcal mol^{-1}) [21]. The formation of the Cu^I-acetylide complexes is also water assisted, since water lowers the pKa of the acetylene C-H by 9.8 pKa units [20]. Formation of the Cu^I-acetylide species allows for subsequent ligand displacement with azide and results in a dimeric copper species (B). Complexation with azide activates it toward nucleophilic attack at the N-3 with the acetylide C-4 (the numbering is presented according to triazole nomenclature). The resulting metallocycle (C) undergoes facile ring contraction via a transannular association of the N-1 lone pair with the C5-Cu π*-orbital to give the copper-triazole complex (D). Protonation of the triazole species, possibly with water and dissociation of the labile copper complex affords the 1,4-disubsituted-1,2,3-triazole (E), thus regenerating catalyst and ending the cycle.

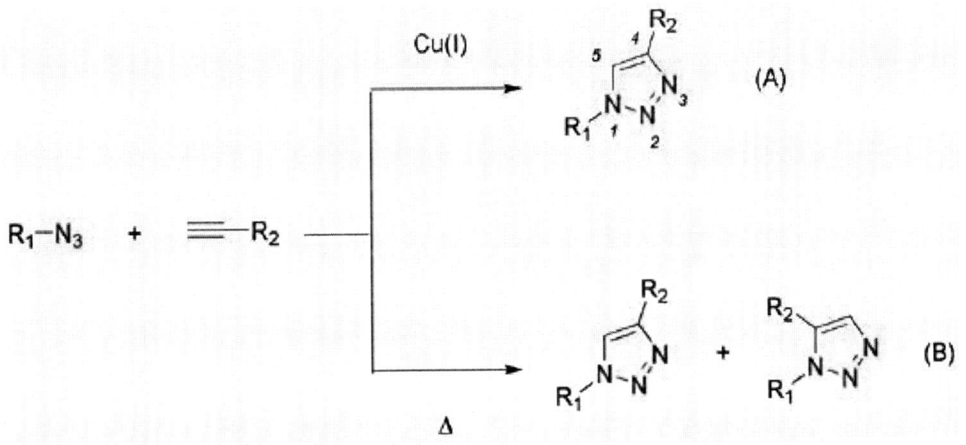

Scheme 1. (A) The Cu^I catalysed 1,3-dipolar cycloaddtion reaction of an organic azide and terminal acetylene affords, exclusively the 1,4-disubstituted-1,2,3-triazole regioisomer. (B) The non-catalysed cycloaddition reaction affords a regioisomeric mixture of 1,4- and 1,5-disubstituted 1,2,3-triazoles.

Within a short time frame, click chemistry has proven to be of remarkable utility and broad scope, not only in organic synthesis, but in chemical biology and drug discovery [22]. Azides and acetylenes are by definition kinetically stable entities possessing high built in energy and are tolerant to a wide range of synthetic conditions [20]. Click chemistry is highly modular and simplifies difficult syntheses, thus enabling a more cost effective and efficient surveillance of structural space. The biocompatibility of the reaction, tolerance towards a broad range of pH and relative inertness of acetylenes and azides within highly functionalised biological milieus has allowed click chemistry to become a viable bioconjugation strategy for labelling biomolecules and for *in situ* lead discovery applications [23]. The 1,2,3-triazole

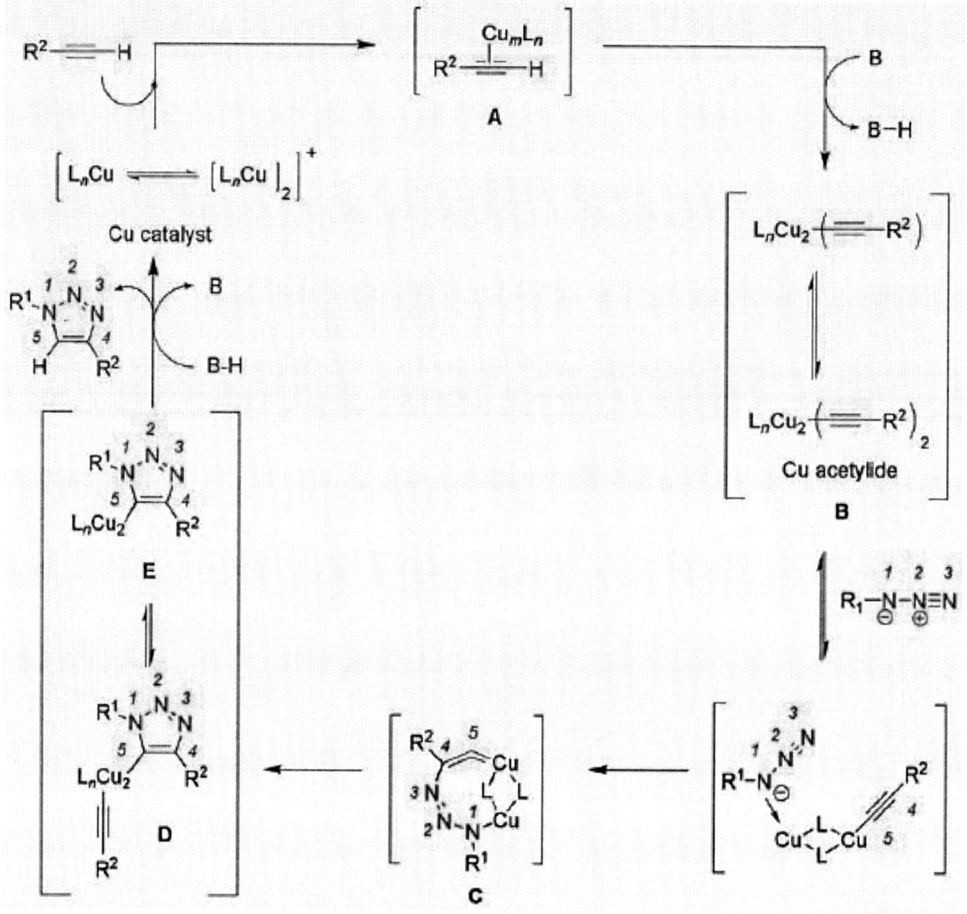

Scheme 2. Proposed stepwise mechanism for the stepwise, Cu^{I} catalysed 1,3-DCR of an azide and acetylene [21].

moiety is a potential pharmacophore owing to its moderate dipole character and rigidity and can therefore be readily incorporated into a design strategy, rather than used as a passive linker between two respective fragments of structural space [24]. Indeed, several examples exist within the literature which describe the biological activity of 1,2,3-triazoles, including antiHIV-1 [25], antibacterial [26], selective β3 adrenergic receptor inhibition [27], antiplatelet activity [28], and antiinflammatory [29] agents to name a few.

More recently, click chemistry has emerged as a powerful conjugation strategy for the preparation of structurally diverse neoglycoconjugates of biomedical interest. This chapter aims to highlight recent developments appearing within the literature concerning the use of click chemistry in carbohydrate based drug discovery and glycobiology. Topics include small molecule probes and drug leads, multivalent neoglycoconjugates acting as lectin inhibitors and potential vaccines, and bioconjugation strategies for labelling of engineered cell surface glycans. This chapter aims to be comprehensive with commentary on future perspective.

2. SMALL MOLECULE PROBES AND DRUG LEADS

2.1. General

Heterocyclic glycoconjugates are widespread in nature. Obvious examples are the nucleosides and nucleotides, the primary building blocks of the nucleic acids. The design of inhibitors of DNA replication based on the mimicry of nucleoside scaffolds has, for decades, been the cornerstone of rational antiinfectives and anticancer drug research [30]. The 1,3-DCR's on carbohydrates, in this case, have proven to be of tremendous utility in such research programmes and have been the subject of previous reviews [31]. Heterocyclic carbohydrates from natural product origin which contain exocyclic nitrogens act as strong inhibitors of certain sugar processing enzymes and are thus valuable antibiotic, antifungal and antiparasitic lead compounds [32]. Examples include Nagstatin, a natural product consisting of a fused tetrahydroimidazo[1,2-*a*]pyridine ring system and a potent inhibitor of β-hexosamidases [33], the Nikkomycin nucleoside-peptide antibiotics isolated from *Streptomyces sp.* which are potent chitin synthase inhibitors [34], the trehazolins [35] and allosamidines containing exocyclic nitrogen heterocycles acts as potent trehalase and chitinase inhibitors, respectively [36]. Intramolecular 1,3-DCR's between azides and alkynes have also been utilised to generate novel fused bicyclic polyhydroxylated alkaloids as potential analogues of bioactive analogues of the indole and indolizodine type [37].

2.2. Glycosidase and Glycosyltransferase Inhibitors

Heterocyclic glycoconjugates are potential mechanistic probes or drug leads owing to their hydrolytic stability towards glycosidases and glycosyltransferases. The strong dipole character and rigidity of the triazole ring can potentially mimic the charge build up and flattening of the carbohydrate ring that is normally associated with the oxocarbenium like ion transition state [32]. In addition, the modular synthesis of carbohydrate triazole analogues by click chemistry lends itself particularly well to the parallel synthesis of small molecule carbohydrate based libraries as potential drug leads.

Recently, Basu and Rossi have demonstrated the β-glycosidase inhibition of 1-β-D-glucosyl-4-phenyl triazole **1** and 1-β-D-galactosyl-4-phenyl triazole **2** as a new class of aglycone modified β-glycosidase inhibitor (Figure 1) [38]. Compounds were screened at 0.24 mM against sweet almond β-glucosidase, *Escherichia coli* β-galactosidase (ECG), and bovine liver β-galactosidase in a preliminary fashion using known glycosidase inhibitors deoxynojirimycin (DNJ) and deoxygalactonojirimycin (DGJ) as comparative reference compounds. The phenyl triazoles were shown to be weak inhibitors of glycosidases and importantly were not substrates for enzyme hydrolysis. In a similar fashion, Périon et al. have employed a mild non-catalysed 1,3-DCR to generate a series of saccharidyl-1,4,5-trisubstituted-1,2,3-triazoles as heterocyclic analogues of the well known glucoamylase inhibitor and antidiabetic drug, acarbose [39].

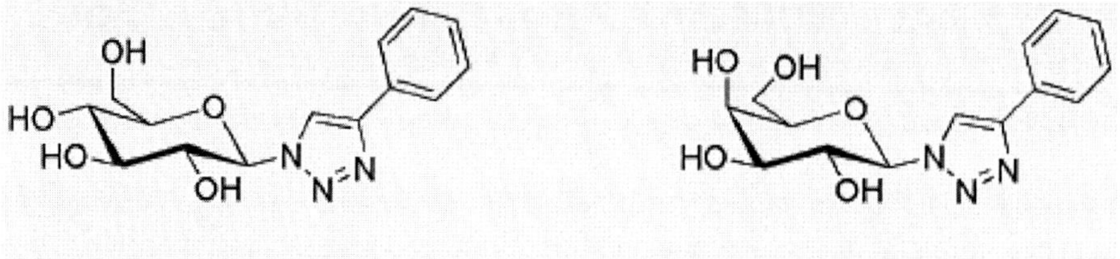

Figure 1. 1-Glycosyl-4-phenyl triazoles **1** and **2** as a new class of aglycone modified β-glycosidase inhibitor [38].

Influenza virus A and B Sialidase (EC 3.2.1.18) is strongly inhibited by 2,3-dianhydro-*N*-acetylneuraminic acid derivatives such as the clinically used drug, Relenza **3** (Zanamivir®) [40]. Motivated by the emergence of avian flu strains (e.g. H5N1) and the need for new frontline antiviral agents, Jiang and colleagues [41] reported the antiinfluenza activity of a library of triazole analogues of Relenza as avian influenza inhibitors. Systemic modification of C-4 ring substituents with amino or guanidino has led to potent and selective inhibition of viral neuraminidase [33, 42]. Thus, the C-4 azido neuraminic acid derivative **4**, which has previously served as a synthetic precursor to **3**, was used as the scaffold for library design (Figure 2). A library of 16 triazole analogues modified in the C-4 position were prepared by click chemistry and were assayed using a neutral red (NR) assay with the data reported as percent protection of MCDK infected cells at 50 μM inhibitor concentration. The protective rate of the triazole analogue **5** against the H5N1 strain was found to be comparable to that of Zanamivir (>64% *cf.* 86%).

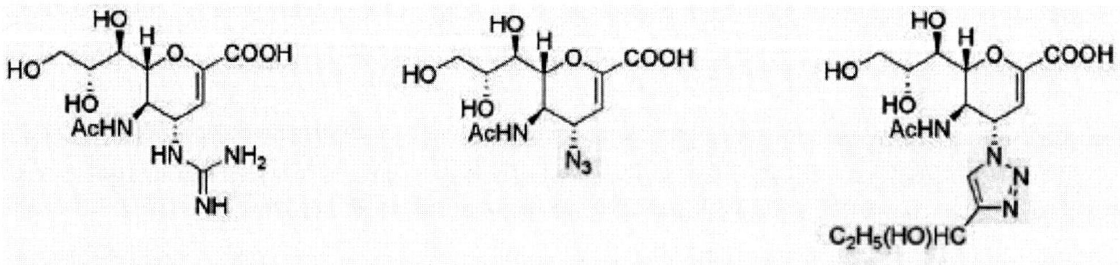

Figure 2. Relenza **3** and analogues **4** (azide) and **5** (the C-4 modified 1,2,3-triazole generated by click chemistry) [41].

There are at least five distinct human α-1,3-fucosyltransferases (α-1,3-FucT III to VIII, EC 2.4.1.152) known which catalyse the transfer of L-fucose from the β-configured GDP-fucose donor to a 3-OH group of an *N*-acetyl lactosamine acceptor to give the Lewis X trisaccharide antigen (Le^X) [43]. Alternatively, some fucosyl transferases (eg αFucT V) utilise the 3'-*O*-sialylated lactosamine acceptor to give the sialyl Lewis X antigen (sialyl Le^X). Both antigenic oligosaccharides are key components implicated in oncogenesis and tumour metastasis, as well as inflammation. Inhibitors of this enzyme are thus important leads in the development of antiinflammatory and antitumour agents [43, 44].

Owing to the complex four partner enzymatic transition state and the low catalytic proficiency of α-1,3-FucT, the successful design of potent inhibitors has been limited. Strongest inhibition lies in the design of substrate donor analogues since most of the binding energy lies in the retention of the GDP moiety [43], yet the strongest inhibitors to date have mainly been in the μM range and too weak to constitute effective leads. Recently, Wong et al. [45] identified a nanomolar inhibitor of α-1,3-FucT from a library of hydrophobic aglycone donor substrate analogues using a novel *in situ* click chemistry screening approach. The library of 85 donor substrate analogues were prepared by click chemistry in microtitre plate format and screened *in situ* within the same well in which they were synthesised (Scheme 3). The GPD donor analogue **6** was identified as the most potent FucT VI inhibitor known to date (K_i = 62 nM).

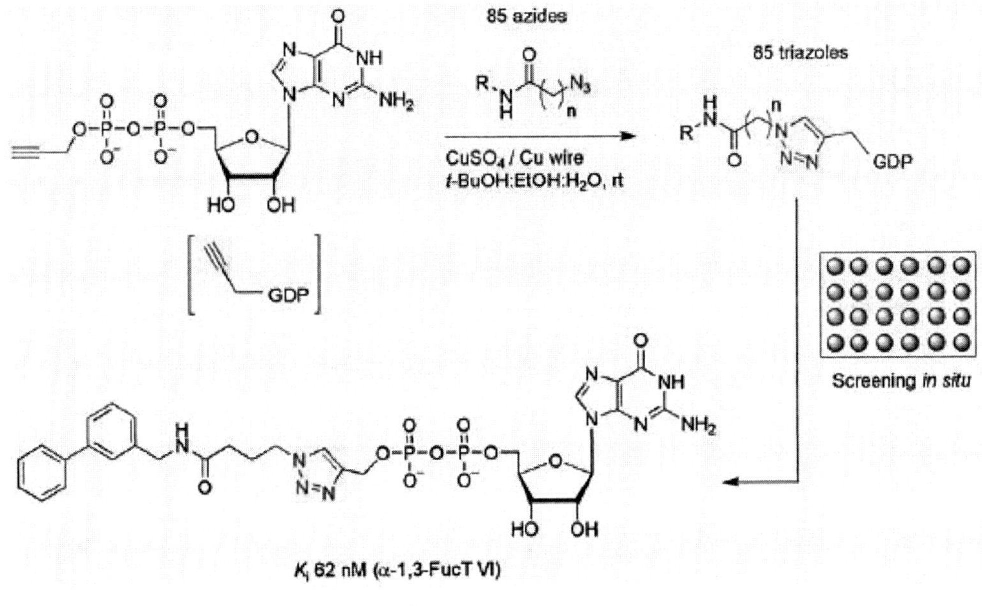

Scheme 3. Hydrophobic donor substrate analogue **6** identified by click chemistry and *in situ* screening as a potent inhibitor α-1,3-FucT VI (K_i = 62 nM) [45].

Leishmania protozoa are the causative agent of Leishmaniasis infection which currently affects 12 million people worldwide [46]. Leishmaniasis is implicated in a host of disease states ranging from subcutaneous lesions to lethal visceral infections. Current therapy of leishmaniasis is far from adequate owing to the emergence of drug resistant strains, as well as the low efficacy, severe side effects, and high toxicity and cost of currently used drugs. The primary carbohydrate reserve of *Leishmania sp.* is a β-1,2 linked mannan (with an average degree of polymerisation of 4–40) which is biosynthesised by a specific β-1,2 mannosyltransferase [44]. It has been shown from previous genetic studies that the β-1,2 linked mannan oligosaccharides are important for parasite survival during intracellular infectivity stages within host macrophages [44, 47]. A detailed investigation into the substrate specificity and active site structure of the β-1,2-mannnosyl transferase is necessary for the development of potent inhibitors of this enzyme and novel Leshmaniasis therapies. Williams and co-workers [48] have recently reported the synthesis of a library of α-mannosides using click chemistry and were assessed as potential substrates for β-1,2-mannosyl transferase (*L. Mexicana*). Thus, two libraries of 14 compounds were prepared by conjugating seven aliphatic and aromatic azides and acetylenes to the structurally conserved α-propynyl mannoside and α-methyl 6-azido-6-deoxy mannoside scaffolds, respectively. Based on HPTLC and fluorography analysis of digest fractions, it was concluded that library members derived from the α-propynyl mannoside scaffold were tolerated as substrates, but not the library derived from the 6-azido mannoside. Also, the enzyme was found to tolerate a broad range of aromatic substituents, an important step in the development of β-1,2-mannosyl transferase inhibitors and new Leishmania therapies.

2.3. Carbonic Anhydrase Inhibitors

The carbonic anhydrase (CA, EC 4.2.1.1) Zn(II)-metalloenzymes are ubiquitous to all kingdoms. These ancient and evolutionary conserved enzymes are efficient catalysts for the reversible hydration of carbon dioxide (CO_2) to give bicarbonate (HCO_3^-) and a proton (H^+), a regulatory and life sustaining reaction underlying physiological processes such as respiration and photosynthesis, the provision of HCO_3^- for crucial biosynthetic pathways (gluconeogenesis, lipogenesis, ureagenesis), the regulation of CO_2 and pH, and electrolyte and fluid secretion [49]. To date, 15 isoforms belonging to α-CA gene family have been discovered in humans (designated hCAs). Their expression dysregulation is implicated in a host of disease states including oncogenesis, elevated intraocular pressure, obesity, memory loss, depression, arteriosclerosis and renal pathology [49]. The modulation of CA activity through inhibition or activation is a potential avenue for therapeutic intervention [50]. The primary pharmacophore for CA recognition and inhibition is the primary aryl sulfonamide (or groups bioisoteric to this functional group, such as sulfamates and sulfamides) [49]. The variable tissue expression and subcellular locations of CA isozymes provides a pharmacological footprint wherein broad specificity CA inhibitors such as acetazolamide exhibit unwanted side effects. This has prompted the development of isozyme selective CA inhibitors as well tissue specific delivery systems. Owing to the strong conservation of amino acid sequence and CA active site architecture, the rational design of isozyme selective inhibitors is a challenging process.

Our group has recently prepared a library of benzene sulfonamides conjugated to monosaccharide and disaccharide "tails" as a dual isozyme-differentiating and solubilising strategy [51, 52, 53]. A total of 50 glycoconjugate benzene sulfonamides were efficiently prepared in parallel fashion using a "click-tailing" strategy developed in our laboratory (Scheme 4). The triazole-linked glycoconjugates were screened using a CO_2 hydration assay against the cytosolic hCA I and II isozymes, as well as the transmembrane hCA IX, XII and XIV isozymes possessing extracellular catalytic domains. Inhibition of these isozymes was in most cases non-clustered indicating that the carbohydrate tails were able to impart selectivity for the isozymes relative to the nonselective parent sulfonamide scaffolds. The methyl β-D-glucuronate triazole **7** was found to be a potent inhibitor of the tumour associated isozyme, hCA IX (K_i = 23 nM), comparable to the nonselective bis-sulfonamide indisulam **8**, which is in phase II clinical trials for the treatment of solid tumours. Compound **7** was 16-fold selective for hCA IX over the physiologically dominant hCA II isozyme, an important finding in the quest for new cancer chemotherapeutics.

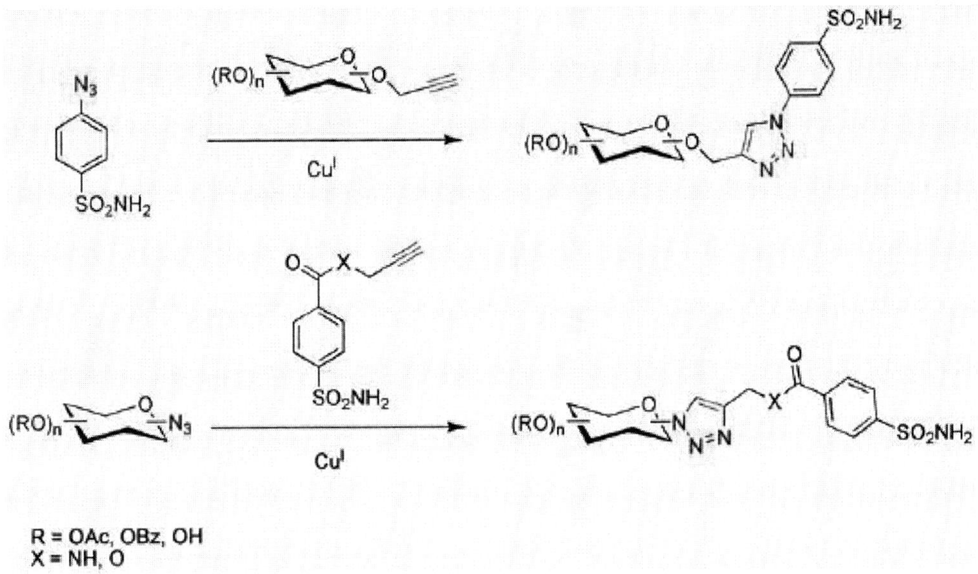

Scheme 4. "Click-tailing" of sugar azides and acetylenes onto benzene sulfonamide scaffolds to generate isozyme selective human carbonic anhydrase inhibitors [51-53].

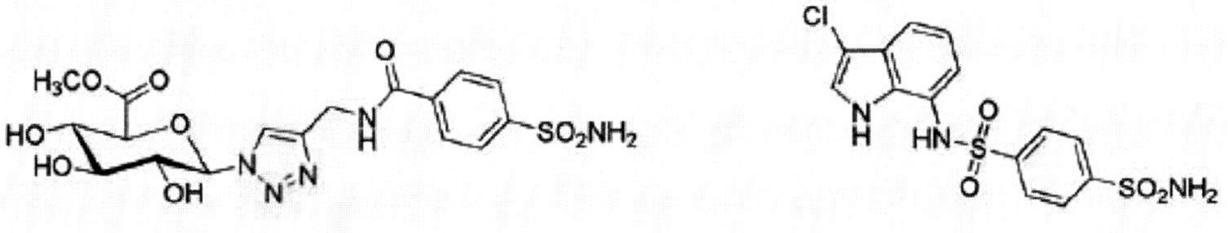

7 | **8**

K_i 23 nM, K_i (hCA II/hCA IX) = 16.4 | K_i 24 nM, K_i (hCA II/hCA IX) = 0.63

Figure 3. Methyl β-D-glucorunate derivative 7 was a potent and selective inhibitor of tumour-associated carbonic anhydrase isozyme IX [51].

2.4. Glycoamino Acid and Glycopeptide Mimics

Glycoprotein based therapeutics constitute more than one third of approved biopharmaceuticals [3]. Glycoproteins and peptides that contain glycans commonly *N*-linked to an asparagine (belonging to the Asn-X-Ser/Thr consensus sequence) or *O*-linked to Ser or Thr residues play pivotal roles in biology. The branched glycan component (as an poly-, oligo-, or monosaccharide) are not only important for protein folding, stability and specificity, but can determine the primary function of the mature glycoprotein as a modulator of intercellular communication [1]. The *de novo* synthesis method is a viable approach for accessing native glycoforms and artificial glycoproteins (neoglycoproteins) of biomedical interest [4, 54]. The conjugation of a saccharide to a protein or peptide often results in the enhancement of an immune response to a carbohydrate antigen or the *in vivo* stability and biological activity of a peptide drug [55].

The theme strongly reverberated in the mimicry of endogenous glycopeptides is the substitution of labile *N*- and *O*-glycosidic bonds with robust linkages that also mimic the geometry and electronics of the native glycopeptide bond [56]. Current impetus is now directed toward the ligation of preformed oligosaccharide and peptide fragments containing mutually reactive functional groups in aqueous solution, a term coined native chemoselective ligation [16]. Click chemistry of preformed carbohydrate and peptide fragments bearing azide and acetylene functionalities has numerous advantages over traditional glycosylation and peptide bond formation chemistry and has emerged as a promising ligation strategy. It allows for the mild, chemoselective and stereospecific conjugation of preformed building blocks/fragments under aqueous conditions and is insensitive to global protecting group effects or synthetic conditions. Azide and acetylenes are also powerful functional groups since they are relatively inert to a wide range of synthetic transformations occurring on the peptide or oligosaccharide component (e.g. glycosylation, protecting group chemistry, peptide coupling etc). Triazole linked glycomimetics are also chemically robust entities towards a wide range of synthetic transformations commonly encountered in carbohydrate chemistry, as has recently been demonstrated by our group [57].

Click chemistry can therefore greatly simplify synthesis and can be introduced at virtually any stage of the glycoprotein/peptide synthetic pathway (convergent or linear strategies). From a mimetic point-of-view, 1,2,3-triazole linked glycoamino acid mimics share similar structural and electronic features to that of the native amide counterparts and are relatively impervious to enzyme or digestive hydrolysis [22a]. 1,2,3-Triazoles have a larger dipole moment compared with amide bonds (~5.0 D by *ab initio* calculation cf. 3.8 D of *N*-methylacetamide) and can potentially enhance amide mimicry due to their weak hydrogen bond acceptor properties [22]. The dipole character can also be fine-tuned depending on the electronic effects of the neighbouring groups. For example, the C-5 bond can function as a potential hydrogen bond donor, much like an amide proton given suitable polarisation exists between the N-2 and N-3 bonds of the triazole ring. Distances between the atoms in the 1 and 4 position of the triazole ring are similar compared to the distance between the α carbon (R) and the carbon attached to the amide (R'), 5.0 Å cf. 3.8 Å (Figure 4) [22a].

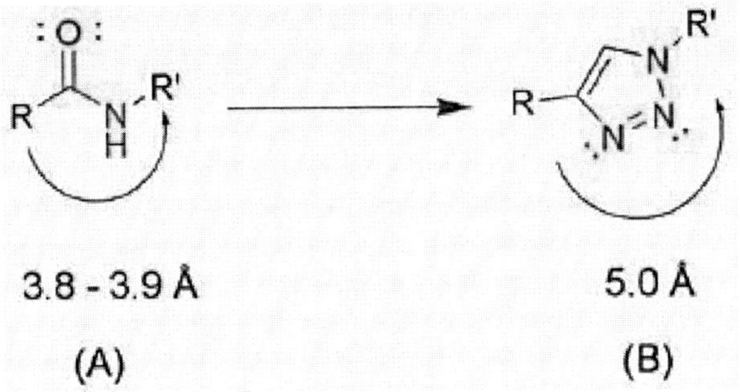

Figure 4. 1,4-Disubstituted-1,2,3-triazoles (B) can function as a potential mimics of glycopeptide bonds (A) due to similar geometric features [22a].

Since the pioneering work of Meldal and colleagues [18b] who described the first use of click chemistry to generate glycopeptides on solid support, there has been a flurry of activity within the literature concerning the development of glycoamino acid and glycopeptides using this synthetic strategy. In an a comprehensive study, Rutjes et al. [58] have recently reported the expedient synthesis of *N*-glycosyl and *C*-glycosyl triazoles as glycoamino acid mimics from glycosyl azide and ethynyl glycosides, respectively (Scheme 5). Click chemistry between acetylene or azide functionalised amino acids and the respective saccharide building blocks in solution resulted in the formation of triazole- glycoamino acids under mild conditions in moderate to good yields. The bifunctional glycoamino acids could potentially then undergo subsequent peptide couplings at either the carboxy or amino terminus. In a preliminary study, the triazole moiety was also shown to be stable to mineral acid and base treatment. Likewise, ethynyl glycosides have also been exploited by Massi and co-workers [59] for the efficient generation of *C*-glycosyl isoxazole and *C*-glycosyl triazaole α-amino acids. In the latter case, the Cu^{I} catalysed cycloaddition of the various ethynyl glycosides with a suite of azide functionalised amino acids proceeded smoothly at room temperature in toluene in good yields.

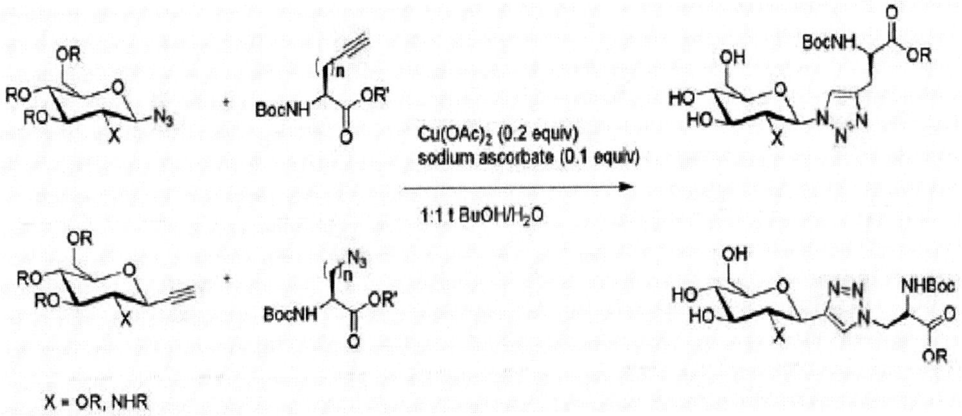

Scheme 5. Synthesis of glycosyl triazole linked glycoamino acid and glycopeptide mimics [58].

In a simple, yet elegant study Macmillan and Blanc [60] have demonstrated the compatibility of click chemistry to a native chemical ligation strategy employing a chemoselective nucleophilic displacement of alkyl bromides with side chain cysteine thiols. The glycoamino acid mimics were based on *N*-acetyl glucosamine, *N*-acetyl lactosamine and chitobiose, all major constituents in endogenous *N*-linked glycoproteins. An important finding revealed the unimportance of the order in which the chemoselective ligation and click chemistry events took place in constructing the triazole modified neoglycopeptides. In this respect, the click reaction itself could potentially function as the chemoselective ligation event, as demonstrated by the model reaction of benzyl mercaptan with 2-bromoacetyl propynyl amide or with the preformed glycosyl triazole glycoamino acid mimic containing the bromoacetamide moiety (Scheme 6).

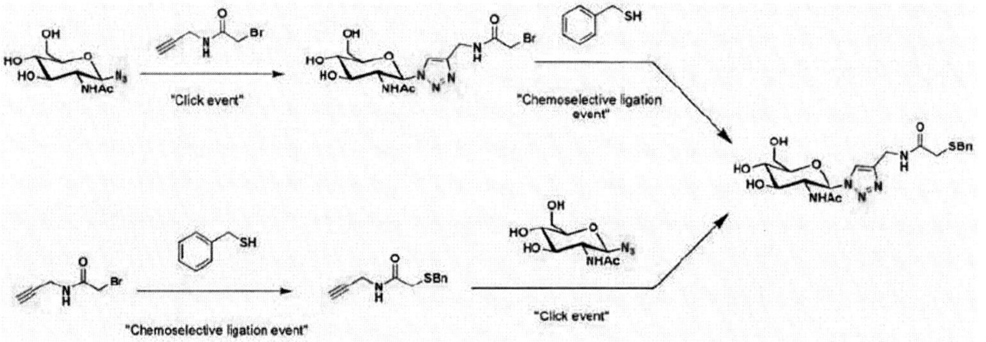

Scheme 6. Synthesis of glycopeptide mimics using a dual chemoselective ligation/click chemistry approach [60].

Click chemistry has recently been used as a glycosylation tool to modify the physicochemical properties of an azabicycloalkane amino acid peptidomimetic of a homoSer-Pro dipeptide [61]. The click reaction between acetylated and free propynyl β-D-glucopyranoside with various azide functionalised peptidomimetic scaffolds afforded the triazole conjugate in good yields (72-84%) under aqueous conditions. In a similar fashion, 1,2,3-triazole analogues of pyrazinone based peptidomimetics have been generated by click chemistry of various glycosyl azides with substituted ethynyl pyrazinone scaffolds [62].

Many carbohydrate macrocycles of natural product origin, including glycosylated cyclic peptides, possess remarkable biological activities and diverse structural characteristics. Indeed, several are used clinically as antibiotics and antifungal drugs [9]. The macrocyclisation and glycosylation are two key structural modifications that occur late in the biosynthesis and confer much of the unique physicochemical properties and biological activity *per se*. Westermann and colleagues [63] have recently utilised click chemistry in combination with ring closing metathesis in the preparation of a series of macrocyclic glycolipid mimics. The synthetic strategy involved a consecutive click ligation and ring closing metathesis using dual functionalised carbohydrate building blocks incorporating azide and olefin groups.

Motivated by the emergence of vancomycin resistant strains of *Enterococcus faecium* (VRE), Thorson and co-workers [64] have reported the synthesis and biological activity of a library of 1,2,3-triazole modified vancomycin analogues. From a library of 15 compounds, they identified a triazole derivative that was twice as effective as vancomycin against antibiotic resistant strains *E. faecium* and *S. aureus*. Walsh and Lin [65] have also exploited click chemistry as part of a chemoenzymatic approach to prepare carbohydrate modified cyclic peptide tyrocidine (tyc) antibiotics with lowered toxicity. A library of 247 glycopeptides was prepared by employing a thioesterase mediated macrocyclisation followed by Cu^I catalysed triazole formation in 96-well microtitre plate format. Following antibacterial and haemolytic assays, two glycopeptides (Tyc4PG-14 and Tyc4PG-15) were identified as retaining the antibacterial potency of the native peptide, whist increasing its therapeutic index by a factor of 6 (Figure 5).

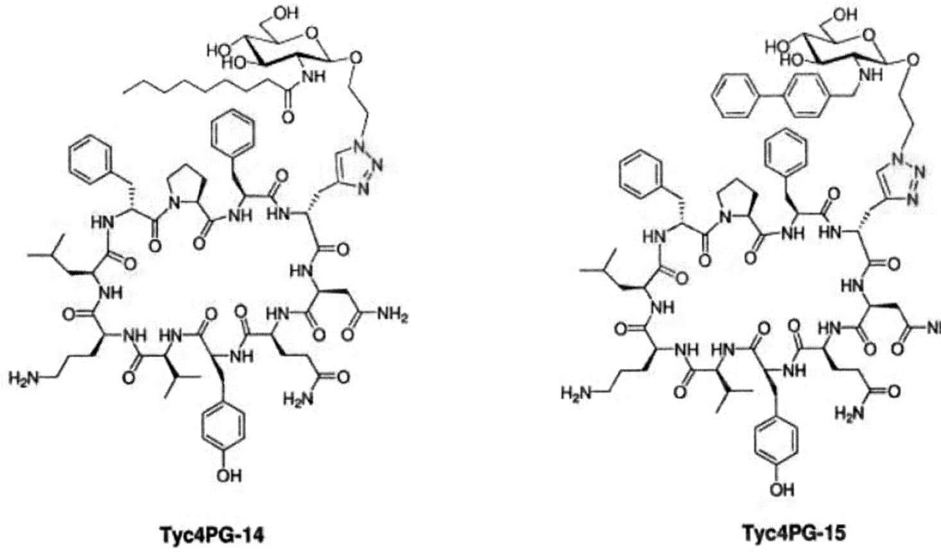

Tyc4PG-14 **Tyc4PG-15**

Figure 5. Glycopeptide antibiotic analogues of tyrocidine (tyc), Tyc4PG-14 and Tyc4PG-15. [65]. Reprinted with permission from Lin, H., Walsh, C. T. *J. Am. Chem. Soc.* 2004, *126* (43), 13998-14003. Copyright ©2004 American Chemical Society.

2.5. Bio-Orthogonal Chemical Reporter Strategies

The development of bioorthogonal chemical methods for profiling disease related glycoproteins expressed on living cell surfaces is an important and recent development in glycomics research and carbohydrate based drug development [66]. In particular, the biosynthetic incorporation of unnatural saccharides into glycan epitopes expressed on living cell surfaces is a recent strategy described by the Bertozzi group and has provided considerable insight into glycosylation structure and function *in vivo* [47, 67]. The principle is called metabolic oligosaccharide engineering and involves the incorporation of synthetically modified saccharides containing an inert but mutually reactive functional group into a host glycoprotein by exploiting a particular metabolic pathway. The favourable size and inertness of azide and acetylene functionalised substrates within living cellular environments as well as the promiscuity of certain glycosyl transferases enables these abiotic substrates it to "slip through" the cellular catalytic machinery and become incorporated into cell surface glycans. The residues then can serve as chemoselective handles for bioconjugation to a fluorescent marker for visualisation, or an affinity tag for proteomic enrichment. In this respect, click chemistry has been instrumental as a bio-orthogonal tagging strategy, owing largely to the biocompatibility of the reaction and the inertness of azides and acetylenes within complex biological environments.

In an early example, Bertozzi and co-workers [68] utilised a strain promoted cycloaddition of a cyclooctane functionalised biotin probe with recombinant glycoproteins and cell surface glycoproteins displaying *N*-azidoacetyl sialic acid (Sia-NAz) residues. The 1,2,3-triazole products were detected in both recombinant systems and on the cell surface of human T-lymphoma Jurkat cells by Western blot analysis and flow cytometry, respectively. In the latter case, fluorescence measurements were shown to be dose dependent and proportional to the density of the Sia-NAz residues on the cell surface glycoproteins as well as the incubation time with the cyclooctane biotin probe.

L-Fucose is a ubiquitous component in sialyl Lewis^y (sialyl Le^y) and sialyl Lewis^x (sialyl Le^x) antigenic selectins found unregulated in certain cancers [8b, 69]. The Bertozzi group recently has recently utilised click reaction as a key bioconjugation step in a fluorescent reporter strategy for the chemical profiling of fucosylated glycoproteins as new cancer biomarkers (Scheme 7) [70]. Utilising the fucose salvage pathway and the promiscuity of fucosyl transferases, three synthetic azido fucose analogues, functionalised with azide at the 2, 4 and 6 positions, respectively, were incorporated into cell surface glycans of human T-lymphoma cell line Jurkat. The azide markers were incorporated into the glycan and presented on the cell surface by culturing the cell culture with 125 μM of each azide analogue. The cells were lysed with an NP-40 detergent which is known to be compatible with click chemistry. The click bioconjugation of the cell lysates with acetylene biotin marker, followed by binding of triazole tagged glycoprotein to α-biotin-horse-radish peroxidise conjugate (HRP) allowed for Western blot analysis. Significant protein labelling (up to 40% of all available fucosylation sites) of only the 6-azido analogue was observed by Western blot analysis of the hydrolysates, although this analogue was deemed cytotoxic at elevated culture concentrations (100 μM). It was postulated that azide analogues substituted at C-2 and C-4 were not acceptable substrates for the fucose salvage pathway.

Very recently, Wong and co-workers [71] have developed a click activated fluorogenic labelling technique to analyse cell surface glycoprotein fucosylation *in vivo* (Scheme 8). In a similar manner to that reported by Bertozzi and colleagues, the artificial 6-azido and 6-ethynyl fucose substrates were incorporated into cell surface glycoproteins of human lymphoma Jurkat cell lines via the fucose salvage pathway. The extent of cell surface glycoprotein fucosylation was analysed *in vivo* by flow cytometry and fluorescent microscopy following the click bioconjugation of the fluorogenic probe to the functionalised L-fucose analogues. Fluorescence was only detected for the 6-azido analogue, following fluorogenic bioconjugation with 4-ethynyl-*N*-ethyl-1,8-naphthalimide, indicating selective uptake and integration of this artificial substrate into cell surface glycans.

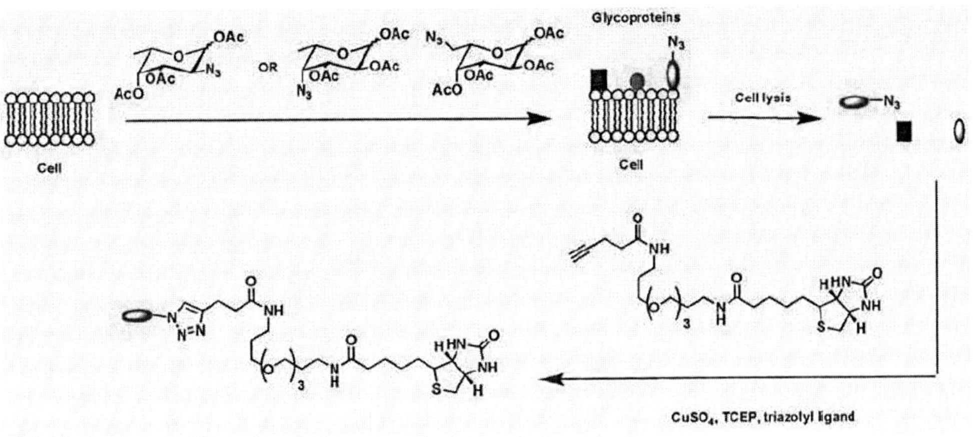

Scheme 7. Metabolic engineering of human T-lymphoma cell line Jurkat and labelling of azido L-fucose derivatives with a biotin reporter probe using click chemistry [70].

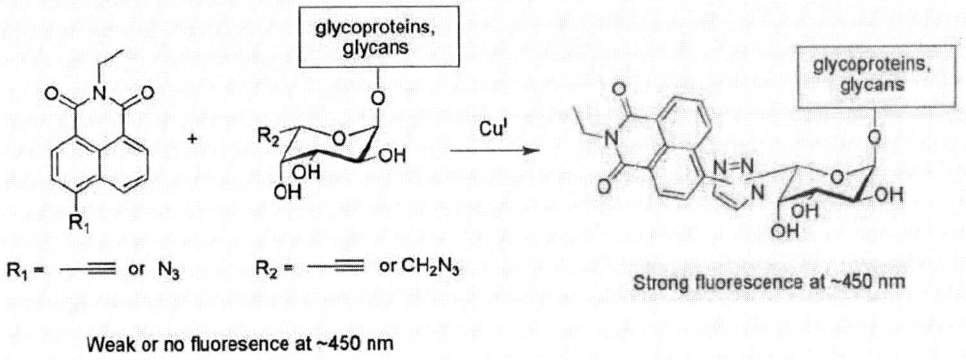

Scheme 8. The bioconjugation of the fluorogenic naphthalene probe to a functionalised L-fucose derivative by click chemistry [71].

3. Multivalent Neoglycoconjugates

3.1. General

Nature compensates for the weak binding affinity between a monovalent carbohydrate ligand and cognate receptor protein (K_a in the ranges of 10^{3-4} M^{-1}, although exceptions are known) by expressing multiple copies of the carbohydrate recognition epitope, or multiple receptor binding sites on a cell surface receptor - a concept known as the "multivalent" or "glycoside cluster effect" [72]. The overall effect is the increase in binding affinity to an extent which exceeds the sum of the constituent binding events in sufficient strength to effect or inhibit a cellular or physiological response. By exploiting one or more of these mechanisms, chemists have assembled a wide range of structurally diverse polymeric artificial carbohydrate polymers (glycopolymers) as effectors or inhibitors of cellular processes and analytical tools for investigating carbohydrate-protein binding events [73]. Artificial glycopolymers hold great promise as inhibitors of bacterial adhesion [74] and anticancer vaccines [75], but also sustainable and economic biomaterials with a wide range of biomedical and drug delivery applications, including biocompatible matrices for tissue engineering [76] and sustained drug release and as polymeric vehicles for targeted drug and gene delivery [77]. Apart from their widespread biomedical applications, glycopolymers can also function as mimics of naturally occurring polysaccharides and display antiinflammatory, anticoagulant and antitumour properties and are thus viable leads for drug development and vaccine development [78,79].

3.2. Glycopolymers and Polysaccharide Mimics

Novel synthetic methods are critical for the development of innovative glycopolymers of biomedical and pharmaceutical importance. Yet, devising novel synthetic pathways for developing biocompatible glycopolymers that are well defined and tailored for a specific purpose is a continuing challenge faced by carbohydrate chemists [79]. Subtle changes in polymer structure and pendant attachments can have profound effects on the mechanical properties and biological activity of the material. Motivated by the requirement for more expedient synthesis of multivalent glycopolymers, Haddleton and co-workers [80] have recently reported the use of a novel, dual click chemistry/transition-metal-mediated-living-radical-polymerisation (TMM-LRP) strategy for the facile preparation of a linear glycopolymer library containing pendantly attached carbohydrates. The acetylene functionalised poly(methylacrylate) polymer was assembled through TMM-LRP and was used as a subsequent scaffold for combinatorial design. A series of glycosyl azides and azido ethyl glycosides were then grafted by click chemistry in a highly efficient manner to the acetylene functionalised polymer backbone providing rapid access to glycopolymer libraries with different bulk properties and potential protein binding affinity. The work demonstrates the powerful utility of click chemistry in preparing pendant-functionalised glycopolymers in a combinatorial fashion.

Reineke et al. [81] have recently utilised click chemistry as a pivotal step in the preparation of glycopolymer-nucleic acid complexes (polyplexes) as gene delivery agents. The synthetic poly(glycoamidoamine) co-polymers were chemically tailored to be non-toxic and biocompatible, whilst improving genetic transfection efficiency by stabilising plasmid DNA (pDNA) complexation, and preventing cellular disassociation, aggregation and biological degradation of the polyplex system prior to delivery at the therapeutic site. On the basis of previous studies implicating heterocyclic-amide co-functionalised polymers as strong DNA binding agents, the triazole-amide moiety (two per repeat unit) was postulated as a strong DNA binding motif via hydrogen bonding and van der Waals interactions. Diazido trehalose units were efficiently polymerised by click chemistry to a variety of dialkyne-amide-terminated oligoamine monomers containing 1, 2 and 3 secondary amine(s) (Scheme 9). *N*-Boc deprotection of the ethyleneamine groups resulted a library of biocompatible and polycationic carbohydrate co-polymers with similar degrees of polymerisation, dispersity, molecular weights and polymer stiffness in aqueous solution. The highly cationic nature rendered the polymers capable of phosphate charge neutralisation and polyplex formation. To elucidate the effect of the 1,2,3-triazole ring and the relative number of ethyleneamine groups in the polymer backbone on pDNA complexation, a series of gel electrophoresis and fluorescent ethidium bromide assays were conducted. Cellular delivery efficacy into HeLa cells was also determined using fluorescent flow cytometry analysis.

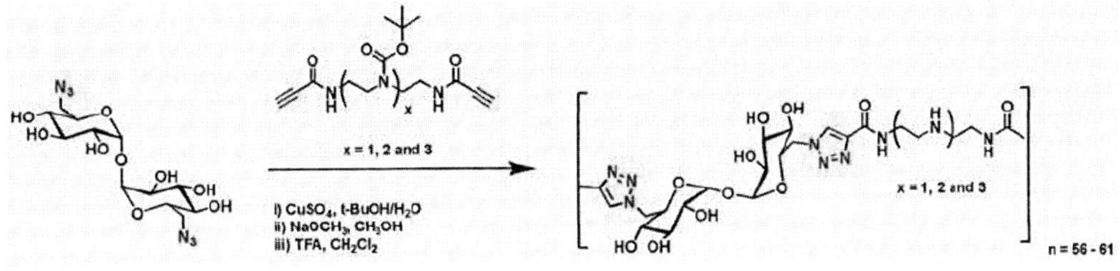

Scheme 9. Click chemistry polymerisation of triazole functionalised, trehalose poly(glycoamidoamine) co-polymers as stabilised polyplex delivery agents [81].

Nature's ability to perform myriad chemical transformations under exquisite regio- and stereocontrol within complex, polyfunctional chemical environments has captured the imagination of synthetic organic chemists for many years. The ability to perform multiple transformations on a single substrate *in vitro* using orthogonal chemistries is a key challenge facing chemists in the quest for novel materials and drugs [82]. Two possible routes have emerged; either conducting multiple transformations simultaneously on separate functional groups within the same molecule, or multiple transformations on the same functional group in a "cascade-type" fashion. Impressive examples have emerged thus far and these are primarily concerned with total synthesis of natural products and peptides [83]. However its adaptation to polymer science has been limited owing to the difficult nature of these polyfunctional macromolecules. The orthogonal functionalisation of polymeric macromolecules requires methods which operate under high fidelity, mild conditions and with strong regio- and stereocontrol [77]. Click chemistry provides appropriate scope in glycopolymer chemistry, as it can be used as a polymerisation tool or to functionalise preformed polymers with equal fidelity under orthogonal reaction conditions.

Hawker and colleagues [84] have recently exploited click chemistry as a mild, regio- and stereoselective strategy for performing cascade functionalisations on glycopolymers. In the simultaneous method, a pendant acetylene functionalised polystyrene terpolymer was reacted with 2,3,4,6-tetra-*O*-β-D-glucopyranosyl azide using click chemistry whilst a hydroxyl pendant side chain was reacted simultaneously with an *N*-hydroxysuccinaimde (NHS) activated ester, affording the functionalised co-polymer containing a triazole and ester pendant sequence (Scheme 10A). On the other hand, the cascade mechanism proceeded via the sequential reaction of a dual ester-functionalised polyacrylate with propargyl amine affording the propynyl amide which was subsequently grafted to the glucosyl azide in a one-pot manner (Scheme 10B).

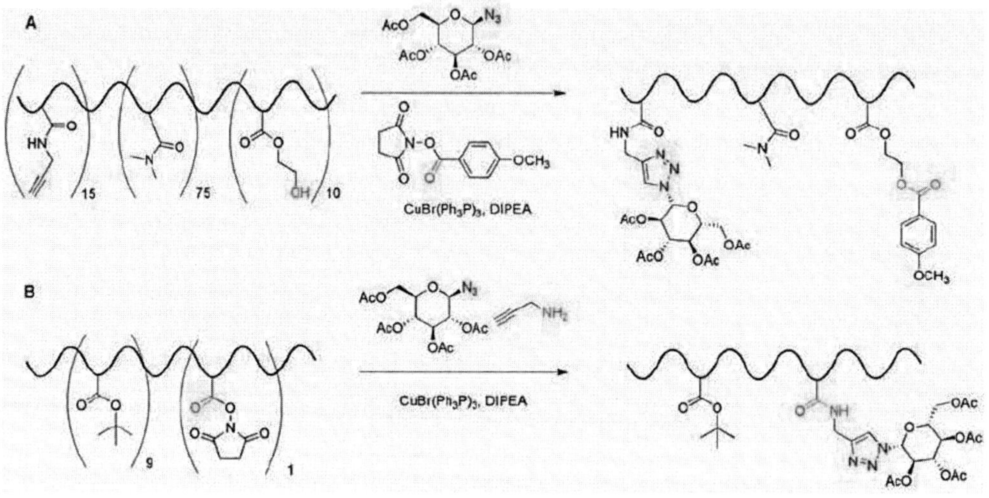

Scheme 10. Preparation of glycopolymers with pendant functionality through (A) the simultaneous functionalisation of a water soluble terpolymer and (B) the cascade functionalisation of an ester functionalised polyacrylamide [84].

Cellulose is an abundant natural biomaterial and its synthetic modification can lead to materials possessing bulk properties that are suitable for a wide range of industrial and biomedical applications [76]. In the pharmaceutical industry, functionalised cellulose homopolymers (cellulosics) have been extensively used as biodegradable enteric coatings for controlled drug release formulations and more recently for bone, cartilage and brain tissue engineering owing to theri excellent biocompatibility [76]. The first example of the chemical modification of cellulose using click chemistry was recently reported by Heinze and co-workers [85]. Click chemistry of the 6-azido-6-deoxy cellulose scaffold afforded triazole spaced carboxyester, thiophene and aniline functionalised cellulosics displaying degree of substitution (DS) values up to 0.9 and good solubility in organic solvents (DMF, DMSO). The click methodology offers hydrolytically stable bonds between the polymer backbone and functional groups and overcomes the formation of polymer crosslinking byproducts that are typically observed with conventional cellulose esterification.

The conjugation of multivalent antigenic oligosaccharides to preformed biomacromolecules, in particular polypeptides, often results in an increase in an immune response and has become a primary strategy in the development of antitumour vaccines [75]. Research efforts in this area are now focused towards the preparation of multiantigenic constructs with the aim of stimulating the immune system towards a broad range of malignant transformations rather than one (i.e. using monomeric antigenic constructs). In this respect, more sophisticated and complex synthetic strategies will be required and are being sought [86]. In a proof-of-concept manner, the Danishefsky group has recently demonstrated the viability of click chemistry as a strategy for linking antigenic oligosaccharides to keyhole limpet hemocyanin (KLH) polypeptide carriers [87]. The authors concluded click chemistry to be a practical and orthogonal linker strategy owing to its high degree of chemoselectivity and tolerance towards other synthetic transformations. In addition, the compatibility of the chemistry under aqueous conditions permits the use of a broad range of carrier proteins/peptides proteins that are otherwise unstable or insoluble within organic solvents.

Many (1→3)-β-D-glucopyranan polymers, such as Lentinan, possess immunostimulating and anticancer properties [88]. Natural and functionalised (1→3)-β-D-glucopyranan polymers have been utilised as novel biomaterials possessing lectin and RNA binding affinity owing to their unique triple stranded helical superstructure [89]. In an effort to generate more regioselective methods for the preparation of functionalised glucan polymers, Shinkai and co-workers [90], have utilised click chemistry to chemically modify the (1→3)-β-D-glucan, 6-azido-6-deoxycurdlan. In a pilot study, the click reactions were performed with a variety of acetylene terminated modules with potential lectin binding, redox active, phytoactive, and fluorescent properties in order to develop novel polysaccharide based materials.

Dondoni and co-workers [91] have recently prepared a series of triazole linked (1,6)-oligomannose analogues as potential mimics of lipoglycan structures and high density oligomannose-type antimicrobial vaccines [92]. The orthogonally substituted ethynyl α-*C*-mannoside building block was conjugated to the 6-azido-α-*C*-mannoside platform to afford the triazole linked dimannoside coupling product. Using an iterative click strategy, involving the necessary installation of an azide group at C-6 and coupling to the ethynyl building block resulted in the facile construction of a structurally rigid triazole linked oligomannose over a total of four consecutive cycles. Whilst the click coupling reactions required up to 30 h for completion at room temperature, microwave irradiation at 100 °C dramatically shortened this (to five minutes), whilst retaining good yields and regioselectivity.

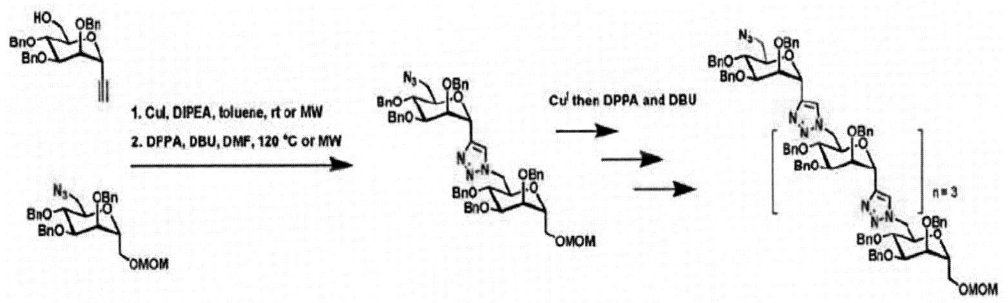

Scheme 11. Iterative assembly by of a triazole linked α-1,6-oligomannose analogue using click chemistry [91].

The first reported example of the use of click chemistry in the modular synthesis of branched oligosaccharide mimics was demonstrated by the work of Field and colleagues [93]. The facile synthesis of amylopectin fragments occurred in a single step between dipropynylated maltoside core fragments and glycosyl azide building blocks based on β-D-glucose, maltose, maltotriose and maltoheptose, respectively. Employing propynyl functionalised maltose templates substituted at C-4/C-6’ and C-6/C-6’, respectively, allowed for the template assisted click synthesis of the respective regioisomeric amylopectin analogues.

3.3. Cyclodextrins and Artificial Receptors

Cyclodextrins (CDs) are water soluble, cyclic malto-oligosaccharides derived from starch which possess a hydrophilic outer surface and a relatively lipophilic central cavity [94]. For quite some time, CDs have been used in drug and cosmetic formulations as solubilising carriers of otherwise hydrophobic and water insoluble molecules through the reversible formation of water soluble inclusion complexes but recently have been gaining considerable attention as topical delivery agents to improve the bioavailability of drugs administered by this route [95]. There has also been much interest in CDs as vehicles for the cell specific delivery of hydrophobic drugs to target sites through multivalent interactions with carbohydrate receptors [71]. CDs are also non-immunogenic and offer enhanced biocompatibility and water solubility compared to other multivalent carbohydrate constructs based on rigid macrocyclic scaffolds, such as azacrown ethers, cyclic peptides, calixarenes and cyclophanes [94, 96].

The first example of an artificial CD analogue incorporating a heterocyclic motif within the binding cavity was demonstrated by Vasella and co-workers [97] during ongoing structural and binding investigations of artificial CD analogues. The intramolecular, thermal Huisgen 1,3-DCR between a dual azide/acetylene functionalised amylose fragment proceeded under thermal conditions to afford the regioisomeric mixture of a heterocyclic β-CD mimic.

Click chemistry offers numerous advantages over existing cycloglycosylation and cyclooligomerisation methods, which are often impeded by long synthetic sequences, low cyclisation yields, as well as difficulties in resolving anomeric mixtures and/or macrocycles of different sizes. Click chemistry was originally proposed by Gin et al [98] as a high yielding, stereo- and regiospecific cyclodimerisation strategy for preparing well defined, modified β-CD analogues. The work reported the convergent click cyclodimerisation of a dual azide/acetylene functionalised trimannoside using a CuI/DBU catalytic system, affording the cyclic dimer in good yields (80%) with the concomitant formation of cyclic trimer (15%) (Scheme 12). The mannose based macrocycle also showed similar propensity to form inclusion complexes with hydrophobic fluorogenic molecules compared to native β-CD, as determined by fluorescence emission spectroscopy. In a recent extension from this work, a series of C_2 and C_3-symmetric oligosaccharide macrocycles were prepared using click chemistry in a convergent manner from a suitably difunctionalised monosaccharide and disaccharide, respectively [99]. The investigation of macrocycle cavity shape and the potential for host-guest interactions was conducted using computational and NMR shift titration methods, revealing a functionalised and sterically congested cavity which may prove useful as novel molecular pores or receptors for studying host-guest interactions.

Click chemistry has been used to prepare C_2-symmetric carbohydrate/amino acid hybrid macrocycles as potential water soluble inclusion complexes [100]. The cyclodimerisation of bifunctional azido-acetylene glycopeptide building blocks proceeded in acetonitrile using a CuI/DIPEA catalyst system in moderate yields. Interestingly, the analogous reaction was attempted using aqueous conditions prescribed by Sharpless and showed minor product conversion at elevated temperatures (70 ^{o}C). The novel carbohydrate/amino acid hybrid macrocycles were investigated as potential water soluble artificial receptors using

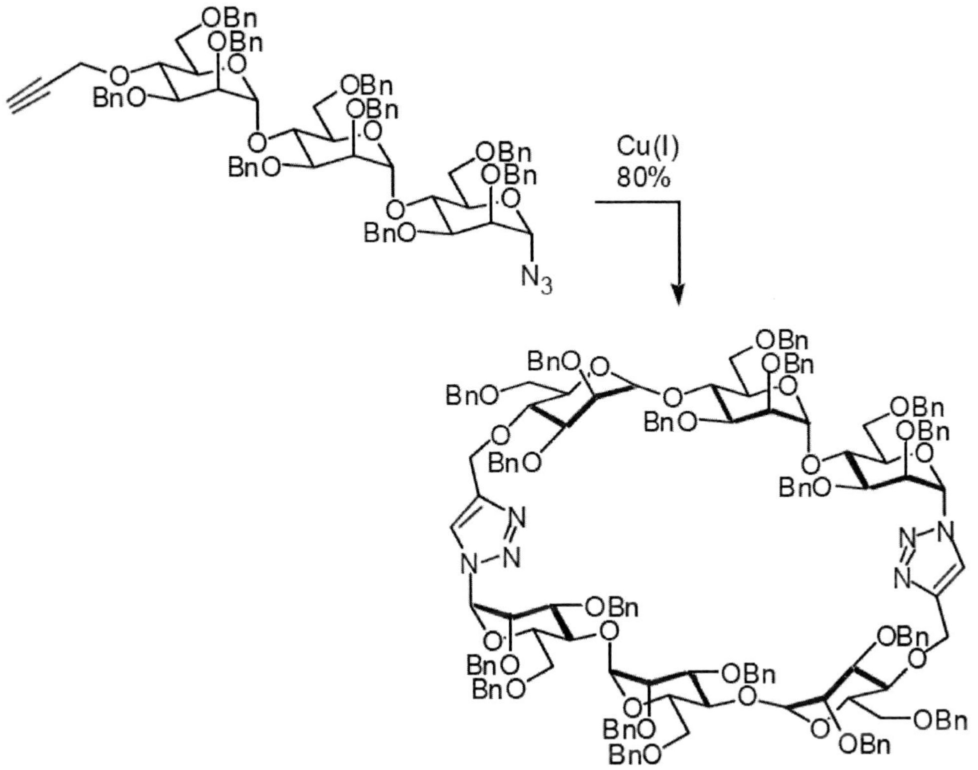

Scheme 12. Convergent synthesis of a manno-β-cyclodextrin using click chemistry [98].

computational methods. It was demonstrated that the 1,2,3-triazole motif imparted increased macrocycle rigidity and cavity size, relative to previously reported nontriazole analogues.

The synthetic modification of preformed CDs provides an alternative route to artificial CDs with particular properties. Exploiting differences in reactivity between primary and secondary hydroxyl groups orientated at the outer and inner faces of CDs, respectively, has allowed for the selective substitution of these groups. Schubert et al. [101] have recently reported the use of click chemistry in the preparation of seven armed star shaped polymer, heptakis-pεCL-β-cyclodextrin. Star shaped polymers have added advantages of over linear polymers since they have lower hydrodynamic volume and a larger number of functionalised end groups for further modification. The acetylene functionalised poly(ε-caprolactone) (pεCL) was prepared via a ring opening polymerisation of ε-caprolactone utilising 5-hexyn-1-ol initiator and tin(II)octanoate catalyst, which was in turn appended to heptakis-azido-β-cyclodextrin via click chemistry and microwave irradiation, affording poly(ε-caprolactone)-β-cyclodextrin in a single step.

3.4. Glycodendrimers, Glycoclusters and Lectin Inhibitors

Glycodendrimers are highly globular, monodisperse, oligo- or polymeric carbohydrate containing nanostructures. Carbohydrates are typically presented at the periphery of a dendritic core in high-densities with well defined, symmetrical branching patterns, and are prepared through a series of iterative steps via convergent or divergent synthetic strategies [102]. The innate structural control, monodispersity and relative size of glycodendrimers - in between that of glycoclusters and polyvalent glycopolymers, endow numerous advantages which have been successfully exploited for a variety of biological and biomedical applications, including multivalent tools for investigating carbohydrate-protein interactions, microbial antiadhesives, cancer vaccines, and as biocompatible scaffolds for drug and gene delivery purposes [78, 103].

The successful synthesis of structurally well defined glycodenrimers and glycoclusters must employ chemistries that maximise atom efficiency. Such chemistries should display a high degree of tolerance towards polyfunctionality and provide access to structurally diverse architectures in the smallest number of steps. Click chemistry has allowed for flexible and rapid synthetic design of a wide range of sophisticated carbohydrate nanostructures of biomedical importance. The first example of the use of Cu^I catalysis in generating triazole linked glycodenrimers was demonstrated Santoyo-González and colleagues [104]. The employment of organic soluble copper catalysts ($(Ph_3P)_3$•CuBr and $(EtO)_3$•CuI) enabled the facile construction of a diverse range of 1,4-disubstituted-1,2,3-triazole linked neoglycoconjugates and glycodendrimers.

Synthetic modifications of the original click chemistry conditions have recently appeared within the literature with the aim of improving the efficiency and scope of click chemistry in multivalent neoglycoconjugate design and synthesis. For example, Van der Eycken and colleagues [105] have reported a highly efficient, one-pot, microwave assisted click chemistry method based on a three component system employing sodium azide, alkyl halide and acetylene. Pieters et al. [106] have reported the facile synthesis of glycodendrimers containing up to nine peripheral carbohydrate units in high yields and shortened reaction times using microwave-assisted click chemistry. Another recent example is provided by Moran and co-workers who have described microwave-assisted click chemistry to rapid label oligonucleotides on solid support with pendant carbohydrate attachments [107] Testament to the impressive scope and compatibility with other synthetic tools, Chang and co-workers have recently explored the utility of sonication (sonochemistry) as a facile means for generating triazole linked neoglycoconjugates under mild, ambient conditions [108]. In addition, a one-pot, orthogonal transformation of glycosyl azides from glycosyl halides followed by click chemistry has also been reported for the efficient preparation of multivalent neoglycoconjugates in solution [109].

Perhaps the greatest advantage of the click methodology in multivalent neoglycoconjugate synthesis is the ability to conduct convergent or divergent syntheses in aqueous solution. This effectively enables the anchoring of deprotected saccharides to the core in a single step, furnishing the target glycodendrimer in high yields and purity, whilst bypassing otherwise necessary activation and/or protection/deprotection steps. Click chemistry has been utilised by Sharpless, Hawker and colleagues [110] as a general and facile strategy for the preparation of a series of bifunctionalised and unsymmetrical dendritic co-polymers incorporating dual purpose binding/detection chain terminals (Scheme 13). Multi-functionalised dendrimers have potential recognition/detection/delivery applications and can be utilised as multivalent ligands for *in situ* monitoring or as novel drug delivery agents. The facile nature of the reaction between the appropriately functionalised dendrons with azide or acetylene substituents allowed for the highly modular, flexible and divergent assembly of the difunctionalised glycodendrimer. The binding element was a polymannose functionalised periphery capable of binding to Concanavalin A (Con A). The fluorescent coumarin functionalised periphery was located at the opposite chain terminal and was employed for potential diagnostic purposes. Multi-functionalised glycodendrimers allow for the introduction of binding elements, reporter elements and/or drug molecules at defined locations within the same nanostructure. The bifunctionalised glycodendrimer was assayed for binding with Con A in a standard hemagglutination assay. From this assay, a relative binding affinity of 240 was observed relative to free mannose, which equated to a per unit affinity of 15.

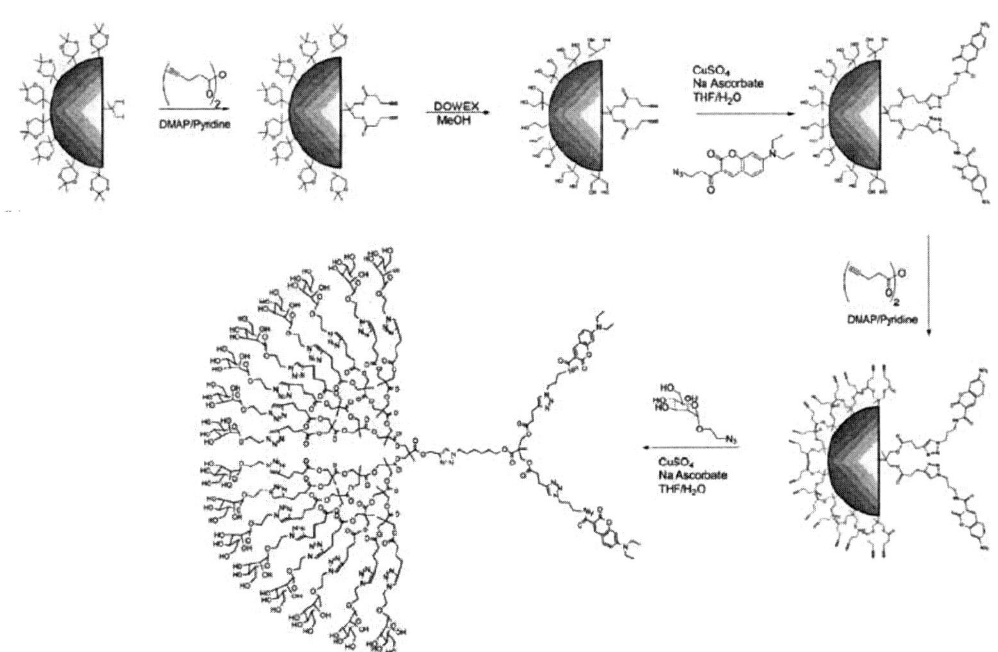

Scheme 13. The synthesis of an unsymmetrical and bifunctionalised dendrimer using click chemistry [110]. From: P. Wu, et al. *Chem. Commun.* 2005, *46*, 5775–5777. Reproduced by permission of The Royal Society of Chemistry ©2005.

Riguera and colleagues [111] have recently reported the facile preparation of unprotected glycodendrimers based on click chemistry. Several *O*-propynyl glycosides based on fucose, mannose, and lactose were appended to three generations of azido terminated gallic acid-triethylene glycol dendrimers in a highly efficient and divergent manner under aqueous conditions and in the presence of catalytic quantities of Cu^I catalyst (1 mol%). Impressive atom efficiency and reported yields (up to 92%) were observed, with the incorporation of up to 27 free propynyl glycosides in a single step. In a separate study by the same group, the applicability of the click methodology was further investigated on other co-polymeric core materials, such as azido terminated PEG-dendritic block co-polymers [112]. Similar atom efficiency and fidelity was also achieved using this versatile and novel co-polymer.

Galectins are a ubiquitous family of cytosolic β-D-galactoside binding proteins, of which 14 are known in mammalian systems [113]. They display broad and diffuse biological activities and their expression during cellular development and oncogenesis is highly variable. In particular, Galectin-1 (Gal-1) and Galectin-3 (Gal-3) are implicated in a wide range of biological and pathological processes including cell apoptosis, innate immune function, oncogenesis as well as HIV-1 infectivity. The chemical modulation of these ambiguous proteins using multivalent inhibitors and effectors is therefore of great interest to glycobiology and clinical research [113b]. Roy and colleagues [114] have prepared a suite of β-D-galactose containing neoglycoconjugates as potential galectin inhibitors from a series of sugar and non-sugar building blocks containing azide or acetylene functionalities. The resulting triazole neoglycoconjugates were rapidly prepared using click chemistry and screened individually against three Galectin proteins; Gal-1, Gal-2 and Gal-3. A C_3-symmetrical triazolo lactose analogue was prepared by a Cu^I catalysed cycloaddition in good yield (83%) and was identified as a strong inhibitor of Gal-1 (20 μM) and a moderate inhibitor of Gal-3 (250 μM). In particular, a 13.3-fold increase in potency for each lactose unit in the trivalent array was observed relative to free lactose against Gal-1, leading to a total relative potency of ~ 40.

In a similar study, Nilsson et al. [115] have examined the binding affinity of a wide range of multivalent lactose-bearing neoglycoconjugates against cancer- and tumour-related galectins, Gal-1, -3, -4N, -4C, -4, -7, -8N and -9N. The multivalent neoglycoconjugates were prepared by a Cu^I catalysed 1,3-DCR of the readily available 2-azidoethyl-β-D-galactopyranosyl-(1→4)-β-D-glucopyranoside and various acetylene functionalised phenyl-bis-alanine and phenyl-tris-alanine building blocks, affording the di- and trivalent neoglycoconjugates, respectively. To investigate the potential effect of the aglycone on binding, the corresponding monovalent homologues were prepared and used as a reference for relative binding affinity along with methyl β-D-lactoside. From a total of eight potential galectin ligands, the divalent carbamate species was found to have a binding affinity for Gal-1 of one order of magnitude stronger than all other ligands studied, with a K_d of 3.2 μM and a cluster effect of 7.7 and 30 relative to the corresponding monomer and methyl-β-D-lactoside, respectively. In contrast, the corresponding monomer displayed markedly weaker affinity for Gal-1, with a K_d of 49 μM.

Deprotected 2-azidoethyl glycosides of lactose and *N*-acetyl glucosamine have been employed by Gao and co-workers [116] as water soluble building blocks in multivalent ligand design. The facile click reaction between the deprotected ω-azido glycosides and the uniformly propynyl alkylated methyl-β-D-glucoside scaffold resulted in the rapid, modular generation of glycoclusters under aqueous conditions (Scheme 14). The two glycoclusters were assayed by a competitive ligand displacement assay (AGP-asialoglycan) using capillary electrophoresis. For both glycoclusters, an impressive 400-fold increase in binding affinity for plant lectin protein RCA120 was observed relative to free lactose, which is indicative of a strong multivalent effect.

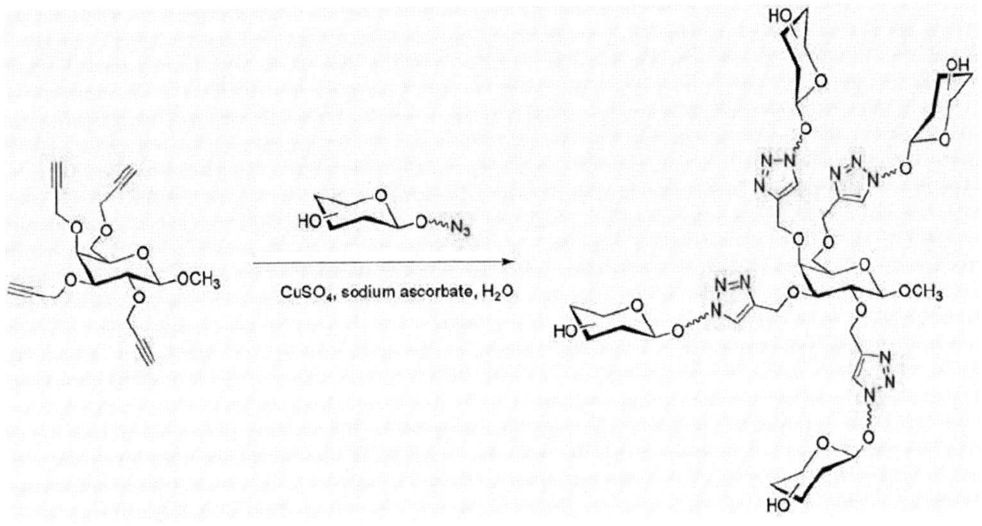

Scheme 14. The preparation of glycoclusters using click chemistry [116].

Du and co-workers [117] have previously reported the mild antitumour and increased white blood cell proliferation activity of linear β-D-(1→6)-*N*-acetyl-β-D-glucosamine hexa- and nonasaccharides. Based on these findings, synthetic efforts were directed towards multivalent models in order to enhance *in vivo* potency. The facile convergent synthesis of a C_3-symmetric (1→6)-*N*-acetyl-β-D-glucosamine octadecasaccharide and the corresponding monovalent hexasaccharide was achieved in good yields using click chemistry. In a preliminary *in vivo* assay, the C_3-symmetic, triantennary octadecasaccharide showed moderate tumour growth inhibition relative to the linear oligosaccharides and the corresponding monovalent hexasaccharide.

4. CARBOHYDRATE MICROARRAYS AND SELF-ASSEMBLED MONOLAYERS

4.1. General

Progress in glycomics research and carbohydrate drug discovery has arisen not only from an impressive array of synthetic and separation technologies, but also the development of robust, high throughput screening tools for the identification of protein-carbohydrate receptor interactions [118]. Carbohydrate microarrays (glycoarrays) have emerged as versatile, high throughput tools for the screening of potential lectin proteins and more recently for investigating the binding of pathogenic bacteria to carbohydrates [119]. Carbohydrate microarrays consist of several hundred or thousands of carbohydrate ligands immobilised on a solid surface. The screening of test proteins for carbohydrate affinity occurs simultaneously in chip format using a variety of design strategies, whilst consuming only minute quantities of test compound and protein, an important prerequisite when using complex oligosaccharides and test compounds which have been painstakingly acquired. Self-assembled monolayers (SAMs) are also gaining attention and consist of homogeneous saccharides packed onto surfaces in a well defined and controllable manner. SAMs technology allows for exquisite control over carbohydrate density and concentration, which allows for real time monitoring and quantitation of binding events [113c]. Owing to the near quantitative product conversion and high fidelity, click chemistry is well poised to be the forerunner biomolecular immobilisation strategy for potential *in situ* screening of carbohydrate-protein interactions.

4.2. Glycoarrays

The earliest example of the construction of glycoarrays using click chemistry was demonstrated by Wong and co-workers [120] in an attempt to develop an accessible, high throughput screening method in microtitre plate format for the identification lectin-like proteins selective for galactose. An azidoethyl galactoside was used as the model carbohydrate for the study and was tethered via a Cu^I/DIPEA catalysed cycloaddition to the hydrophobic propionamide which, inturn was attached noncovalently to the polystyrene support of the well surface (Scheme 15).

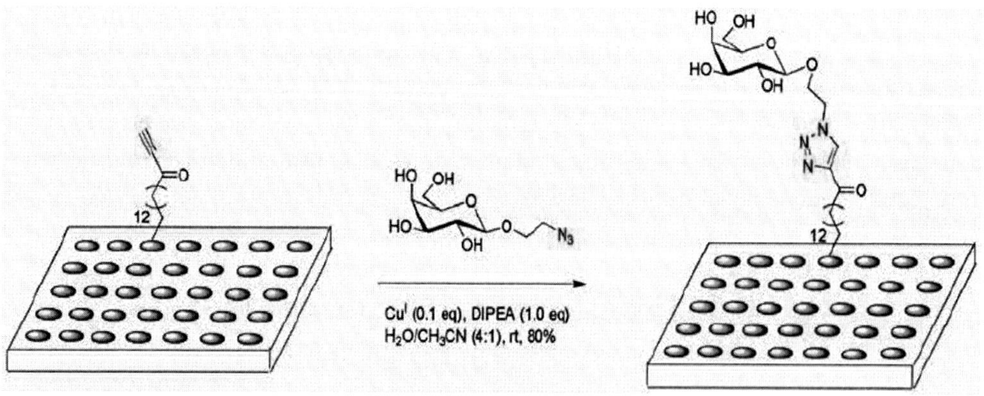

Scheme 15. The immobilisation of an azido ethyl galactoside to a noncovalent, polystyrene supported acetylene linker using click chemistry [120].

The click reaction optimised to allow conjugation on micromolar scales (4-80 μM), and was found complete under all conditions by ESI-MS in excellent yields (up to 92%), according to a previously reported sulfuric acid-phenol (SAP) assay. The Ricin B lectin protein isolated from *Ricinus communus* (castor bean), which selectively binds β-linked D-galactosides, was screened against the galactolipid array. Using the same click chemistry immobilisation strategy, a suite of 11 oligosaccharides displaying variable terminal branching patterns and degrees of terminal sialylation and/or fucosylation were then screened against potential lectin proteins. The enzyme catalysed fucosylation of the immobilised trisaccharide allowed for the *in situ* generation of immobilised sialyl Le^X ligands. The *in situ* generated sialyl Le^X ligand bound selectively to a fucose specific lectin from *Tetragonolobus purpureas* which did not show any affinity for the nonfucosylated trisaccharide precursor.

Wong and colleagues have also reported a similar noncovalent glycoarray utilising click chemistry as a key immobilisation strategy for the high throughput screening of Fuc-T inhibitors [121]. *N*-Acetyl lactosamine was immobilised to the solid support by click chemistry and was incubated with the triazole inhibitor, Fuc-T and the substrate, GDP-Fuc. The *in situ* fucosylation of the displayed *N*-acetyl lactosamine array was quantified by binding to *T. purpureas* lectin conjugated to a peroxidase. In this manner, four compounds were detected with nanomolar inhibition constants, three of which contained a biphenyl substituent which is consistent with earlier findings [45]. Mei et al. [122] have also employed click chemistry as an immobilisation strategy in conjunction with quantum dots (QDs) as a fluorescent label for the detection and quantitation of carbohydrate-protein binding.

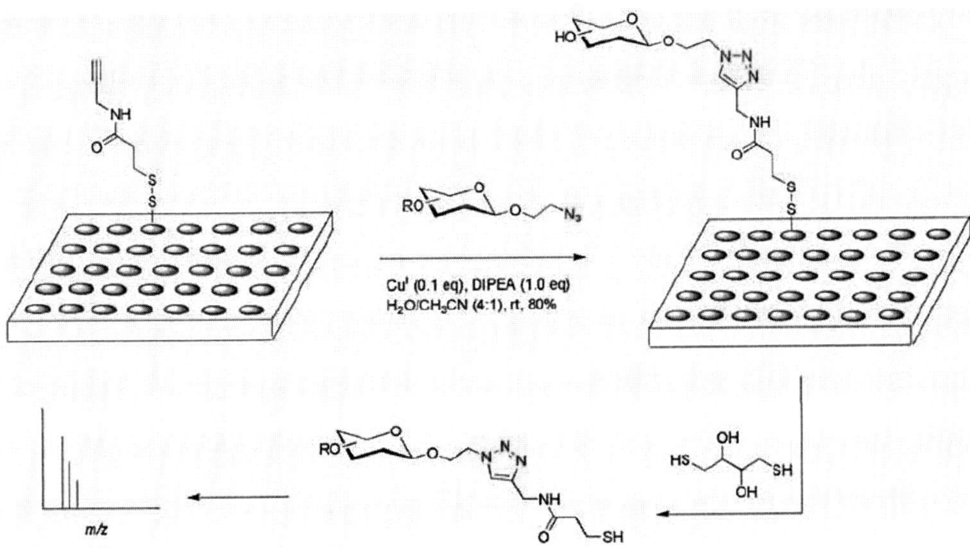

Scheme 16. Triazole linked glycoarrays formed by a Cu^I catalysed 1,3-DCR to a covalently supported propionamide linker [123].

In an effort to generate glycoarrays with enhanced stability and robustness towards high throughput screening protocols, Wong and co-workers [123] have exploited click chemistry to immobilise carbohydrate arrays to *covalently* attached solid supported linkers via thiourea and amide bonds. The immobilisation of azido ethyl glycosides to an alkyne terminated propionamide linker proceeded with high fidelity and efficiency using click chemistry (Scheme 16). Analysis of the immobilised ligand by ESI-MS could be achieved by the mild, reductive cleavage of the disulfide linker from the covalent anchor and subsequent dissolution. In a similar approach to the previous study, two glycoarrays were screened against lectins selective for α-fucose (*Lotus tetragonolobus*, LTL) and galactose (*Erythrina cristagalli,* EC) residues. The tetrasaccharide breast cancer antigenic determinant, Globo-H was also screened against LTL and was found to bind to this lectin in a concentration dependent manner (K_d = 22.8 μM), this was comparable with previous findings. In the same study, the immobilisation of carbohydrate ligands using click chemistry was also shown to be applicable to ELISA type formats. The covalently attached glycoarrays were also assessed for their ability to bind a GP-120-antibody (2G12), which recognises dense oligomannose clusters presented on the surface of HIV-1 gp120 ($Man_9GlcNAc_2$). Several triazole tethered oligomannose arrays were reported to display very strong affinity for the antibody (K_d = 0.1-1.0 μM).

4.3. Carbohydrate Self-Assembled Monolayers

Click chemistry between azide functionalised sugar linkers with solid-supported alkynes has recently been exploited by Wang and co-workers [124] for the fabrication of carbohydrate containing SAMs, using an approach similar to that developed by Chidsey and co-workers [125] for the fabrication of oligonucleotide SAMs. In this case, traditional alkane thiol linkers were employed, owing to their clean and spontaneous adsorption to the solid gold surface. Further functionalisation at the linker terminus with an alkyne group allowed for the fabrication of the carbohydrate SAM by the chemoselective delivery of various azido(triethylene)glycol-*O*-glycosides, based on α-D-mannose, α-D-galactose and β-D-lactose via click chemistry (Scheme 17). The carbohydrate ligands were successively interspersed with azido(triethylene)glycol forming a stable, rigid and well dispersed SAM. The relative binding affinities (reported as affinity constants, K_A) of various lectin proteins selective for these carbohydrate epitopes were determined by real time analytical techniques – surface plasmon resonance (SPR) and quartz crystal microbalance (QCM). Relative binding affinities to the lectin proteins, in particular that of the triazole-mannose SAM with Con A lectin ($K_A = 3.9 \pm 0.2 \times 10^6$ M^{-1}), were found to be in good agreement with previously reported values.

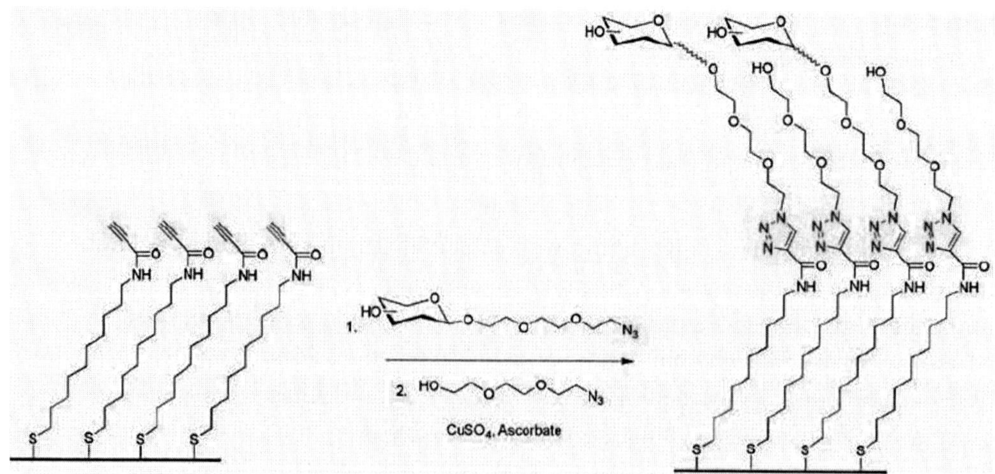

Scheme 17. Fabrication of carbohydrate SAM using Cu^I catalysed 1,3-DCR [124].

Chaikof and co-workers [126] have recently demonstrated the potential of click chemistry to immobilise carbohydrates to glass surfaces using flexible, biocompatible polyethylene glycol (PEG) linkers. The attachment of the PEG linker to the maleimide terminated solid support was achieved using a PEG chain functionalised at both termini with cyclopentadiene and an alkyne. Successive immobilisation of the azide functionalised carbohydrate was then achieved using click chemistry under buffered aqueous conditions (PBS, pH 7.0) with impressive fidelity and homogeneity.

5. Conclusions

Click chemistry has had a considerable impact on drug discovery and biotechnology since its conception some 7 years ago. However, the application to glycomics research and carbohydrate based drug development is more recent, yet the sheer volume of literature which has appeared the past 3 to 4 years is testament to its broad scope and versatility. Click chemistry holds great promise as a premier bioconjugation strategy for a wide variety of preparative and analytical applications. It has become a powerful synthetic tool for preparing a wide range of structurally diverse neoglycoconjugates, either as probes for elucidation of carbohydrate-protein binding or as drug leads. It has also shown potential as a bioorthogonal strategy for the labelling of chemically engineered cell surface glycans. These tools serve to correlate glycan structure and function in physiological states of health and disease, leading to enhanced understanding of the nature of protein and lipid glycosylation and the identification of novel targets for clinical intervention and vaccine development. Click chemistry has also become a robust immobilisation strategy for the fabrication of surface based carbohydrate arrays and monolayers that facilitate high throughput lectin carbohydrate investigations for the identification of lectin proteins of particular relevance in pathology. Owing to its impressive scope and utility in many aspects of carbohydrate based drug design processes, it is reasonable to envisage that click chemistry within carbohydrate based drug research will flourish.

6. Update

The premier click reaction, the Cu-catalyzed azide-alkyne cycloaddition [127], has seen further widespread application in carbohydrate chemistry and glycobiology in the short span of time since the original publication of this chapter [128]. The authors have had additional success in application of this chemistry to the identification of new antitubercular compounds [129] and carbonic anhydrase inhibitors [130]. Kumar has reported the direct one-pot conversion of glycosyl halides into glycosyl triazoles [131], bypassing purification of the glycosyl azide [57], chemistry that can be accelerated by microwave irradiation [105]. Bertozzi [132] and Boons [133] have each developed copper-free Huisgen cycloaddition chemistry for the visualization of metabolically labeled glycoconjugates of living cells by taking advantage of the increased reactivity of strained, cyclic alkynes. This represents a significant advance for the characterization and manipulation of cell surfaces by chemical means. Bertozzi and coworkers have monitored incorporation of a GalNAc azide into glycan expression of deveopling zebrafish with precise spatiotemporal analysis [132]. The Boons group has visualized incorporation of ManNAc azide into sialic acid cell-surface structures of Chinese hamster ovary cells using confocal microscopy [133]. Wang reported a novel tetramer of the tetrasaccharide of high-mannose N-glycans as synthetic immunogen for generation of HIV-neutralizing antibodies [134]. The ability to rapidly assemble oligomers of carbohydrate ligands has made click chemistry a frequent method of choice for the study of lectins as these proteins generally recognize a cluster of their respective ligands [135]. Finally, Click chemistry has been recently reviewed in the context of its medicinal and carbohydrate chemistry applications [136]. The low steric demand of both azide and alkyne allow coupling between molecules of all shapes and sizes under both biological and abiologic conditions. It is chemistry that is here to stay.

7. References

[1] (a) Varki, A., Cummings, R., Esko, J., Freeze, H., Hart, G., Marth, J. Eds. In *Essentials of Glycobiology*. Cold Spring Harbor Laboratory Press, Cold Spring Harbor, NY, 1999; (b) Haltiwanger, R. S., Lowe, J. B. Role of glycosylation in development. *Annu. Rev. Biochem.* 2004, *73*, 491-537; (c) Rudd, P. M., Elliott, T., Cresswell, P., Wilson, I. A., Dwek, R. A. Glycosylation and the Immune System. *Science* 2001, *291* (5512), 2370-2375.

[2] Ohtsubo, K., Marth, J. D., Glycosylation in Cellular Mechanisms of Health and Disease. *Cell* 2006, *126*, 855-867; (b) Ley, K. The role of selectins in inflammation and disease. *Trends Mol. Med.* 2003, *9* (6), 263-268; (c) Hölemann, A., Seeberger, P. H., Carbohydrate diversity: synthesis of glycoconjugates and complex carbohydrates. *Chem. Biotech.* 2004, *15*, 615-622.

[3] (a) Walsh, G., Roy Jefferis, R. Post-translational modifications in the context of therapeutic proteins. *Nature Biotechnol.* 2006, *24* (10), 1241-1252; (b) Shriver, Z., Raguram, S., and Sasisekharan, R. Glycomics: A pathway to a class if new and improved therapeutics. *Nat. Rev. Drug Discov.* 2004, *3*, 863–873; (c) Doores, K. J., Gamblin, D. P., Davis, B. G. Exploring and exploiting the therapeutic potential of glycoconjugates. *Chem. Eur. J.* 2006, *12* (3), 656-665; (d) Dove, A. The bittersweet promise of glycobiology. *Nat. Biotechnol.* 2001, *19*, 913-917.

[4] (a) Davis, B. G., Synthesis of Glycoproteins. *Chem. Rev.* 2002, *102*, 579-601. (b) Spiro, R. G. Protein glycosylation: nature, distribution, enzymatic formation, and disease implications of glycopeptide bonds. *Glycobiology* 2002, *12*, 43R-50R.

[5] (a) Bertozzi, C., R., Kiessling, L. L. Chemical Glycobiology. *Science* 2001, *291* (5512), *2357-2364*; (b) Buskas, T., Ingale, S., Booms, G.–J., Glycopeptides as versatile tools for glycobiology. *Glycobiology* 2006, *16* (8), 113R-136R; (c) Seitz, O. Glycopeptides and Glycoproteins: Synthetic Chemistry and Biology. In *Carbohydrate-based Drug Discovery*. Wong, C.–H. ed., Wiley VCH, Weinheim, Germany, 2003. pp 169-209.

[6] (a) Saotome, C., Kanie, O. Synthetic Methodologies. In *Carbohydrate-based Drug Discovery*. Wong, C.–H. ed., Wiley VCH, Weinheim, Germany, 2003. pp 1-33; (b)

Seeberger, P. H., Solid-Phase Oligosaccharide Synthesis. In *Carbohydrate-based Drug Discovery*. Wong, C.–H. ed., Wiley VCH, Weinheim, Germany, 2003. pp 103-107; (c) Dudkin, V. Y., Miller, J. S., Danishefsky, S. J. Chemical Synthesis of Normal and Transformed PSA Glycopeptides. *J. Am. Chem. Soc.* 2004, *126* (3), 736-738; (d) Zhang, J., Shao, J., Kowal, P., Wang, P. G. Enzymatic Synthesis of Oligosaccharides. In *Carbohydrate-based Drug Discovery*. Wong, C.–H. ed., Wiley VCH, Weinheim, Germany, 2003. pp 129-162.

[7] (a) Sears, P., Wong, C.–H., Carbohydrate Mimetics: A New Strategy for Tackling the Problem of Carbohydrate-Mediated Biological Recognition. *Angew. Chem. Int. Ed.* 1999, *38* (16), 2300-2324; (b) Ernst, B., Kolb, H. C., Schwardt, O. Carbohydrate mimetics in drug discovery. In *Organic Chemistry of Sugars*. Levy, D. E. Ed., CRC Press LLC, Boca Raton, FL, USA, 2006, pp. 803-861.

[8] (a) Marcaurelle, L. A., Seeberger, P. H. Combinatorial carbohydrate chemistry. *Curr. Opin. Chem. Biol.* 2002, *6* (3), 289-296; (b) Velter, I., La Ferla, B., Nicotra, F., Carbohydrate-Based Molecular Scaffolding. *J. Carbohydr. Chem.* 2006, *25* (2-3), 97-138; (c) Gabius, H.–J., Siebert, H.–C., Andry, S., Jimenez-Barbero, J., Rüdiger, H. Chemical Biology of the Sugar Code. *ChemBiochem.* 2004, *5* (6), 741-764.

[9] (a) Macmillan, D., Daines, A. M. Recent Developments in the Synthesis and Discovery of Oligosaccharides and Glycoconjugates for the Treatment of Disease. *Curr. Med. Chem.,* 2003, *10* (24), 2733-2773; b) Kannagi, R. Carbohydrate-Based Treatment of Cancer Metastasis. In *Carbohydrate-based Drug Discovery*. Wong, C.–H. ed., Wiley VCH, Weinheim, Germany, 2003. pp 803-829; (c) Drinnan, N. Vari, F. Aspects of the Stability and Bioavailability of carbohydrates and Carbohydrate Derivatives. *Mini-Rev. Med. Chem.* 2003, *3* (7), 633-649; (d) Dube, D. H., Bertozzi, C. R., Glycans in cancer and inflammation - potential for therapeutics and diagnostics. *Nat. Rev. Drug Disc.* 2005, *4* (6), 477-488.

[10] (a)Thorson, J. S., Vogt, T. Glycosylated Natural Products. In *Carbohydrate-based Drug Discovery*. Wong, C.–H. ed., Wiley VCH, Weinheim, Germany, 2003. pp.803-829; (b) Wessjohann, L. A., Ruijter, E., Garcia-Rivera, D., Brandt, W. What can a chemist learn from nature's macrocycles? – A brief, conceptual view. *Mol. Div.* 2005, *9* (1-3), 171–186; (c) Quader, S.; Boyd, S. E.; Jenkins, I. D.; Houston, T. A. Multisite Modification of Neomycin B: A Combined Mitsunobu and Click Chemistry Approach. *J. Org. Chem.* 2007, *72* (6), 1962-1979.

[11] For examples of the recent failures of carbohydrate-based drugs during clinical trials see: Michael, A. Carbohydrates down, but not out. *Nature Biotechnol.* 1997, *15* (12),1233–1234.

[12] Holzer, C. T., von Itzstein, M., Jin, B., Pegg, M. S., Stewart, W. P., Wu, W. Y. Inhibition of sialidases from viral, bacterial and mammalian sources by analogues of 2-deoxy-2,3-didehydro-*N*-acetylneuraminic acid modified at the C-4 position. *Glycoconjugate J.* 1993, *10* (1), 40-44.

[13] Islam, T., Linhardt, R. J. Chemistry, Biochemistry, and Pharmaceutical potentials of Glycosaminoglycans and Related Saccharides. In *Carbohydrate-based Drug Discovery*. Wong, C.–H. ed., Wiley VCH, Weinheim, Germany, 2003. pp. 407-440.

[14] (a) Asano, N., Nash, R. J., Molyneux, R. J., Fleet, G. W. J. Sugar-mimic glycosidase inhibitors: natural occurrence, biological activity and prospects for therapeutic application. *Tetrahedron Asymm.* 2000, *11* (8), 1645-1680; (b) Tietze, L. F., Lieb, M., Herzig, T., Haunert, F., Schubert, I. A Strategy for Tumor-Selective Chemotherapy by Enzymatic Liberation of seco-duocarmycin SA-derivatives from Non-toxic Prodrugs. *Bioorg. Med. Chem.* 2001, *9* (7), 1929-1939.

[15] Masereel, B., Rolin, S., Abbate, F., Scozzafava, A., Supuran, C. T. Carbonic Anhydrase Inhibitors: Anticonvulsant Sulfonamides Incorporating Valproyl and Other Lipophilic Moieties. *J. Med. Chem.* 2002, 45, 312-320.

[16] For reviews on chemoselective ligation in glycopeptide synthesis, see: a) Hang, H.C.; Bertozzi, C.R. Chemoselective approaches to glycoprotein assembly. *Acc. Chem. Res.* 2001, *34* (9), 727-736; b) Peri, F., Nicotra, F., Chemoselective ligation in glycochemistry. *Chem. Commun.* 2004, (6), 623-627.

[17] (a) Huisgen, R. 1,3-Dipolar cycloadditions. *Angew. Chem.* 1963, *75* (13), 604-637; (b) Huisgen, R. 1,3-Dipolar cycloadditions. 76. Concerted nature of 1,3-dipolar cycloadditions and the question of diradical intermediates. *J. Org. Chem.* 1976, *41* (3), 403-419; (c) Gothelf, K. V., Jørgensen, K. A. Asymmetric 1,3-Dipolar Cycloaddition Reactions. *Chem. Rev.* 1998, *98* (2), 863-909.

[18] (a) Rostovtsev, V. V., Green, L. G., Fokin, V. V., Sharpless, K. B. A Stepwise Huisgen Cycloaddition Process: Copper(I)-Catalyzed Regioselective Ligation of Azides and Terminal Alkynes. *Angew. Chem. Int. Ed.* 2002, *41* (14), 2596-2599; (b) Tornøe, C. W., Christensen, C., Meldal, M. Peptidotriazoles on Solid Phase: [1,2,3]-Triazoles by Regiospecific Copper(I)-Catalyzed 1,3-Dipolar Cycloadditions of Terminal Alkynes to Azides. *J. Org. Chem.* 2002, *67* (9), 3057-3064.

[19] Kolb, H. C., Finn, M. G., Sharpless, K. B. Click Chemistry: Diverse Chemical Function from a Few Good Reactions. *Angew. Chem. Int. Ed.* 2001, *40* (11), 2004-2021.

[20] (a) Sustmann, R. Simple model for substituent effects in cycloaddition reactions. I. 1,3 - Dipolar cycloadditions. *Tetrahedron Lett.* 1971, *29*, 2717-2720; (b) Bräse, S., Gil, C., Knepper, K., and Zimmermann, V. Organic Azides: An Exploding Diversity of a Unique Class of Compounds. *Angew. Chem. Int. Ed.* 2005, *44* (33), 5188–5240.

[21] (a) Himo, F., Lovell, T., Hilgraf, R., Rostovtsev, V. V., Noodleman, L., Sharpless, K. B., Fokin, V. V. Copper(I)-Catalyzed Synthesis of Azoles. DFT Study Predicts Unprecedented Reactivity and Intermediates. *J. Am. Chem. Soc.* 2005, *127* (41), 210-216; (b) Bock, V. D., Hiemstra, H., van Maarseveen, J. H. CuI-Catalyzed Alkyne–Azide "Click" Cycloadditions from a Mechanistic and Synthetic Perspective *Eur. J. Org. Chem.* 2006, 51–68.

[22] (a) Kolb, H. C., Sharpless, K. B. The growing impact of click chemistry on drug discovery. *Drug Disc. Today* 2003, *8* (24), 1128-1136; (b) Röper, S., Kolb, H. C. Click Chemistry for Drug Discovery. In *Fragment-based Approaches in Drug Discovery.* Jahnke, W., Erlanson, D. A. Eds. Wiley-VCH Verlag GmbH & Co. KGaA, Weinheim, Germany, 2006, pp. 113-139.

[23] Weber, L. In vitro combinatorial chemistry to create drug candidates. *Drug Disc. Today* 2004, *1* (3), 261-267.

[24] (a) Oh, K., Guan, Z. A convergent synthesis of new β-turn mimics by click chemistry. *Chem. Commun.* 2006, (29), 3069–3071. b) Horne, W. S., Yadav, M. K., Stout, C. D., Ghadiri, M. R. Heterocyclic Peptide Backbone Modifications in an r-Helical Coiled Coil. *J. Am. Chem. Soc.* 2004, *126* (47), 15366-15367.

[25] For examples of the anti HIV-1 activity of 1,2,3-triazole containing compounds, see: (a) Whiting, M., Tripp, J. C., Lin, Y.–C., Lindstrom, W., Olson, A. J. Elder, J. H., Sharpless, K. B., Fokin, V. V. Rapid Discovery and Structure-Activity Profiling of Novel Inhibitors of Human Immunodeficiency Virus Type 1 Protease Enabled by the Copper(I)-Catalyzed Synthesis of 1,2,3-Triazoles and Their Further Functionalization. *J. Med. Chem.* 2006, *49* (26), 7697-7710; (b) Hosahudya N. Gopi, H. N., Tirupula, K. C., Baxter, S., Ajith, S., Chaiken, I. W. Click Chemistry on Azidoproline: High-Affinity Dual Antagonist for HIV-1 Envelope Glycoprotein gp120. *Chem MedChem.* 2006, *1* (1), 54–57; (c) Whiting, M., Muldoon, J., Lin, Y.–C., Silverman, S. M., Lindstrom, W., Olson, A. J., Kolb, H. C., Finn, M. G., Sharpless, K. B., Elder, J. H., Fokin, V. V. Inhibitors of HIV-1 protease by using in situ click chemistry. *Angew. Chem. Int. Ed.* 2006, *45* (9), 1435-1439.

[26] For examples of the antimicrobial properties of 1,2,3-triazole containing compounds, see: (a) Genin, M. J., Allwine, D. A., Anderson, D. J., Barbachyn, M. R., Emmert, D. E., Garmon, S. A., Graber, D. R., Grega, K. C., Hester, J. B., Hutchinson, D. K., Morris, J., Reischer, R. J., Ford, C. W., Zurenko, G. E., Hamel, J. C., Schaadt, R. D., Stapert, D., Yagi, B. H. Substituent Effects on the Antibacterial Activity of Nitrogen-Carbon-Linked (Azolylphenyl) oxazolidinones with Expanded Activity Against the Fastidious Gram-Negative Organisms *Haemophilus influenzae* and *Moraxella catarrhalis*. *J. Med. Chem.* 2000, *43* (5), 953–970; (b) Dabak, K., Sezer, Ö., Akar, A., Anaç, O. Synthesis and investigation of tuberculosis inhibition activities of some 1,2,3-triazole derivatives. *Eur. J. Med. Chem.* 2003, *38* (2), 215-218.

[27] Brockunier, L. L., Parmee, E. R., Ok, H. O., Candelore, M. R., Cascieri, M. A., Colwell, L. F., Deng, L., Feeney, W. P., Forrest, M. J., Hom, G. J., MacIntyre, D. E., Tota, L., Wyvratt, M. J., Fisher, M. H., Weber, A. E. Human β3-Adrenergic Receptor Agonists Containing 1,2,3-Triazole-Substituted Benzenesulfonamides. *Bioorg. Med. Chem. Lett.* 2000, *10* (20), 2111–2114.

[28] Cunha, A. C., Figueiredo, J. M., Tributino, J. L. M., Miranda, A. L. P., Castro, H. C., Zingali, R. B., Fraga, C. A. M., de Souza, M. C. B. V., Ferreirac, V. F., Barreiroa, E. J. Antiplatelet Properties of Novel *N*-Substituted-phenyl-1,2,3-triazole-4-acylhydrazone Derivatives. *Bioorg. Med. Chem.* 2003, *11* (9), 2051–2059.

[29] For examples of the anti-inflammatory activity of 1,2,3-triazole containing compounds, see: of (a) Tullis, J. S., Van-Rens, J. C., Natchus, M. G., Clark, M. P., De, B., Hsieh, L. C., Janusz, M. J. The Development of New Triazole Based Inhibitors of Tumor Necrosis Factor-α (TNF-α) Production. *Bioorg. Med. Chem. Lett.* 2003, *13* (10), 1665–1668; (b) Kim, D. K.; Kim, J.; Park, H. J. Synthesis and biological evaluation of novel 2-pyridinyl-[1,2,3]triazoles as inhibitors of transforming growth factor *b*1 type 1 receptor. *Bioorg. Med. Chem. Lett.* 2004, *14* (10), 2401–2405.

[30] (a) Matsuda, A., Sasaki, T. Antitumor activity of sugar-modified cytosine nucleosides. *Cancer Science* 2004, *95* (2), 105-111; (b) Al-Madhoun, A. S., Tjarks, W., Eriksson, S.

The role of thymidine kinases in the activation of pyrimidine nucleoside analogues. *Mini-Rev. Med. Chem.* 2004, *4* (4), 341-350; (c) Ichikawa, E., Kato, K. Sugar-modified nucleosides in past 10 years, a review. *Curr. Med. Chem.* 2001, *8* (4), 385-423; (d) Gumina, G., Chong, Y., Choo, H., Song, G.–Y., Chu, C. K. L-nucleosides: antiviral activity and molecular mechanism. *Curr. Top. Med. Chem.* 2002, *2* (10), 1065-1086.

[31] For reviews concerning 1,3-dipolar cycloadditions on carbohydrates, see: (a) Gallos, J. K., Koumbis, A. E. 1,3-Dipolar cycloadditions in the Synthesis of Carbohydrate Mimics. Part 1: Nitrile Oxides and Nitronates. *Curr. Org. Chem.* 2003, *7* (5), 397-425; (b) Gallos, J. K., Koumbis, A. E., 1,3-Dipolar cycloadditions in the Synthesis of Carbohydrate Mimics. Part 2: Nitrones and Oximes. *Curr. Org. Chem.* 2003, *7* (6), 585-628; (c) Gallos, J. K., Koumbis, A. E., 1,3-Dipolar cycloadditions in the Synthesis of Carbohydrate Mimics. Part 3: Azides, Diazo Compounds and Other Dipoles. *Curr. Org. Chem.* 2003, *7* (8), 771-793.

[32] (a) Lillelund, V. H., Jensen, H. H., Liang, X., Bols, M., Recent Developments of Transition-State Analogue Glycosidase Inhibitors of Non-Natural Product Origin. *Chem. Rev.* 2002, *102* (2), 515-553. (b) Houston, T. A.; Blanchfield, J. T. Back to (non)-Basics: Recent Developments in Neutral and Charge-Balanced Glycosidase Inhibitors. *Mini-Rev. Med. Chem.* 2003, *3*, 669-678.

[33] Nagstatin and heterocyclic mimics as a potent hexosamidase inhibitors, see: (a) Shanmugasundaram, B., Debowski, A. W., Dennis, R. J., Davies, G. J., Vocadlo, D. J., Vasella, A. Inhibition of *O*-GlcNAcase by a gluco-configured nagstatin and a PUGNAc-imidazole hybrid inhibitor. *Chem. Commun.* 2006, (42), 4372-4374; (b) Miroslav, T., Andrea, V. Synthesis of *N*-acetylglucosamine-derived nagstatin analogues and their evaluation as glycosidase inhibitors. *Helv. Chim. Acta.* 2005, *88* (1), 10-22. (c) Tatsuta, K., Shozo, M., Ohta, S., Gunji, H. Syntheses and glycosidase inhibiting activities of nagstatin analogs. *J. Antibiotics.* 1995, *48* (3), 286-288; (d) Tatsuta, K., Ikeda, Y., Miura, S. Synthesis and glycosidase inhibitory activities of nagstatin triazole analogs. *J. Antibiotics* 1996, *49* (8), 836-838.

[34] Li, R. K., Rinaldi, M. J. In Vitro Antifungal Activity of Nikkomycin Z in Combination with Fluconazole or Itraconazole. *Antimicrob. Agents Chemother.* 1999, *43* (6), 1401–1405.

[35] Kobayashi, Y. Chemistry and biology of trehazolins. *Carbohydr. Res.* 1999, *315* (1-2), 3-15.

[36] Rao, F. V., Douglas R. Houston, D. R., Boot, R. G., Aerts, J. M. F. G., Sakuda, S., van Aalten, D. M. F. Crystal Structures of Allosamidin Derivatives in Complex with Human Macrophage Chitinase. *J. Biol. Chem.* 2003, *278* (22), 20110-20116.

[37] (a) Dolhem, F., Al Tahli, F., Lièvre, C., Demailly, G. Efficient Synthesis of 1,2,3-Triazole-Fused Bicyclic Compounds from Aldoses. *Eur. J. Org. Chem.* 2005, *23*, 5019-5023; (b) Pearson, W. H., Lovering, F. E. Assembly of 3α-Arylperhydroindoles by the Intramolecular Cycloaddition of 2-Azaallyl Anions with Alkenes. Total Syntheses of (±)-Crinine, (±)-6-Epicrinine, (-)-Amabiline, and (-)-Augustamine. *J. Org. Chem.* 1998, *63* (26), 3607-3617.

[38] Rossi, L. L., Basu, A. Glycosidase inhibition by 1-glycosyl-4-phenyl triazoles. *Bioorg. Med. Chem. Lett.* 2005, *15* (15), 3596–3599.

[39] Périon, R., Ferrières, V., M. Isabel García-Moreno, M. I., Mellet, C. O. Duval, R., José M. Fernández, G., Plusquelleca, D. 1,2,3-Triazoles and related glycoconjugates as new glycosidase inhibitors. *Tetrahedron.* 2005, *61*, 9118-9128.

[40] Thomson, R., von Itzstein, M. *N*-Acetylneuraminic acid Derivatives and Mimetics as Anti-influenza Agents. In *Carbohydrate-based Drug Discovery.* Wong, C.–H., Ed.; Wiley VCH press, Weinheim, Germany, 2003, pp 831-861.

[41] Li, J., Zheng, M., Tang, W., He, P. –L., Zhu, W., Li, T. Syntheses of triazole-modified zanamivir analogues via click chemistry and anti-HIV activities. *Bioorg. Med. Chem. Lett.* 2006, *16* (21), 5009-5013.

[42] Smith, P. W., Whittington, A. R., Sollis, S. L., Howes, P. D., Taylor, N. R. Novel inhibitors of influenza sialidases related to zanamivir. Heterocyclic replacements of the glycerol sidechain. *Bioorg. Med. Chem. Lett.* 1997, *7* (17), 2239-2242.

[43] (a) Schmidt, R. R., Jung, K, -H., Glycosyl transferase inhibitors, in *Carbohydrate based drug discovery.* Vol. 2, Wong, C.–H., ed. Wiley-VCH press, Weinheim, Germany, 2003, pp. 609–659; (b) Palcic, M. M., Qian, X., Glycosyl transferase inhibitors, in *Carbohydrates in chemistry and biology.* Vol. 3, Ernst, B., Hart, G. W., Sinay, P. ed. Wiley-VCH press, Weinheim, Germany, 2000, pp. 293–312.

[44] Mitchell, M. L., Tian, F., Lee, L. V., Wong, C.–H. Synthesis and evaluation of transition-state analogue inhibitors of α-1,3- fucosyltransferase. *Angew. Chem. Int. Ed.* 2002, *41* (16), 3041-3044.

[45] Lee, L. V., Mitchell, M. L., Huang, S, -J., Fokin, V. V., Sharpless, K. B., Wong, C.–H. A Potent and Highly Selective Inhibitor of Human α-1,3-Fucosyltransferase via Click Chemistry. *J. Am. Chem. Soc.* 2003, *125* (32), 9588-9589.

[46] Sernee, M. F., Ralton, J. E., Dinev, Z., Khairallah, G. N., O'Hair, R. A., Williams, S. J., McConville, M. J. Leishmania β-1,2-mannan is assembled on a mannose-cyclic phosphate primer. *Proc. Nat. Acad. Sci. USA.* 2006, *103* (25), 9458-9463.

[47] (a) Naderer, T., McConville, M. J. Evidence that intracellular β1-2 mannan is a virulence factor in *Leishmania* parasites. *Mol. Biochem. Parasitol.* 2002, *125* (1-2), 147-161; (b) Garami, A., Ilg, T. Disruption of mannose activation in *Leishmania Mexica a*: GDP-mannose phosphorylase is required for virulence, but not viability. *EMBO J.* 2001, *20* (14), 3657-3666 (c) Garami, A. Mehlert, T. Ilg, *Mol. Cell. Biol.* 2001, *21* (23), 8168.

[48] van der Peet, P., Gannon, C. T., Ian Walker, I., Dinev, Z., Angelin, M., Tam, S., Ralton, J. E., McConville, M. J., Williams, S. J. Use of Click Chemistry to Define the Substrate Specificity of Leishmania β-1,2-Mannosyltransferases. *ChemBiochem.* 2006, *7* (9), 1384-1391.

[49] Supuran, C. T. Carbonic anhydrases: Catalytic and inhibition mechanism, distribution and physiological roles. In *Carbonic Anhydrase: Its Inhibitors and Activators.* Supuran, C. T., Scozzafava, A., Conway, J., Eds.; CRC Press: Boca Raton, FL, 2004; pp 1-24.

[50] For reviews on CA inhibitors see: (a) Pastorekova, S., Parkkila, S., Pastorek, J., Supuran, C. T. Carbonic Anhydrases: Current State of the Art, Therapeutic Applications and Future Prospects. *J. Enzyme Inhib. Med. Chem.* 2004, *19*(3), 199-229; (b) Scozzafava, A., Mastrolorenzo, A., Supuran, C. T. Carbonic anhydrase inhibitors

and activators and their use in therapy. *Expert Opin. Ther. Patents* 2006, *16* (12), 1627-1664; (c) Clare, B. W., Supuran, C. T., A perspective on quantitative structure-activity relationships and carbonic anhydrase inhibitors. *Expert Opin. Drug Metabol. Toxicol.* 2006, *2* (1), 113-137; (d) Thiry, A., Dogne, J. –A., Masereel, B., Supuran, C. T. Targeting tumor-associated carbonic anhydrase IX in cancer therapy. *Trends Pharm. Sci.* 2006, *27* (11), 564-573.

[51] Wilkinson, B. L., Bornaghi, L. F., Houston, T. A., Innocenti, A., Supuran, C. T., Poulsen, S.–A. A Novel Class of Carbonic Anhydrase Inhibitors: Glycoconjugate Benzene Sulfonamides Prepared by "Click-Tailing" *J. Med. Chem.* 2006, *49* (22), 6539-6548.

[52] (a) Wilkinson, B. L., Bornaghi, L. F., Houston, T. A., Innocenti, A., Vullo, D., Supuran, C. T., Poulsen, S.–A. Inhibition of membrane-associated carbonic anhydrase isozymes IX, XII and XIV with a library of glycoconjugate benzenesulfonamides. *Bioorg. Med. Chem. Lett.* 2007, *17* (4), 987-992; (b) Wilkinson, B. L., Bornaghi, L. F., Wright, A. D., Houston, T. A., Poulsen, S.–A. Antimycobacterial Activity of a Bis-Sulfonamide. *Bioorg. Med. Chem. Lett.* 2007, *17*(5), 1355-1357.

[53] Wilkinson, B. L., Bornaghi, L. F., Houston, T. A., Innocenti, A., Vullo, D., Supuran, C. T., Poulsen, S.–A. Carbonic anhydrase inhibitors: inhibition of isozymes I, II and IX with triazole linked *O*-glycosides of benzene sulfonamides. *J. Med. Chem.* 2007, *50*, 1651-1657.

[54] (a) Seitz, O. Glycopeptides and Glycoproteins: Synthetic Chemistry and Biology, in *Carbohydrate based drug discovery*. Vol. 2, Wong, C.–H., ed. 2003, Wiley-VCH press, Weinheim, Germany, pp. 169-209; (b) Buskas, T., Ingale, S., Boons, G.–J., Glycopeptides as versatile tools for glycobiology. *Glycobiology.* 2006, *16* (8), 113-136.

[55] For examples of enhanced biological activitiy of neoglycopeptides and glycol-engineering of peptide drugs, see: (a) Polt, R., Dhanasekaran, M., Keyari, C. M. Glycosylated neuropeptides: a new vista for neuropsychopharmacology? *Med. Res. Rev.* 2005, *25* (5), 557-585; (b) Susaki, H., Suzuki, K., Ikeda, M., Yamada, H., Watanabe, H. K. Synthesis and antidiuretic activities of novel glycoconjugates of arginine-vasopressin. *Chem. Pharm. Bull.* 1998, *46* (10), 1530; (c) Elliot, S., Lorenzini, T., Asher, S., Aoki, K., Brankow, D., Buck, L., Busse, L., Chang, D., Fuller, J., Grant, J., Hernday, N., Hokum, M., Hu, S., Knudten, A., Levin, N., Komorowski, R., Martin, F., Navarro, R., Osslund, T., Rogers, G., Rogers, N., Trail, G., Egrie, J. Enhancement of therapeutic protein *in vivo* activities through glycoengineering. *Nature Biotechnol.* 2003, *21* (4), 414-421.

[56] Dondoni, A., Marra, A. Methods for Anomeric Carbon-Linked and Fused Sugar Amino Acid Synthesis: The Gateway to Artificial Glycopeptides. *Chem. Rev.* 2000, *100* (12), 4395–4422.

[57] Wilkinson, B. L., Bornaghi, L. F., Poulsen, S.–A., Houston, T. A. Synthetic Utility of Glycosyltriazoles in Carbohydrate Chemistry. *Tetrahedron.* 2006, *62* (34), 8115–8125.

[58] Kuijpers, B. H. M., Groothuys, S., Keereweer, A., R., Quaedflieg, P. J. L. M., Blaauw, R. H., van Delft, F. L., Rutjes, F. P. J. T. Expedient Synthesis of Triazole linked Glycosyl Amino Acids and Peptides. *Org. Lett.* 2004, *6* (18), 3123-3126.

[59] Dondoni, A., Giovannini, P. P., Massi, A. Assembling Heterocycle-Tethered *C*-Glycosyl and α-Amino Acid Residues via 1,3-Dipolar Cycloaddition Reactions. *Org. Lett.* 2004, *6* (17), 2929-2932.

[60] Macmillan, D., Blanc, J. A novel neoglycopeptide linkage compatible with native chemical ligation. *Org. Biomol. Chem.* 2006, *4* (15), 2847-2850.

[61] Arosio, D., Bertoli, M., Manzonib, L., Scolastico, C. Click chemistry to functionalise peptidomimetics. *Tetrahedron Lett.* 2006, *47* (22), 3697–3700.

[62] Ermolat'ev, D. Dehaen, W., Van der Eycken, E. Indirect coupling of the 2(1H)-pyrazinone scaffold with various (oligo)-saccharides via "click chemistry": En route towards glycopeptidomimetics. *QSAR & Comb. Sci.* 2004, *23* (10), 915-918.

[63] Dörner, S., Westermann, B. A short route for the synthesis of ''sweet'' macrocycles via a click-dimerization–ring-closing metathesis approach. *Chem. Commun.* 2005, (22), 2852–2854.

[64] Fu, X., Albermann, C., Jiang, J., Liao, J., Zhang, C., Thorson, J. S. Antibiotic Optimisation via *in vitro* glycorandomisation. *Nature Biotechnol.* 2003, *21* (12), 1467-1469.

[65] Lin, H., Walsh, C. T. A Chemoenzymatic Approach to Glycopeptide Antibiotics. *J. Am. Chem. Soc.* 2004, *126* (43), 13998-14003.

[66] (a) Prescher, J. A., Bertozzi, C. R. Chemistry in living systems. *Nature Chem. Biol.* 2005, *1* (1), 13-21; (b) Prescher, J. A., Bertozzi, C. R. Chemical Technologies for Probing Glycans. *Cell.* 2006, *126* (5), 851-854.

[67] Prescher, J. A., Bertozzi, C. R. Chemical remodelling of cell surfaces in living animals. *Nature* 2004, *430* (7002), 873-877.

[68] Agard, N. J., Prescher, J. A., Bertozzi, C. R. A Strain-Promoted [3 + 2] Azide-Alkyne Cycloaddition for Covalent Modification of Biomolecules in Living Systems. *J. Am. Chem. Soc.* 2004, *126* (46), 15046-15047.

[69] (a) Kannagi, R. Molecular mechanism for cancer-associated induction of sialyl Lewis X and sialyl Lewis A expression-The Warburg effect revisited. *Glycoconjugate J.* 2004, *20* (5), 353-64; (b) Dube, D. H., Bertozzi, C. R. Glycans in cancer and inflammation - potential for therapeutics and diagnostics. *Nature Rev. Drug Disc.* 2005, *4* (6), 477-488; (c) Ohyama, C., Tsuboi, S., Fukuda, M. Dual roles of sialyl Lewis X oligosaccharides in tumor metastasis and rejection by natural killer cells. *EMBO J.* 1999, *18* (6), 1516-1525; (d) Fernandes B., Sagman U., Auger M., Demetrio M., Dennis J. W. Beta 1-6 branched oligosaccharides as a marker of tumor progression in human breast and colon neoplasia. *Cancer Res*. 1991, *51* (2), 718-723.

[70] Rabuka, D., Hubbard, S. C., Laughlin, S. T., Argade, S. P., Bertozzi, C. R. A Chemical Reporter Strategy to Probe Glycoprotein Fucosylation. *J. Am. Chem. Soc.* 2006, *128* (37), 12078-12079.

[71] Sawa, M., Hsu, T.–L., Itoh, T., Sugiyama, M., Hanson, S. R., Vogt, P. K., Wong, C. – H. Glycoproteomic probes for fluorescent imaging of fucosylated glycans in vivo. *Proc. Natl. Acad. Sci. USA.* 2006, *103* (33), 12371–12376.

[72] For reviews on the multivalent or glycoside cluster effect, see: (a) Kiessling, L., Pontrello, J. K., Schuster, M. C. Synthetic Multivalent Ligands as Effectors or Inhibitors of Biological Processes. In *Carbohydrate-based Drug Discovery*. Wong, C.–

H., Ed.; Wiley-VCH, Weinheim, Germany, 2003, pp. 575-607.b) Kiessling, L. L., Gestwicki, J. E., Strong, L. E. Synthetic Multivalent Ligands as Probes of Signal Transduction. *Angew. Chem. Int. Ed.* 2006, *45* (15), 2348–2368. c) Kitov, P. I., Bundle, D. R. Thermodynamic Models of the Multivalency Effect. In *Carbohydrate-based Drug Discovery*; Wong, C.–H., Ed.; Wiley-VCH, Weinheim, Germany, 2003, pp. 541-573; d) Lundquist, J. J., Toone, E. J. The Cluster Glycoside Effect. *Chem. Rev.* 2002, *102* (2), 555-578.

[73] Collins, B. E., Paulson., J. C. Cell surface biology mediated by low affinity multivalent protein–glycan interactions. *Curr. Opin. Chem. Biol.* 2004, *8* (6), 617-625.

[74] For reviews detailing multivalent neoglycoconjugates as bacterial anti-adhesives: (a) Pieters, R. J. Intervention with Bacterial Adhesion by Multivalent Carbohydrates. *Med. Res. Rev.* 2006, 1-14; (b) Pieters, R. J. Interference with lectin binding and bacterial adhesion by multivalent carbohydrates and peptidic carbohydrate mimics. *Trends Glycosci. Glycotech.* 2004, *16* (90), 243-254; (c) Schengrund, C. "Multivalent" saccharides: development of new approaches for inhibiting the effects of glycosphingolipid-binding pathogens. *Biochem. Pharmacol.* 2003, *65* (5), 699-707.

[75] For reviews detailing multivalent carbohydrates as cancer vaccines: (a) Ragupathi, G., Coltart, D. M., Williams, L. J., Koide, F., Kagan, E., Allen, J.; Harris, C., Glunz, P. W., Livingston, P. O., Danishefsky, S. J. On the power of chemical synthesis: immunological evaluation of models for multiantigenic carbohydrate-based cancer vaccines. *Proc. Natl. Acad. Sci. USA.* 2002, *99* (21), 13699-13704; (b) Danishefsky, S. J. Synthetic Carbohydrate-Based Vaccines. In *Carbohydrate-based Drug Discovery.* Wong, C.–H., Ed.; Wiley-VCH, Weinheim, Germany, 2003, pp. 381-406.

[76] For reviews on glycopolymers as biomatierals see: (a) Klemm, D., Heublein, B., Fink, H, -P., Bohn, A. Cellulose: Fascinating Biopolymer and Sustainable Raw Material. *Angew. Chem. Int. Ed.* 2005, *44* (22), 3358-3393. (b) Heinze, T., Liebert, T., Unconventional methods in cellulose functionalisation. *Prog. Polym. Sci.* 2001, 1698-1762; (c) Müllera, F. A., Müllera, L., Hofmanna, I., Greila, P., Wenzelb, M. M., Staudenmaierb, R. Cellulose-based scaffold materials for cartilage tissue engineering. *Biomaterials.* 2006, *27*, 3955-3963; (d) Langer, R., David A. Tirrell, D. A. Designing materials for biology and medicine. *Nature* 2004, *428* (6982), 487-492.

[77] For reviews on glycopolymers as drug delivery and gene delivery, see: (a) Ladmiral, V., Melia, E., Haddleton, D. M. Synthetic glycopolymers: an overview. *Eur. Polymer J.* 2004, *40*, 431–449; (b) D'Souza, A. J. M., Topp, E. M. Release from Polymeric Prodrugs: Linkages and Their Degradation. *J. Pharm. Sci.* 2004, *93* (8), 1962-1979; (c) Hoste, K., De Winne, K., Schacht, E. Polymeric prodrugs. *Int. J. Pharm.* 2004, *277*, 119–131; (d) Wang, W., Liu, X., Xie, Y., Zhang, H., Yu, W., Xiong, Y., Xie, W., Ma, X. Microencapsulation using natural polysaccharides for drug delivery and cell implantation. *J. Mater. Chem.* 2006, *16* (32), 3252-3267. (e) Green, D. W., Mann S., Oreffo, R. O. C. Mineralized polysaccharide capsules as biomimetic micro-environments for cell, gene and growth factor delivery in tissue engineering. *Soft Matter* 2006, *2* (9), 732-737.

[78] Duncan, R. The Dawning Era of Polymer Therapeutics. *Nat. Rev. Drug Disc.* 2003, *2* (5), 347-360.

[79] Wang, Q., Dordick, J. S., Linhardt, R. J. Synthesis and Application of Carbohydrate-Containing Polymers. *Chem. Mater.* 2002, *14* (8), 3232-3244.

[80] Ladmiral, V., Mantovani, G., Clarkson, G. J., Cauet, S., Irwin, J. L., Haddleton, D. M. Synthesis of Neoglycopolymers by a Combination of "Click Chemistry" and Living Radical Polymerization. *J. Am. Chem. Soc.* 2006, *128* (14), 4823-4830.

[81] Srinivasachari, S., Liu, Y., Zhang, G., Prevette, L., Reineke, T. M. Trehalose Click Polymers Inhibit Nanoparticle Aggregation and Promote pDNA Delivery in Serum. *J. Am. Chem. Soc.* 2006, *128* (25), 8176-8184.

[82] Hawker, C. J., Wooley, K. L. The Convergence of Synthetic Organic and Polymer Chemistries. *Science* 2005, *309* (5738), 1200-1205.

[83] (a) Oaksmith, J. M., Peters, U., Ganem, B. Three-Component Condensation Leading to β-Amino Acid Diamides: Convergent Assembly of β-Peptide Analogues *J. Am. Chem. Soc.* 2004, *126* (42), 13606-13607; (b) Wallace, G. A., Heathcock, C. H. Further Studies of the *Daphniphyllum* Alkaloid Polycyclization Cascade. *J. Org. Chem.* 2001, *66* (2), 450-454.

[84] Malkoch, M., Thibault, R. J., Drockenmuller, E., Messerschmidt, M., Voit, B., Russell, T. P., Hawker, C. J. Orthogonal Approaches to the Simultaneous and Cascade Functionalization of Macromolecules Using Click Chemistry. *J. Am. Chem. Soc.* 2005, *127* (42), 14942-14949.

[85] Liebert, T., Hänsch, C., Heinze, T. Click Chemistry with Polysaccharides. *Macromol. Rapid Commun.* 2006, *27* (3), 208–213.

[86] Keding, S. J. Danishefsky, S. J. Natural Product Synthesis Special Feature: Prospects for total synthesis: A vision for a totally synthetic vaccine targeting epithelial tumors. *Proc. Natl. Acad. Sci. USA.* 2004, *101* (33), 11937-11942.

[87] Wan, Q., Chen, J., Chen, G., Danishefsky, S. J. A Potentially Valuable Advance in the Synthesis of Carbohydrate-Based Anticancer Vaccines through Extended Cycloaddition Chemistry. *J. Org. Chem.* 2006, *71* (21), 8244-8249.

[88] Bao, X.; Duan, J.; Fang, X.; Fang, J. Chemical modifications of the (1→3)-β-D-glucan from spores of *Ganoderma lucidum* and investigation of their physicochemical properties and immunological activity. *Carbohydr. Res.* 2001, *336* (2), 127-140.

[89] (a) Sakurai, K., Shinkai, S., Molecular Recognition of Adenine, Cytosine, and Uracil in a Single-Stranded RNA by a Natural Polysaccharide: Schizophyllan. *J. Am. Chem. Soc.* 2000, *122* (36), 4520-4521; (b) Hasegawa, T., Fujisawa, T., Numata, M., Umeda, M., Matsumoto, T., Kimura, T., Okumura, S., Sakurai, K., Shinkai, S. Single-walled carbon nanotubes acquire a specific lectin-affinity through supramolecular wrapping with lactose-appended schizophyllan. *Chem. Commun.* 2004, (19), 2150-2151.

[90] Hasegawa, T., Umeda, M., Numata, M., Li, C., Bae, A. -H., Fujisawa, T., Haraguchi, S., Sakuraib, K., Shinkai, S. 'Click chemistry' on polysaccharides: a convenient, general, and monitorable approach to develop (1→3)-β-D-glucans with various functional appendages. *Carbohydr. Res.* 2006, *341* (1), 35–40.

[91] Cheshev, P., Marra, A., Dondoni, A. First synthesis of 1,2,3-triazolo-linked (1,6)-α-D-oligomannoses (triazolomannoses) by iterative Cu(I)-catalyzed alkyne–azide cycloaddition. *Org. Biomol. Chem.* 2006, *4* (17), 3225–3227.

[92] (a) Wu, X. Y., Bundle, D. R. Synthesis of Glycoconjugate Vaccines for *Candida albicans* Using Novel Linker Methodology. *J. Org. Chem.* 2005, *70* (18), 7381-7388; (b) Ni, J., Song, H., Wang, Y., Stamatos, N. M., Wang, L.–X. Toward a Carbohydrate-Based HIV-1 Vaccine: Synthesis and Immunological Studies of Oligomannose-Containing Glycoconjugates. *Bioconjugate. Chem.* 2006, *17* (2), 493-500; (c) Watt, J. A., Williams, S. J. Rapid, iterative assembly of octyl α-1,6-oligomannosides and their 6-deoxy equivalents. *Org. Biomol. Chem.* 2005, *3* (10), 1982-1992.

[93] Marmuse, L., Nepogodiev, S. A., Field, R. A. "Click chemistry" *en route* to pseudo-starch. *Org. Biomol. Chem.* 2005, *3* (12), 2225-2227.

[94] Mellet, C. O., Defaye, J., Fernándéz, J., M. G. Multivalent Cyclooligosaccharides: Versatile Carbohydrate Clusters with Dual Role as Molecular Receptors and Lectin Ligands. *Chem. Eur. J.* 2002, *8* (9), 1983-1990.

[95] (a) Loftsson, T., Masson, M. Cyclodextrins in topical drug formulations: theory and practice. *Int. J. Pharm.* 2001, *225*, 15-30; (b) Rao, V. M., Stella, V. J. When can cyclodextrins be considered for solubilization purposes? *J. Pharm. Sci.* 2003, *92* (5), 927-932; (c) Chourasia, M. K.; Jain, S. K. Polysaccharides for Colon Targeted Drug Delivery. *Drug Delivery* 2004, *11*(2), 129-148.

[96] Conteron, J. M., Vincent, C., Bosso, C., Penades, S. Glycophanes, cyclodextrin-cyclophane hybrid receptors for apolar binding in aqueous solutions. A stereoselective carbohydrate-carbohydrate interaction in water. *J. Am. Chem. Soc.* 1993, *115* (22), 10066-10076; See also reference 94 and references cited therein.

[97] Hoffmann, B., Bernet, B., Vasella, A. Oligosaccharide Analogues of Polysaccharides (Part 241): Synthesis of Cyclodextrin Analogues Containing a Substituted Buta-1,3-diyne or a 1,2,3-Triazole Unit and Analysis of Intramolecular Hydrogen Bonds. *Helv. Chim. Acta.* 2002, *85* (3), 265-286.

[98] Bodine, K. D., Gin, D. Y., Gin, M. S. Synthesis of Readily Modifiable Cyclodextrin Analogues via Cyclodimerization of an Alkynyl-Azido Trisaccharide. *J. Am. Chem. Soc.* 2004, *126* (6), 1638-1639.

[99] Bodine, K. D., Gin, D. Y., Gin, M. S. Highly Convergent Synthesis of C_3- or C_2-Symmetric Carbohydrate Macrocycles. *Org. Lett.* 2005, *7* (20), 4479-4482.

[100] Billing, J. F., Nilsson, U. J., C_2-Symmetric Macrocyclic Carbohydrate/ Amino Acid Hybrids through Copper (I)-Catalyzed Formation of 1,2,3-Triazoles. *J. Org. Chem.* 2005, *70* (12), 4847-4850.

[101] Hoogenboom, R., Moore, B. C., Schubert, U. S. Synthesis of star-shaped poly(ε-caprolactone) via 'click' chemistry and 'supramolecular click' chemistry. *Chem. Commun.* 2006, (38), 4010–4012.

[102] For reviews on the biological applications of glycodendrimers, see: (a) Cloninger, M. J. Biological applications of dendrimers. *Curr. Opin. Chem. Biol.* 2002, *6* (6), 742–748. (b) Gillies, E. R., Fréchet, J. M. J. Dendrimers and dendritic polymers in drug discovery. *Drug Disc. Today.* 2005, *10* (1), 35-43. (c) Bezouška, K. Design, functional evaluation and biomedical applications of carbohydrate dendrimers (glycodendrimers). *Rev. Mol. Biotechnol.* 2002, *90* (3-4), 269-290. (d) Benito, J. M., Gomez-Garcia, M., Ortiz Mellet, C., Baussanne, I., Defaye, J., Fernandez, J. M., Garcia. J. M. Optimizing saccharide-directed molecular delivery to biological receptors: design, synthesis, and

biological evaluation of glycodendrimer-cyclodextrin conjugates. *J. Am. Chem. Soc.* 2004, *126* (33), 10355-10363. Bhadra, D., Yadav, A. K., Bhadra, S., Jain, N. K. Glycodendrimeric nanoparticulate carriers of primaquine phosphate for liver targeting. *Int. J. Pharm.* 2005, *295* (1-2), 221-233.

[103] (a) Vannucci, L., Fiserova, A., Sadalapure, K., Lindhorst, T. K., Kuldova, M., Rossmann, P., Horvath, O., Kren, V., Krist, P., Bezouska, K., Luptovcova, M., Mosca, F., Pospisil, M. Effects of N-acetyl-glucosamine-coated glycodendrimers as biological modulators in the B16F10 melanoma model in vivo. *Int. J. Oncol.* 2003, *23* (2), 285-296; (b) Roy R., Baek M.–G. Glycodendrimers: novel glycotope isosteres unmasking sugar coding. Case study with T-antigen markers from breast cancer MUC1 glycoprotein. *J. Biotechnol.* 2002, *90* (3-4), 291-309.

[104] (a) Pérez-Balderas, F., Ortega-Muñoz, M., Morales-Sanfrutos, J. Hernández-Mateo, F., Calvo-Flores, F. G., Calvo-Asín, J. A., Isac-García, J. Francisco Santoyo-González, F. Multivalent Neoglycoconjugates by Regiospecific Cycloaddition of Alkynes and Azides Using Organic-Soluble Copper Catalysts. *Org. Lett.* 2003, *5* (11), 1951-1954. (b) Calvo-Flores, F. G., Isac-García, J., Hernández-Mateo,F., Pérez-Balderas, F., Calvo-Asín, J. A., Sanchéz-Vaquero, E., Santoyo-González, F. 1,3-Dipolar Cycloadditions as a Tool for the Preparation of Multivalent Structures. *Org. Lett.* 2000, *2* (16), 2499-2502.

[105] Appukkuttan, P., Dehaen, W., Fokin, V. V., Van der Eycken, E. A Microwave-Assisted Click Chemistry Synthesis of 1,4-Disubstituted 1,2,3-Triazoles via a Copper(I)-Catalyzed Three-Component Reaction. *Org. Lett.* 2004, *6* (23), 4223-4225.

[106] Joosten, J. A. F., Niels T. H. Tholen, N. T. H., El Maate, F., Brouwer, A. J., van Esse, G. W., Rijkers, D. T. S., Liskamp, R. M. J., Pieters, R. J. High-Yielding Microwave-Assisted Synthesis of Triazole linked Glycodendrimers by Copper-Catalyzed [3+2] Cycloaddition. *Eur. J. Chem.* 2005, *15*, 3182–3185.

[107] Bouillon, C., Meyer, A., Vidal, S., Jochum, A., Chevolot, Y., Cloarec, J.-P., Praly, J.–P., Vasseur, J.–J., Morvan, F. Microwave Assisted "Click" Chemistry for the Synthesis of Multiple Labeled-Carbohydrate Oligonucleotides on Solid Support. *J. Org. Chem.* 2006, *71* (12), 4700-4702.

[108] Deng, S. L., Gangadharmath, U., Chang, C.-W. Sonochemistry: A Powerful Way of Enhancing the Efficiency of Carbohydrate Synthesis. *J. Org. Chem.* 2006, *71* (14), 5179-5185.

[109] Chittaboina, S., Xie, F., Wang, Q. One-pot synthesis of triazole linked glycoconjugates. *Tetrahedron Lett.* 2005, *46* (13), 2331-2336.

[110] Wu, P., Malkoch, M., Hunt, J. N., Vestberg, R., Kaltgrad, E., Finn, M. G., Fokin, V. V., Sharpless, K. B., Hawker, C. J. Multivalent, bifunctional dendrimers prepared by click chemistry. *Chem. Commun.* 2005, (46), 5775–5777.

[111] Fernandez-Megia, E., Correa, J., Rodríguez-Meizoso, I., Riguera, R. A Click Approach to Unprotected Glycodendrimers. *Macromolecules.* 2006, *39* (6), 2113-2120.

[112] Fernandez-Megia, E., Correa, J., Riguera, R. Clickable PEG-Dendritic Block Copolymers. *Biomacromolecules.* 2006, *7* (11), 3104-3111.

[113] For reviews on galectins, see: (a) Leffler, H., Carlsson, S., Hedlund, M., Qian, Y., Poirier, F. Introduction to galectins *Glycoconjugate J.* 2004, *19* (7/8/9), 433–440; (b) Pieters, R. J. Inhibition and detection of galectins. *ChemBiochem.* 2006, *7* (5), 721-728.

[114] Giguère, D., Patnam, R., Bellefleur, M.–A., St-Pierre, C., Satob, S., Roy, R. Carbohydrate triazoles and isoxazoles as inhibitors of galectins-1 and -3. *Chem. Commun.* 2006, (22), 2379–2381.

[115] Tejler, J., Tullberg, E., Frejd, T., Lefflerb, H., Nilsson, U. J., Synthesis of multivalent lactose derivatives by 1,3-dipolar cycloadditions: selective galectin-1 inhibition. *Carbohydr. Res.* 2006, *341* (10), 1353–1362.

[116] Gao, Y. J., Eguchi, A., Kakehib, K., Lee, Y. C. Synthesis and molecular recognition of carbohydrate-centered multivalent glycoclusters by a plant lectin RCA_{120}. *Bioorg. Med. Chem.* 2005, *13* (22), 6151-6157.

[117] Chen, Q., Yang, F., Du, Y. Synthesis of a C_3-symmetric (1→6)-N-acetyl-β-D-glucosamine octadecasaccharide using click chemistry. *Carbohydr. Res.* 2005, *340* (16), 2476–2482.

[118] (a) Ratner, D. M., Adams, E. W., Disney, M. D., Seeberger, P. H. Tools for Glycomics: Mapping Interactions of Carbohydrates in Biological Systems. *ChemBiochem.* 2004, *5* (10), 1375-1383; b) Dyukova V. I., Shilova N. V., Galanina O. E., Rubina A. Y., Bovin N. V. Design of carbohydrate multiarrays. *Biochim. Biophys. Acta* 2006, *1760* (4), 603-609; c) Smith, E. A. Thomas, W. D., Kiessling, L. L., Corn, R. M. Surface Plasmon Resonance Imaging Studies of Protein-Carbohydrate Interactions. *J. Am. Chem. Soc.* 2003, *125* (20), 6140-6148.

[119] Disney, M. D., Seeberger, P. H. The Use of Carbohydrate Microarrays to Study Carbohydrate-Cell Interactions and to Detect Pathogens. *Chem. Biol.* 2004, *11* (12), 1701-1707.

[120] Fazio, F., Bryan, M. C., Blixt, O., Paulson, J. C., Wong, C.-H. Synthesis of Sugar Arrays in Microtiter Plates. *J. Am. Chem. Soc.* 2002, *124* (48), 14397-14402.

[121] Bryan, M. C., Lee, L. V., Wong, C.–H. High throughput identification of fucosyltransferase inhibitors using carbohydrate microarrays. *Bioorg. Med. Chem. Lett.* 2004, *14* (12), 3185–3188.

[122] Huang, G-L., Liu, T.-C., Liu, M.-X., Mei, X.-Y. The application of quantum dots as fluorescent label to glycoarray. *Anal. Biochem,* 2005, *340* (1), 52-56.

[123] Bryan, M. C., Fazio, F., Lee, H.-K., Huang, C.–Y., Chang, A., Best, M. D., Calarese, D. A., Blixt, O., Paulson, J. C., Burton, D., Wilson, I. A., and Wong, C.–H., Covalent Display of Oligosaccharide Arrays in Microtiter Plates. *J. Am. Chem. Soc.* 2004, *126* (28), 8640-8641.

[124] Zhang, Y., Luo, S., Tang, Y., Yu, L., Hou, K.–Y., Cheng, J.–P., Zeng, X., and Wang, P. G. Carbohydrate-Protein Interactions by "Clicked" Carbohydrate Self-Assembled Monolayers. *Anal. Chem.* 2006, *78* (6), 2001-2008.

[125] Devaraj, N. K., Miller, G. P., Ebina, W., Kakaradov, B., Collman, J. P., Kool, E. T., Chidsey, C. E. D. Chemoselective Covalent Coupling of Oligonucleotide Probes to Self-Assembled Monolayers. *J. Am. Chem. Soc.* 2005, *127* (24), 8600-8601.

[126] Sun, X.–L., Stabler, C. L., Cazalis, C. S., and Chaikof, E. L. Carbohydrate and Protein Immobilization onto Solid Surfaces by Sequential Diels-Alder and Azide-Alkyne Cycloadditions. *Bioconjugate Chem.* 2006, *17* (1), 52-57.

[127] Meldal, M., Tornoe, C. W. Cu-catalyzed azide-alkyne cycloaddition. *Chem. Rev.* 2008, *108* (8), 2952-3015.

[128] Wilkinson, B. L.; Bornaghi, L. F.; Houston, T. A.; Poulsen, S.-A. "Click Chemistry in Carbohydrate Based Drug Development and Glycobiology" in *Drug Design Research Perspectives*, S. P. Kaplan, Ed., Nova Publishers, 2007, pp. 57-102.

[129] (a) Wilkinson B. L., Long H., Sim E., Fairbanks A. J. Synthesis of Arabino glycosyl triazoles as potential inhibitors of mycobacterial cell wall biosynthesis. *Bioorg. Med. Chem. Lett.* 2008, *18* (23), 6265-6267; b) Singh B. K., Yadav A. K., Kumar B., Gaikwad A., Sinha S. K., Chaturvedi V., Tripathi R. P. Preparation and reactions of sugar azides with alkynes: synthesis of sugar triazoles as antitubercular agents. *Carbohydr. Res.* 2008, *343* (7), 1153-1162.

[130] (a) Poulsen S.-A., Wilkinson B. L., Innocenti A., Vullo D., Supuran, C. T. Inhibition of human mitochondrial carbonic anhydrases VA and VB with para-(4-phenyltriazole-1-yl)-benzenesulfonamide derivatives. *Bioorg. Med. Chem. Lett.* 2008, *18* (16), 4624-4627; b) Wilkinson B. L., Innocenti A., Vullo D., Supuran, C. T. Poulsen S.-A., Inhibition of carbonic anhydrases with glycosyltriazole benzene sulfonamides. *J. Med. Chem.* 2008, *51* (6), 1945-1953.

[131] Kumar R., Maulik P. R., Misra A. K. Significant rate accelerated synthesis of glycosyl azides and glycosyl 1,2,3-triazole conjugates. *Glycoconjugate J.* 2008, *25* (7), 595-602.

[132] Laughlin S. T., Baskin J. M., Amacher S. L., Bertozzi, C. R. In vivo imaging of membrane-associated glycans in developing zebrafish. *Science* 2008, *320* (5876), 664-667.

[133] Ning XH, Guo J, Wolfert MA, Boons GJ. Visualizing metabolically labeled glycoconjugates of living cells by copper-free and fast Huisgen cycloadditions. *Angew. Chem. Int. Ed.* 2008, *47* (12),2253-2255.

[134] Wang JS, Li HG, Zou GZ, Wang LX. Novel template-assembled oligosaccharide clusters as epitope mimics for HIV-neutralizing antibody 2G12. Design, synthesis, and antibody binding study. *Org. Biomol. Chem.* 2007, *5* (10), 1529-1540.

[135] (a) Touaibia M., Wellens A., Shiao T. C., Wang Q., Sirois S., Bouckaert J., Roy R. Mannosylated G(0) dendrimers with nanomolar affinities to Escherichia coli FimH. *ChemMedChem.* 2007, *2* (8), 1190-1201; b) Gouin S. G., Vanquelef E., Fernandez J. M. G., Mellet C. O., Dupradeau F. Y., Kovensky J. Multi-mannosides based on a carbohydrate scaffold: Synthesis, force field development, molecular dynamics studies, and binding affinities for lectin Con A. *J. Org. Chem.* 2007, *72* (24), 9032-9045; c) Palomo C., Aizpurua J. M., Balentova E., Azcune I., Santos J. I., Jimenez-Barbero J., Canada J., Miranda J. I. "Click" saccharide/beta-lactam hybrids for lectin inhibition. *Org. Lett.* 2008, *10* (11), 2227-2230; d) Branderhorst H. M., Ruijtenbeek R., Liskamp R. M. J., Pieters R. J. Multivalent carbohydrate recognition on a glycodendrimer-functionalized flow-through chip. *ChemBiochem.* 2008, *9* (11), 1836-1844; e) Vecchi A., Melai B., Marra A., Chiappe C., Dondoni A. Microwave-enhanced ionothermal CuAAC for the synthesis of glycoclusters on a calix[4]arene platform. *J. Org. Chem.*

2008, *73* (23), 6437-6440; f) Papp, I., Dernedde, J., Enders S., Haag, R. Modular synthesis of multivalent glycoarchitectures and their unique selectin binding behavior. *Chem. Commun.* 2008, (44), 5851-5853; g) Touaibia M., Roy R. First Synthesis of "Majoral-Type" Glycodendrimers Bearing Covalently Bound alpha-D-Mannopyranoside Residues onto a Hexachlocyclotriphosphazene Core. *J. Org. Chem.* 2008, *73* (23),9292-9302.

[136] (a) Moorhouse A. D., Moses J. E. Click chemistry and medicinal chemistry: A case of "Cyclo-Addiction". *ChemMedChem* 2008, *3* (5), 715-723; b) Dedola S., Nepogodiev S. A., Field R. A. Recent applications of the Cu-I-catalysed Huisgen azide-alkyne 1,3-dipolar cycloaddition reaction in carbohydrate chemistry. *Org. Biomol. Chem.* 2007, *5* (7), 1006-1017; c) Dondoni, A. Triazole: the keystone in glycosylated molecular architectures constructed by a click reaction. *Chem.-Asian J.* 2007, *2* (6), 700-708.

In: Glycobiology Research Trends
Editors: G. Powell and O. McCabe
ISBN: 978-1-60692-841-7

Chapter VIII

Functional Enzyme Complex Involved in HNK-1 Carbohydrate Biosynthesis

Yasuhiko Kizuka[1], Yasuhiro Tonoyama[2], and Shogo Oka[2]
[1]Department of Biological Chemistry, Graduate School of Pharmaceutical Sciences, Kyoto University, Kyoto 606-8501, Japan
[2]Department of Biological Chemistry, Human Health Sciences, Graduate School of Medicine, Kyoto University, Kyoto 606-8507, Japan

Abstract

The HNK-1 carbohydrate epitope, also known as CD57, has a unique structural feature comprising a negatively charged trisaccharide (HSO_3-3GlcAβ1-3Galβ1-4GlcNAc-). This carbohydrate was first found to be an epitope on the human natural killer cell surface, but later experiments revealed that the HNK-1 carbohydrate was mainly expressed in the nervous system in various species. As key enzymes for its biosynthesis, we and others have cloned two glucuronyltransferases (GlcAT-P and GlcAT-S) and a sulfotransferase (HNK-1ST). Gene targeting analyses of these enzymes clearly revealed that the HNK-1 carbohydrate plays important roles in synaptic plasticity and memory formation. Although the key molecules related to its biosynthesis have been cloned, the detailed expression mechanism remains unclear probably due to the low expression levels of these transferases and to the complexity of the cellular glycosylation system in the ER and Golgi apparatus. During the course of a study to clarify the complicated HNK-1 biosynthesis system, we discovered a functional enzyme complex consisting of GlcAT-P (or –S) and HNK-1ST. Since the specific activity of HNK-1ST was up-regulated with the co-existence of GlcAT-P (or –S), the complex was found to raise the efficiency of HNK-1 biosynthesis. Moreover, recent research on the diverse glycan biosynthesis mechanism indicated that several enzyme complexes exist in cells to control the well-regulated glycan expression. It has been evident that to these complexes, non-transferase proteins are also related such as sugar-nucleotide transporter, glycosidase, phosphatase, and so on. These results suggest that not only the expression level of a single glycosyltransferase but also its surrounding environment serve as regulators of the expression of diverse glycoconjugates.

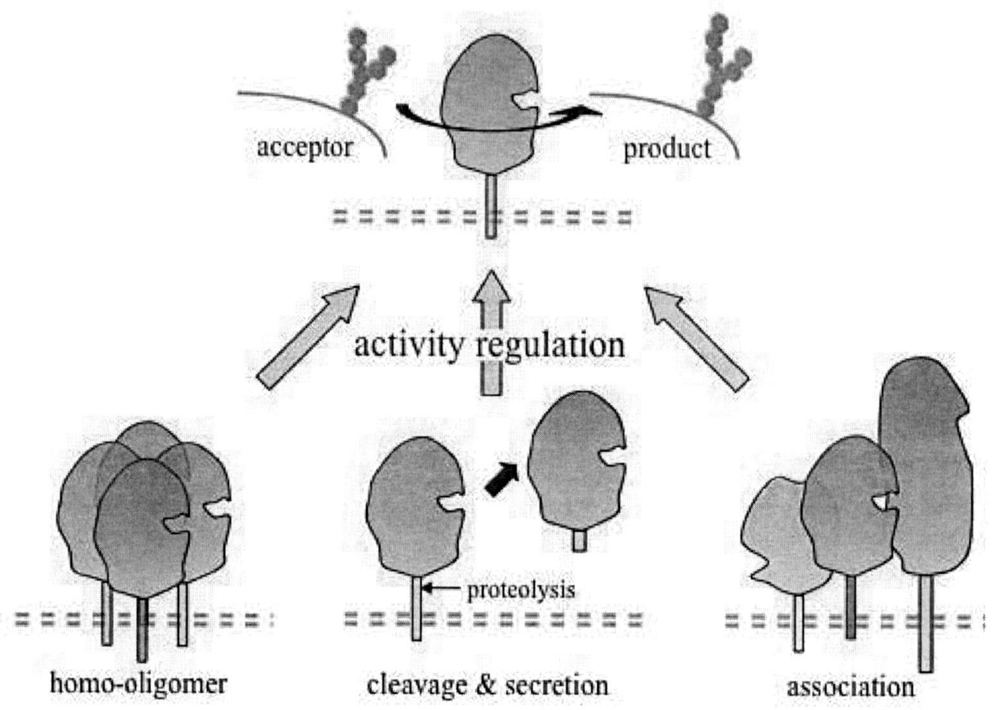

Figure 1. Regulation of glycosyltransferases.
Although glycosylation is completed by several glycosyltransferases, the expression level of the product glycan does not always reflect the expression level of its biosynthetic transferase. Several factors like oligomerization, secretion and association control its activity in cells.

INTRODUCTION

An increasing number of findings has proven that glycosylation participates in various biological events, and many efforts have been made to elucidate the functions and biosynthetic mechanisms of glycans [1, 2]. Cells express on their surface various glycans depending on the situation such as during differentiation or disease progression, and these diverse glycans are biosynthesized through the concerted actions of many glycosyltransferases and related proteins in the ER and Golgi apparatus [3]. So far, most glycosyltransferase genes have been cloned. However, in a certain condition, only the existence or absence of a glycosyltransferase cannot account for the expression of the product glycan [4]. These results indicate that unknown regulatory mechanisms other than the presence or expression level of a given glycosyltransferase, control its activity in cells. Over the past few years, in fact, several experimental results supporting this idea have been reported. For example, oligomerization, cleavage and secretion, or protein-protein complex formation of a certain glycosyltransferase may regulate glycan expression via control of its activity or intracellular transport [Figure 1, and refs. 5, 6, 7]. In addition, it is hard to see native aspects of an endogenously expressed glycosyltransferase in cells or organs, since the expression level is relatively low [8]. Therefore, we often come to some conclusions based on the results obtained only in overexpression experiments. Judging from these results, although the biosynthetic mechanism of glycans is extremely complicated and is difficult to elucidate entirely, we think there are intriguing and fascinating phenomena remaining to be uncovered.

Glycosylation in the Nervous System

In the nervous system, specific glycan epitopes, such as polysialic acid (PSA) or human natural killer-1 (HNK-1) carbohydrate, are also expressed to form the complicated neural network [9]. These carbohydrate epitopes are mainly carried by cell adhesion molecules, such as neural cell adhesion molecule (NCAM), and are considered to regulate cell-cell recognition or adhesion by modulating the adhesion properties of carrier glycoproteins [9]. Moreover, recent gene targeting techniques in mice allowed us to discover that each glycan plays an important role in maintenance of precise neural functions at the individual level [9], and it was shown that a complete loss of PSA in mice resulted in early death [10]. Although the enzymes responsible for the biosynthesis of these glycans biosynthesis were identified [11, 12, 13, 14], the mechanism controlling these enzymes' activities in cells remains to be elucidated. Since these neural specific glycans have been shown to function in precise high order brain function, as just described, research on the expression mechanism of these glycans will provide us with meaningful information about glycan functions in the nervous system. In particular, we have been interested in the functional role and expression mechanism of the HNK-1 carbohydrate. In this chapter, we described the formation of an enzyme complex involving the HNK-1 carbohydrate that regulates the expression.

Structure and Biosynthesis of the HNK-1 Carbohydrate

As indicated by its name, the HNK-1 carbohydrate was first found as an epitope on the human natural killer cell surface [15], and it is also referred to as CD57 in the field of immunology. However, later experiments revealed that the HNK-1 carbohydrate is predominantly expressed in the nervous system [16], and its neural expression is conserved in various species such as Drosophila, zebrafish and mammals [17, 18, 19]. This carbohydrate comprises a unique sulfated trisaccharide (HSO_3-3GlcAβ1-3Galβ1-4GlcNAc-) [Figure 2A], and is attached to both glycoproteins (NCAM, L1, P0, etc.) and glycolipids (SGGL1 and SGGL2) [20, 21]. Temporally and spatially regulated expression of the HNK-1 carbohydrate is observed in migrating neural crest cells [22] and rhombomeres [23], and it was suggested that it is associated with neural crest cell migration and neuron to glial cell adhesion [24]. The key enzymes for its biosynthesis were successfully cloned one after another, that is, two glucuronyltransferases (GlcAT-P and GlcAT-S) and a sulfotransferase (HNK-1ST) [Figure 2A, and refs. 13, 14, 25]. Since in the brain, higher and wider expression of the GlcAT-P transcript than that of GlcAT-S is found [26], GlcAT-P was considered to be a predominant glucuronyltransferase responsible for the HNK-1 biosynthesis. Actually, GlcAT-P gene-deficient mice exhibited great loss of HNK-1 in the brain [27], indicating that GlcAT-P is probably the main enzyme *in vivo*. Targeted deletion of the HNK-1ST gene in mice was also performed by Senn *et al.*, and they revealed that HNK-1ST is the sole sulfotransferase for the HNK-1 carbohydrate in the mouse brain [28]. Moreover, these mutant mice showed reduced synaptic plasticity in the hippocampus and impaired spatial learning task performance [Figure 2B, and refs. 27, 28], indicating that the HNK-1 carbohydrate has an important role in brain high order functions like learning and memory. On the other hand, we recently found that GlcAT-S mRNA is expressed in mouse kidney as well as in the brain to biosynthesize a non-sulfated form of the HNK-1 carbohydrate because of the lack of HNK-1ST expression in the mouse kidney [29]. Moreover, while HNK-1 carrier molecules in the brain are often cell adhesion molecules, non-sulfated HNK-1 carrier proteins are completely different such as metallo-protease or basement membrane component [29, 30]. These results suggest that GlcAT-P and GlcAT-S have distinct substrate specificities and functions *in vivo*.

Glucuronyltransferases Involved in HNK-1 Biosynthesis

GlcAT-P and GlcAT-S are Golgi-resident type II membrane proteins like many other glycosyltransferases, consisting of an N-terminal short cytoplasmic tail, a single spanning transmembrane domain, a stem region and a C-terminal catalytic domain [13, 14]. Overexpression of GlcAT-P or GlcAT-S gave rise to cell surface expression of the HNK-1 carbohydrate even in a non-neural cell line such as COS-1 cells [13, 14]. In addition, the recombinant catalytic domains of both GlcAT-P and GlcAT-S showed glucuronyltransferase activities toward glycoproteins, glycolipids and oligosaccharide acceptors *in vitro* [31]. Although their detailed acceptor substrate specificities are somewhat different, they fundamentally recognize the *N*-acetyllactosamine structure (Galβ1-4GlcNAc) and transfer glucuronic acid to the 3-position of galactose residues using UDP-GlcA as a donor substrate. Intriguingly, however, the specific activities of the two enzymes were increased by phospholipids in a distinct way [31]. For instance, sphingomyelin enhanced the GlcAT-P activity toward a glycoprotein acceptor but not that of GlcAT-S, while both enzymes required a phospholipid like phosphatidylinositol for transfer to a glycolipid acceptor. These results indicate that the microenvironment around these enzymes in the Golgi apparatus affects their enzymatic activities in cells.

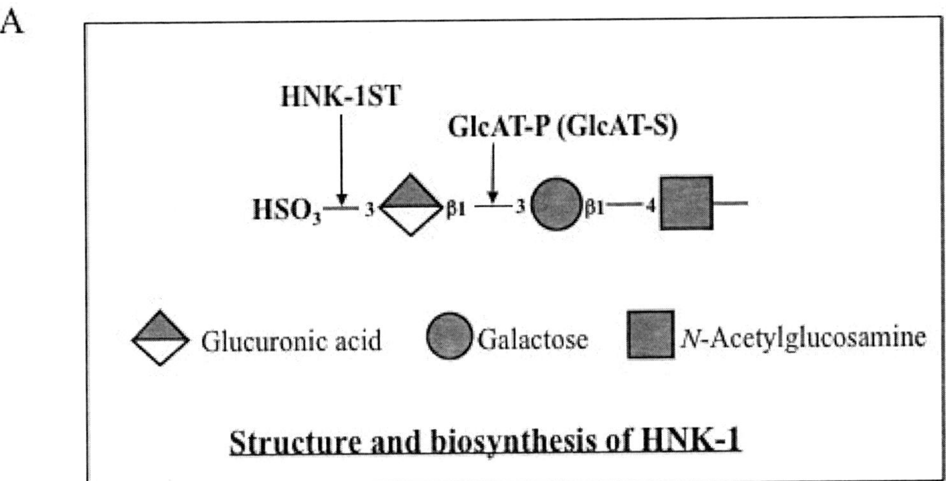

B

	Phenotype of GlcAT-P deficient mice
HNK-1 expression	Almost complete loss (*remaining expression)
Synaptic plasticity	Reduced LTP in hippocampal CA1 region
Morris mater maze test	Increased escape latency (impaired spatial learning)
Y-maze test	Increased entry into arms (greater exploratory activity)
Elevated plus maze test	Longer time in open arms (reduced anxiety)

* ref 27

Figure 2. Biosynthesis and function of the HNK-1 carbohydrate.
(A) The HNK-1 carbohydrate comprises a sulfated trisaccharide and is sequentially biosynthesized by GlcAT-P (or –S) and HNK-1ST. Glucuronic acid is transferred to the non-reducing end of an *N*-acetyllactosamine residue. (B) The phenotypes of GlcAT-P gene-deficient mice are summarized. These mutant mice exhibit reduced synaptic plasticity and impaired learning with almost complete loss of HNK-1 expression in the brain.

Another glucuronyltransferase related to glycan biosynthesis, GlcAT-I, was cloned according to the sequence similarity with GlcAT-P [32]. *In vitro* substrate specificity analysis revealed that the enzyme catalyzes the glucuronic acid transfer to Galβ1-3Galβ1-4Xylβ-Ser to complete the linkage tetrasaccharide of proteoglycans required for glycosaminoglycan elongation rather than serves as an HNK-1 biosynthetic enzyme. However, exogenous expression of GlcAT-I in Lec2 cells induced HNK-1 carbohydrate expression on its *N*-

glycans [33]. On the other hand, GlcAT-P as well as GlcAT-I restored the glycosaminoglycan synthesis of a GlcAT-I defective cell line [33]. These results suggest that these glucuronyltransferases could compensate for each other under some conditions, but more detailed analysis is required to understand this issue precisely.

Functional Enzyme Complex Between GlcATs and HNK-1ST

Physical Interaction of GlcATs and HNK-1ST

The structure of the HNK-1 carbohydrate epitope on NCAM, P0 and SGGLs, as a sulfated trisaccharide (HSO_3-3GlcAβ1-3Galβ1-4GlcNAc-), was determined [34, 35, 21]. It was conceivable that a non-sulfated form (GlcAβ1-3Galβ1-4GlcNAc-) was also expressed on these carrier molecules because the glucuronylation and sulfation steps are catalyzed sequentially by distinct enzymes, but so far such a glycan structure has not been reported except for in the mouse kidney, as described above. In addition, GlcAT-P-deficient mice and HNK-1ST-deficient ones showed similar neural phenotypes, *i.e.* impaired synaptic plasticity and memory formation, indicating the importance of the sulfate group in the HNK-1 carbohydrate for its function in the nervous system. These experimental results allowed us to hypothesize that GlcATs (GlcAT-P and GlcAT-S) and HNK-1ST form an enzyme complex and that the sulfate group is transferred by HNK-1ST soon after glucuronylation by GlcATs, leading to efficient biosynthesis of the HNK-1 carbohydrate. To exemplify this idea, we investigated the interaction between GlcATs and HNK-1ST in cells using the co-immunoprecipitation technique. As a result, in CHO cells, transiently expressed full-length GlcAT-P (or –S) was found to be co-immunoprecipitated with full-length HNK-1ST and *vice versa* [Figure 3A and ref. 36]. Next, we investigated the specificity of the complex formation by similar co-immunoprecipitation analysis using other transferases. As expected, it was found that a sialyltransferase, ST3GalIV, was not associated with HNK-1ST in cells and that glycosylation-related sulfotransferases, C4ST1 and GalNAc4ST1, were also not co-precipitated with either GlcAT-P or –S, despite their apparent co-localization in the Golgi apparatus, indicating a specific interaction between GlcATs and HNK-1ST. To our surprise, however, an interaction between GlcAT-I and HNK-1ST was clearly observed [36]. Even now, the biological significance of this interaction is unclear, but it might be a clue for determining the reason for the presence of HNK-1ST in HNK-1-negative cells and organs [37].

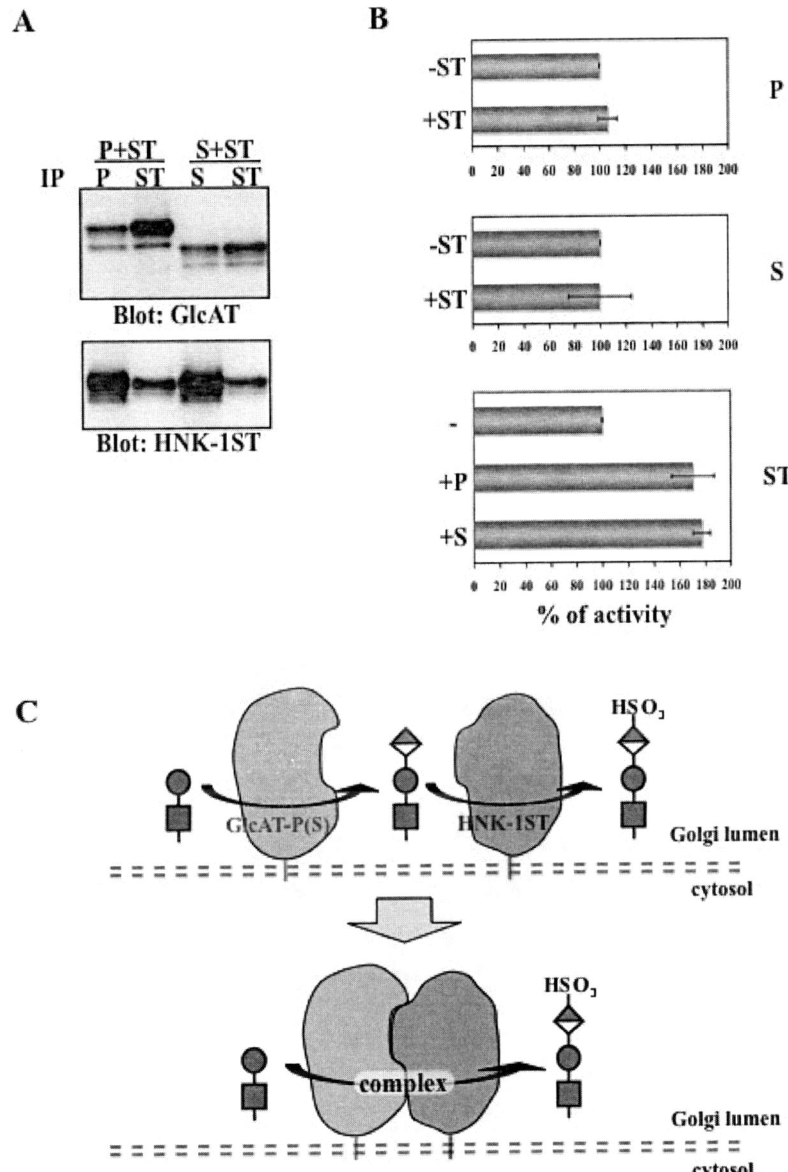

Figure 3. Complex between GlcATs and HNK-1ST.
(A) Co-immunoprecipitation analysis of 3xFLAG tagged GlcATs and EGFP fused to HNK-1ST. They were transiently expressed with the binding counterpart in CHO cells. The expressed GlcATs and HNK-1ST were immunoprecipitated with anti-FLAG or anti-EGFP pAb, respectively, followed by immunoblotting with anti-FLAG mAb (upper panel) or anti-EGFP mAb (lower panel). (B) The *in vitro* transferase activity of GlcAT-P (upper graph), GlcAT-S (middle graph), or HNK-1ST (lower graph) was measured with or without the binding partner. Three independent experiments were performed and the bars indicate standard deviations. (C) Schematic diagram of the GlcAT-HNK-1ST complex. The complex is considered to allow efficient biosynthesis of the HNK-1 carbohydrate.

Direct Interaction through Catalytic Domains

GlcAT-I exhibits high sequence similarity with GlcAT-P and GlcAT-S, especially in the C-terminal catalytic region. Since GlcAT-I also interacted with HNK-1ST, we hypothesized that GlcATs bind with HNK-1ST through their catalytic domains. To examine this, pull-down assays were carried out using the purified recombinant catalytic domains of GlcATs and HNK-1ST [36]. FLAG-tagged GlcAT-P and GlcAT-S were expressed and purified from *E. coli* cells, and protein A fused to the HNK-1ST catalytic domain was expressed and purified from COS-1 cells, and they all maintained their transferase activities *in vitro* [38, 36]. Pull-down assays were performed using them, and we clearly showed direct interaction between GlcATs and the HNK-1ST catalytic regions [36]. Although the *N*-teminal domains could be also involved in the interaction, only their catalytic domains are at least sufficient for the binding between these enzymes.

Biological Significance of the GlcAT and HNK-1ST Interaction

As described above, we hypothesized that interaction between GlcATs and HNK-1ST enables them to biosynthesize the HNK-1 carbohydrate efficiently. In addition, we showed the interaction was mediated by their catalytic domains. Therefore, it was expected that the complex formation enhanced their enzymatic activities. To determine whether or not the interaction affects the specific activity, we measured the *in vitro* transferase activities toward glycoprotein acceptor using purified enzymes with or without the binding counterpart. As a result, while the activities of GlcAT-P and GlcAT-S were not influenced by binding with HNK-1ST, the sulfotransferase activity of HNK-1ST was increased about two-fold on binding with both GlcAT-P and GlcAT-S [Figure 3B and ref. 36]. These results indicated that HNK-1ST efficiently transferred the sulfate group by associating with GlcATs in cells, which was consistent with our hypothesis described above [Figure 3C]. Moreover, although the specific activities of the GlcATs were not affected by HNK-1ST *in vitro*, it was found that the expression level of carbohydrate antigen produced by GlcATs in cells was apparently up-regulated with the co-existence of HNK-1ST [36]. The mechanism underlying this up-regulation was unclear, but it is possible that the enzyme interaction positively regulates the activities of GlcATs as well as HNK-1ST *in vivo*.

Other Protein-protein Interactions during Glycan Biosynthesis

Besides the enzymes involved in HNK-1 biosynthesis, it is evident that other glycan-related proteins also interact *in vivo* and *in vitro*. The first report of an enzyme complex involving a glycosyltransferase was about GnT-I and mannosidase-II [39]. At that time, that finding reinforced the validity of a mechanism called the kin-recognition model, which could explain the Golgi retention of glycosyltransferase via enzyme complex [39]. Even now, however, the Golgi retention mechanism governing glycosyltransferase remains controversial [40]. Maccioni's group revealed that glycolipid glycosyltransferases GalT2 and GalNAcT form heterocomplex through their N-terminal regions [41]. Moreover, they reported that a small GTPase, Sar1, recognizes and binds with the basic amino acid motif (called the di-basic motif) in the cytoplasmic tails of these glycosyltransferases adjacent to the transmembrane border, and that this interaction enables several glycosyltransferases to exit from the ER to the Golgi [42]. More recently, they found, using a yeast two-hybrid system, that a calcium-binding protein, calsenilin or CALP, binds with the GalT2 cytoplasmic tail [43]. The binding affected the subcellular distribution of GalT2. In this way, as for specific glycosyltransferases, it has been proven that several mechanisms exist to control their intracellular localization. Meanwhile, interactions to regulate the transferase activity have also been reported. The enzymes responsible for heparan sulfate biosythesis, EXT1 and EXT2, showed enzymatic activity in cells only when they were co-expressed [44]. Furthermore, Izumikawa *et al.* reported that chondroitin sulfate polymerization was differently achieved with multiple enzyme combinations consisting of ChSy-1, ChSy-2, CSGlcA-T and ChPF [45]. In addition, it was reported that POMT1 and POMT2, *O*-mannosyltransferases which catalyze the first step of *O*-mannosylation in mammals, require their association in cells for *O*-mannosyltransferase activity [46]. Taken together, these accumulating findings point out the significance of an enzyme complex for understanding the glycan expression mechanism.

Complexes involving non-transferase proteins have also been reported. Hart's group found an interaction between *O*-GlcNAc transferase and serine/threonine phosphatase [47]. *O*-GlcNAc modification, carried by many nuclear and cytosolic proteins, is well-known to function in signal transduction and protein degradation, and appears to have a reciprocal relationship with protein phosphorylation [48]. They revealed a dynamic relationship between *O*-GlcNAcylation and phosphorylation from the aspect of enzyme interaction. In addition, it was reported that a nucleotide-sugar transporter also associates with glycosyltransferases. Nucleotide-sugars such as UDP-galactose and CMP-sialic acid serve as donor substrates of glycosyltransferases, and their transporters in the ER or Golgi are essential for glycosyltransferases because nucleotide-sugars are synthesized in the cytosol or nucleus [49]. Sprong *et al.* found that UDP-galactose transporter interacts with UDP-galactose:ceramide galactosyltransferase, and that the transproter localization significantly changed from the Golgi to the ER upon interaction with the galactosyltransferase to enable import of UDP-galactose into the ER [50]. Judging from these results, a protein-protein complex involving not only a glycosyltransferase but also another protein is an important factor modulating glycan biosynthesis [Figure 4].

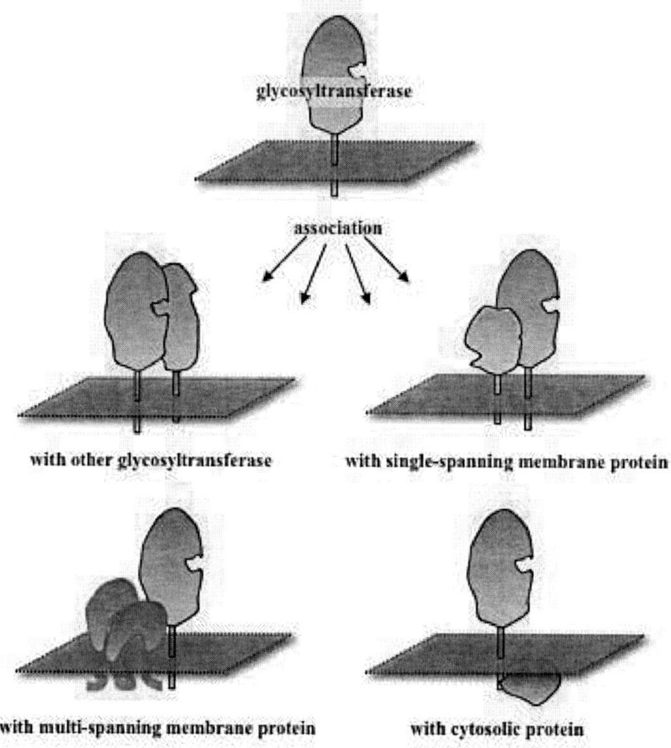

Figure 4. Interaction of glycosyltransferases with several types of proteins.
Glycosyltransferases associate with several proteins in cells, including other glycosyltransferases and non-glycosyltransferase enzymes or proteins. These interactions probably control their activity or subcellular localization.

Contribution of Enzyme Complexes to the Entire Glycosylation System in Cells

As just described, several enzyme complexes have been reported. Considering the results together, almost all complexes consist of proteins highly related in their functions. For example, our finding concerned HNK-1 related enzymes, and others concerned heparan sulfate biosynthetic enzymes, chondroitin sulfate synthases, *O*-mannosyltransferases, and so on. On the other hand, we wonder why only limited kinds of carrier molecules are modified with specific glycans during biosynthesis, such as PSA on NCAM [9]. Based on these results, we presume that each enzyme complex (including a substrate protein of a glycosyltransferase) for a specific glycan biosynthesis accumulates at a different position in the Golgi apparatus and forms a hetero-complex to biosynthesize a specific glycan cooperatively and efficiently. Indeed, recently, Goto's group reported that stacked fragments of the Golgi complex in Drosophila cells are functionally diverse and they called these fragments "Golgi units" [51]. In immunostaining and immunoisolation experiments, they found that Fringe connection and Rhomboid, having distinct functions in the Golgi from each other, resided in distinct Golgi units, and that the units could be separated biochemically. On the other hand, Fringe and Fringe connection, both of which serve to glycosylate Notch receptor, were shown to reside in the same Golgi unit. They finally proposed that the different Golgi units of glycan-related molecules underlie the diversity of glycan modification in cells. Although such a Golgi unit was not observed in mammalian cells because of the single entity of a Golgi complex, unlike in Drosophila cells, it is possible that highly-related enzymes involved in the biosynthesis of a specific glycan reside and accumulate in a distinct compartment in a single Golgi in mammalian cells to biosynthesize the specific glycan on a selected acceptor substrate. Therefore, the protein-protein complexes described here may contribute to this machinery in mammalian cells.

Conclusion

As to HNK-1 carbohydrate expression, we revealed a novel enzyme complex consisting of a glucuronyltransferase and a sulfotransferase, as described above. We also elucidated that the interaction is mediated by their *C*-terminal catalytic domains, leading to the up-regulation of HNK-1ST specific activity and the efficient sequential biosynthesis of the HNK-1 carbohydrate. Previously, we succeeded in crystallization of the GlcAT-P and GlcAT-S catalytic domains, and revealed the structural bases of their catalytic reaction mechanisms at the atomic level [52, 53]. Now we are performing a similar study on HNK-1ST, and if we are successful in co-crystalization with GlcATs, we will probably be able to elucidate the interaction mechanism from the aspect of structural biology. Meanwhile, as indicated by the other glycosyltransferase complexes, it is sufficiently possible that the *N*-terminal domain of GlcATs or HNK-1ST binds with other proteins to regulate HNK-1 carbohydrate biosynthesis. In addition, the physiological meaning of the GlcAT-I and HNK-1ST interaction remains to be determined. But our work and others' indicate that only the expression level of a single glycosyltransferase can no longer explain a specific glycan expression level and that there appears to be many unknown enzyme complexes involved. Besides, it is unclear what role each enzyme complex plays in the total glycosylation system in cells and whether or not a "Golgi unit"-like machinery exists for diverse glycan expression in mammalian cells. We believe future work will shed light on these issues and possibilities.

References

[1] Ohtsubo, K. and Marth, J. D. (2006). Glycosylation in cellular mechanisms of health and disease. *Cell*, *126*, 855-867.

[2] Lowe, J. B. and Marth, J. D. (2003). A genetic approach to Mammalian glycan function. *Annu. Rev. Biochem.*, *72*, 643-691.

[3] Freeze, H. H. (2006). Genetic defects in the human glycome. *Nat. Rev. Genet.*, *7*, 537-551.

[4] Baum, L. G., Derbin, K., Perillo, N. L., Wu, T., Pang, M. and Uittenbogaart, C. (1996). Characterization of terminal sialic acid linkages on human thymocytes. Correlation between lectin-binding phenotype and sialyltransferase expression. *J. Biol. Chem.*, *271*, 10793-10799.

[5] Ma, J. and Colly, K. J. (1996). A disulfide-bonded dimer of the Golgi beta-galactoside alpha2,6-sialyltransferase is catalytically inactive yet still retains the ability to bind galactose. *J. Biol. Chem.*, *271*, 7758-7766.

[6] Sugimoto, I., Futakawa, S., Oka, R., Ogawa, K., Marth, J. D., Miyoshi, E., Taniguchi, N., Hashimoto, Y. and Kitazume, S. (2007). Beta-galactoside alpha2,6-sialyltransferase I cleavage by BACE1 enhances the sialylation of soluble glycoproteins. A novel regulatory mechanism for alpha2,6-sialylation. *J. Biol. Chem.*, *282*, 34896-34903.

[7] de Graffenried, C. L. and Bertozzi, C. R. (2004). The roles of enzyme localization and complex formation in glycan assembly within the Golgi apparatus. *Curr. Opin. Cell Biol.*, *16*, 356-363.

[8] Larsen, R. D., Rajan, V. P., Ruff, M. M., Kukowska-Latallo, J. and Cummings, R. D. (1989). Isolation of a cDNA encoding a murine UDPgalactose:beta-D-galactosyl-1,4-N-acetyl-D-glucosaminide alpha-1,3-galactosyltransferase: expression cloning by gene transfer. *Proc. Natl. Acad. Sci. U. S. A.*, *86*, 8227-8231.

[9] Kleene, R. and Schachner, M. (2004). Glycans and neural cell interactions. *Nat. Rev. Neurosci.*, *5*, 195-208.

[10] Weinhold, B., Seidenfaden, R., Rockle, I., Muhlenhoff, M., Schertzinger, F., Conzelmann, S., Marth, J. D., Gerardy-Schahn, R. and Hildebrandt, H. (2005). Genetic ablation of polysialic acid causes severe neurodevelopmental defects rescued by deletion of the neural cell adhesion molecule. *J. Biol. Chem.*, *280*, 42971-42977.

[11] Eckhardt, M., Muhlenhoff, M., Bethe, A., Koopman, J., Frosch, M. and Gerardy-Schahn, R. (1995). Molecular characterization of eukaryotic polysialyltransferase-1. *Nature*, *373*, 715-718.

[12] Kojima, N., Yoshida, Y. and Tsuji, S. (1995). A developmentally regulated member of the sialyltransferase family (ST8Sial II, STX) is a polysialic acid synthase. *FEBS Lett.*, *373*, 119-122.

[13] Terayama, K., Oka, S., Seiki, T., Miki, Y., Nakamura, A., Kozutsumi, Y., Takio, K. and Kawasaki, T. (1997). Cloning and functional expression of a novel glucuronyltransferase involved in the biosynthesis of the carbohydrate epitope HNK-1. *Proc. Natl. Acad. Sci. U. S. A.*, *94*, 6093-6098.

[14] Seiki, T., Oka, S., Terayama, K., Imiya, K. and Kawasaki, T. (1999). Molecular cloning and expression of a second glucuronyltransferase involved in the biosynthesis of the HNK-1 carbohydrate epitope. *Biochem. Biophys. Res. Commun.*, *255*, 182-187.

[15] Abo, T. and Balch, C. M. (1981). A differentiation antigen of human NK and K cells identified by a monoclonal antibody (HNK-1). *J. Immunol.*, *127*, 1024-1029.

[16] Schuller-Petrovic, S., Gebhart, W., Lassmann, H., Rumpold, H. and Kraft, D. (1983). A shared antigenic determinant between natural killer cells and nervous tissue. *Nature*, *306*, 179-181.

[17] Dennis, R. D., Martini, R. and Schachner, M. (1991). Expression of carbohydrate epitopes L2/HNK-1 and L3 in the larva and imago of Drosophila melanogaster and Calliphora vicina. *Cell Tissue Res.*, *265*, 589-600.

[18] Metcalfe, W. K., Myers, P. Z., Trevarrow, B., Bass, M. B. and Kimmel, C. B. (1990). Primary neurons that express the L2/HNK-1 carbohydrate during early development in the zebrafish. *Development*, *110*, 491-504.

[19] Schwarting, G. A., Jungalwala, F. B., Chou, D. K., Boyer, A. M. and Yamamoto, M. (1987). Sulfated glucuronic acid-containing glycoconjugates are temporally and spatially regulated antigens in the developing mammalian nervous system. *Dev. Biol.*, *120*, 65-76.

[20] Nieke, J. and Schachner, M. (1985). Expression of the neural cell adhesion molecules L1 and N-CAM and their common carbohydrate epitope L2/HNK-1 during development and after transection of the mouse sciatic nerve. *Differentiation*, *30*, 141-151.

[21] Chou, D. K., Ilyas, A. A., Evans, J. E., Costello, C., Quarles, R. H. and Jungalwala, F. B. (1986). Structure of sulfated glucuronyl glycolipids in the nervous system reacting with HNK-1 antibody and some IgM paraproteins in neuropathy. *J. Biol. Chem.*, *261*, 11717-11725.

[22] Bronner-Fraser, M. (1987). Perturbation of cranial neural crest migration by the HNK-1 antibody. *Dev. Biol.*, *123*, 321-331.

[23] Kuratani, S. C. (1991). Alternate expression of the HNK-1 epitope in rhombomeres of the chick embryo. *Dev. Biol.*, *144*, 215-219.

[24] Keihauer, G., Faissner, A. and Schachner, M. (1985). Differential inhibition of neurone-neurone, neurone-astrocyte and astrocyte-astrocyte adhesion by L1, L2 and N-CAM antibodies. *Nature*, *316*, 728-730.

[25] Bakker, H., Friedmann, I., Oka, S., Kawasaki, T., Nifant'ev, N., Schachner, M. and Mantei, N. (1997). Expression cloning of a cDNA encoding a sulfotransferase involved in the biosynthesis of the HNK-1 carbohydrate epitope. *J. Biol. Chem.*, *272*, 29942-29946.

[26] Inoue, M., Kato, K., Matsuhashi, H., Kizuka, Y., Kawasaki, T. and Oka, S. (2007). Distributions of glucuronyltransferases, GlcAT-P and GlcAT-S, and their target substrate, the HNK-1 carbohydrate epitope in the adult mouse brain with or without a targeted deletion of the GlcAT-P gene. *Brain Res.*, *1179*, 1-15.

[27] Yamamoto, S., Oka, S., Inoue, M., Shimuta, M., Manabe, T., Takahashi, H., Miyamoto, M., Asano, M., Sakagami, J., Sudo, K., Iwakura, Y., Ono, K. and Kawasaki, T. (2002). Mice deficient in nervous system-specific carbohydrate epitope HNK-1 exhibit impaired synaptic plasticity and spatial learning. *J. Biol. Chem.*, *277*, 27227-27231.

[28] Senn, C., Kutsche, M., Saghatelyan, A., Bosl, M. R., Lohler, J., Bartsch, U., Morellini, F. and Schachner, M. (2002). Mice deficient for the HNK-1 sulfotransferase show alterations in synaptic efficacy and spatial learning and memory. *Mol. Cell Neurosci.*, *20*, 712-729.

[29] Tagawa, H., Kizuka, Y., Ikeda, T., Itoh, S., Kawasaki, N., Kurihara, H., Onozato, M. L., Tojo, A., Sakai, T., Kawasaki, T. and Oka, S. (2005). A non-sulfated form of the HNK-1 carbohydrate is expressed in mouse kidney. *J. Biol. Chem.*, *280*, 23876-23883.

[30] Kizuka, Y., Kobayashi, K., Kakuda, S., Nakajima, Y., Itoh, S., Kawasaki, N. and Oka, S. (2008). Laminin-1 is a novel carrier glycoprotein for the nonsulfated HNK-1 epitope in mouse kidney. *Glycobiology*, *18*, 331-338.

[31] Kakuda, S., Sato, Y., Tonoyama, Y., Oka, S. and Kawasaki, T. (2005). Different acceptor specificities of two glucuronyltransferases involved in the biosynthesis of HNK-1 carbohydrate. *Glycobiology*, *15*, 203-210.

[32] Kitagawa, H., Tone, Y., Tamura, J., Neumann, K. W., Ogawa, T., Oka, S., Kawasaki, T. and Sugahara, K. (1998). Molecular cloning and expression of glucuronyltransferase I involved in the biosynthesis of the glycosaminoglycan-protein linkage region of proteoglycans. *J. Biol. Chem.*, *273*, 6615-6618.

[33] Wei, G., Bai, X., Sarkar, A. K. and Esko, J. D. (1999). Formation of HNK-1 determinants and the glycosaminoglycan tetrasaccharide linkage region by UDP-GlcUA:Galactose beta1,3-glucuronosyltransferases. *J. Biol. Chem.*, *274*, 7857-7864.

[34] Liedtke, S., Geyer, H., Wuhrer, M., Geyer, R., Frank, G., Gerardy-Schahn, R., Zahringer, U. and Schachner, M. (2001). Characterization of N-glycans from mouse brain neural cell adhesion molecule. *Glycobiology*, *11*, 373-384.

[35] Gallego, R. G., Blanco, J. L., Thijssen-van Zuylen, C. W., Gotfredsen, C. H., Voshol, H., Duus, J. O., Schachner, M. and Vliegenthart, J. F. (2001). Epitope diversity of N-glycans from bovine peripheral myelin glycoprotein P0 revealed by mass spectrometry and nano probe magic angle spinning 1H NMR spectroscopy. *J. Biol. Chem.*, *276*, 30834-30844.

[36] Kizuka, Y., Matsui, T., Takematsu, H., Kozutsumi, Y., Kawasaki, T. and Oka, S. (2006). Physical and functional association of glucuronyltransferases and sulfotransferase involved in HNK-1 biosynthesis. *J. Biol. Chem.*, *281*, 13644-13651.

[37] Ong, E., Yeh, J. C., Ding, Y., Hindsgaul, O. and Fukuda, M. (1998). Expression cloning of a human sulfotransferase that directs the synthesis of the HNK-1 glycan on the neural cell adhesion molecule and glycolipids. *J. Biol. Chem.*, *273*, 5190-5195.

[38] Kakuda, S., Oka, S. and Kawasaki, T. (2004). Purification and characterization of two recombinant human glucuronyltransferases involved in the biosynthesis of HNK-1 carbohydrate in Escherichia coli. *Protein Expr. Purif.*, *35*, 111-119.

[39] Nilsson, T., Rabouille, C., Hui, N., Watson, R. and Warren, G. (1996). The role of the membrane-spanning domain and stalk region of N-acetylglucosaminyl-transferase I in retention, kin recognition and structural maintenance of the Golgi apparatus in HeLa cells. *J. Cell Sci.*, *109*, 1975-1989.

[40] Opat, A. S., van Vliet, C. and Gleeson, P. A. (2001). Trafficking and localization of resident Golgi glycosylation enzymes. *Biochimie*, *83*, 763-773.

[41] Giraudo, C. G. and Maccioni, H. J. (2003). Ganglioside glycosyltransferases organize in distinct multienzyme complexes in CHO-K1 cells. *J. Biol. Chem.*, *278*, 40262-40271.

[42] Giraudo, C. G. and Maccioni, H. J. (2003). Endoplasmic reticulum export of glycosyltransferases depends on interaction of a cytoplasmic dibasic motif with Sar1. *Mol. Cell Biol.*, *14*, 3753-3766.

[43] Quintero, C. A., Valdez-Taubas, J., Ferrari, M. L., Haedo, S. D. and Maccioni, H. J. (2008). Calsenilin and CALP interact with the cytoplasmic tail of UDP-Gal:GA2/GM2/GD2 beta-1,3-galactosyltransferase. *Biochem. J.*, *412*, 19-26.

[44] McCormick, C., Duncan, G., Goutsos, K. T. and Tufaro, F. (2000). The putative tumor suppressors EXT1 and EXT2 form a stable complex that accumulates in the Golgi apparatus and catalyzes the synthesis of heparan sulfate. *Proc. Natl. Acad. Sci. U. S. A.*, *97*, 668-673.

[45] Izumikawa, T., Koike, T., Shiozawa, S., Sugahara, K., Tamura, J. and Kitagawa, H. (2008). Identification of chondroitin sulfate glucuronyltransferase as chondroitin synthase-3 involved in chondroitin polymerization: chondroitin polymerization is achieved by multiple enzyme complexes consisting of chondroitin synthase family members. *J. Biol. Chem.*, *283*, 11396-11406.

[46] Akasaka-Manya, K., Manya, H., Nakajima, A., Kawakita, M. and Endo, T. (2006). Physical and functional association of human protein O-mannosyltransferases 1 and 2. *J. Biol. Chem.*, *281*, 19339-19345.

[47] Wells, L., Kreppel, L. K., Comer, F. I., Wadzinski, B. E. and Hart, G. W. (2004). O-GlcNAc transferase is in a functional complex with protein phosphatase 1 catalytic subunits. *J. Biol. Chem.*, *279*, 38466-38470.

[48] Hart, G. W., Housley, M. P. and Slawson, C. (2007). Cycling of O-linked beta-N-acetylglucosamine on nucleocytoplasmic proteins. *Nature*, *446*, 1017-1022.

[49] Gerardy-Schahn, R., Oelmann, S. and Bakker, H. (2001). Nucleotide sugar transporters: biological and functional aspects. *Biochimie*, *83*, 775-782.

[50] Sprong, H., Degroote, S., Nilsson, T., Kawakita, M., Ishida, N., van der Sluijs, P. and van Meer, G. (2003). Association of the Golgi UDP-galactose transporter with UDP-galactose:ceramide galactosyltransferase allows UDP-galactose import in the endoplasmic reticulum. *Mol. Biol. Cell*, *14*, 3482-3493.

[51] Yano, H., Yamamoto-Hino, M., Abe, M., Kuwahara, R., Haraguchi, S., Kusaka, I., Awano, W., Kinoshita-Toyoda, A., Toyoda, H. and Goto, S. (2005). Distinct functional units of the Golgi complex in Drosophila cells. *Proc. Natl. Acad. Sci. U. S. A.*, *102*, 13467-13472.

[52] Kakuda, S., Shiba, T., Ishiguro, M., Tagawa, H., Oka, S., Kajihara, Y., Kawasaki, T., Wakatsuki, S. and Kato, R. (2004). Structural basis for acceptor substrate recognition of a human glucuronyltransferase, GlcAT-P, an enzyme critical in the biosynthesis of the carbohydrate epitope HNK-1. *J. Biol. Chem.*, *279*, 22693-22703.

[53] Shiba, T., Kakuda, S., Ishiguro, M., Morita, I., Oka, S., Kawasaki, T., Wakatsuki, S. and Kato, R. (2006). Crystal structure of GlcAT-S, a human glucuronyltransferase, involved in the biosynthesis of the HNK-1 carbohydrate epitope. *Proteins*, *65*, 499-508.

In: Glycobiology Research Trends
Editors: G. Powell and O. McCabe
ISBN: 978-1-60692-841-7

Short Communication A

PROTEIN-SPECIFIC GLYCOSYLATION: CIS-CONTROLLING PEPTIDIC ELEMENTS

Franz-Georg Hanisch
Institute of Biochemistry II, Medical Faculty, and Center for Molecular Medicine Cologne, University of Cologne, Joseph-Stelzmann-Str. 52, 50931 Köln, Germany

ABSTRACT

The term "Protein-specific glycosylation" points to important functional implications of a subset of glycosylation types that are under direct control of recognition determinants on the protein. An example of this type is found in the β4-GalNAc modification of N-linked glycans on a selected panel of proteins, like carbonic anhydrase or glycodelin, which was demonstrated recently to require specific protein (sequence) determinants proximal to the glycosylation site functioning as cis-regulating elements. Another example of such a cis-regulating element was described for the control of mammalian O-mannosylation. In this case the structural features of the substrate sites (located within the mucin domain of human α-dystroglycan) are necessary, but not sufficient for determining the transfer of mannose to Ser/Thr. Evidence was provided that an upstream-located peptide was also essential. This short overview will highlight recent work on the topic "Protein-specific Glycosylation" focussing on the above-cited examples, where the primary structure of a protein directly exerts a cis-control of the glycosylation event and will not refer to other examples, where conformational epitopes serve as signal patches, as revealed for the modification of lysosomal hydrolases with mannose-6-phosphate or for the polysialylation of NCAM.

The Initiation of Protein Glycosylation is Controlled by the Primary Structure

The initial transfer of sugars into certain positions of a protein is catalyzed by peptidyl glycosyltransferases and controlled by structural features of the substrate positions and their local peptide environment. Accordingly, the initiation of glycosylation exhibits expectedly a certain degree of protein-specificity. This is most evident in those cases where the first sugar is transferred into a "consensus" sequence. To such consensus motif-dependent forms of glycosylation belong the Asn-N-linked oligosaccharide chains, which are transferred from a dolicholphosphate-anchor into the *Asn*-Xaa-Thr/Ser/Cys (where Xaa can be any amino acid except proline) (Bause and Legler, 1981). Further examples of a stringent sequence control *via* a sequon are found in the rare glycosylation forms initiated by transfer of fucose (Cys-Xaa-Xaa-Gly-Gly-*Thr*-Cys) and glucose (Cys-Xaa-*Ser*-Xaa-Pro-Cys) into Ser/Thr positions located within EGF repeats (Harris and Spellman, 1993). Other forms of O-linked glycosylation are less stringently controlled, like the mucin-type chains, which are initiated by transfer of O-GalNAc, and the modification of nucleocytosolic proteins with single O-GlcNAc. About 20 polypeptide GalNAc-transferases with cell-specific expression patterns transfer the O-GalNAc modification sequon-independent into serine or threonine positions of proteins, but show some degree of selectivity with overlapping site preferences (Tarp and Clausen, 2008; Ten Hagen et al. 2003). Also the acceptor site for O-GlcNAc does not display a definite consensus sequence (Hart et al., 2007). However, the fuzzy motif is characterized by the close vicinity of proline residues, valines and a downstream tract of serines, while other amino acids, like leucines and glutamines are unfavourable.

Double Control of Glycosylation Confers Higher Degree of Specificity

While initiation of glycosylation appears to be generally determined by structural properties of the substrate position, evidence was revealed very recently for the existence of new control mechanisms on the initiation level, which refers to one of the rare types of protein-specific O-glycosylation, i.e. mammalian O-mannosylation (Breloy et al., 2008). Also very recently, a similar type of double control was shown to be exerted by the peptide core on peripheral modifications with LacdiNAc of N-linked chains (Miller et al., 2008). Both observations have in common that a peptide stretch located distant from the putative glycosylation site represents a necessary and sufficient prerequisite for the protein-specific glycosylation. These cis-controlling peptidic elements provide another level of protein specificity to the glycosylation event as outlined below.

Cis-controlling Peptide Elements in the Protein-specific Modification of N-glycans with LacdiNAc

A peripheral modification of N-linked chains, the addition of β4-linked GalNAc to terminal GlcNAc with formation of the LacdiNAc dihexosamine, represents a unique protein-specific glycosylation found on a selected panel of glycoproteins. The peripheral modification appears to play crucial roles in the regulation of circulatory half-life of hormones (Manzella et al., 1996) and in cell recognition (Lowe and Marth, 2003). The modification was detected on N-linked glycans in a series of mammalian glycoproteins including the glycoprotein hormone LH (luteinizing hormone) (Green et al., 1985), glycodelin (Dell et al., 1995), prolactin-like proteins (Manzella et al., 1997), proopiomelanocortin (Skelton et al., 1992), SorLA/LR11 (Fiete et al., 2007), sialoadhesin (Martinez-Pomares et al., 1992), tenascin-R (Woodworth et al., 2001), and carbonic anhydrase VI (Hooper et al., 1995). As a peripheral modification of O-linked chains it was found on murine *zona pellucida* glycoprotein (ZP3), which plays a role in initial sperm-egg binding (Dell et al., 2003).

The group of Jacques Baenziger could recently provide evidence that a 19-meric peptide sequence within the target protein (LRRFIEQKITKRKKEKYWP) is necessary and sufficient for the modification of N-linked glycans with LacdiNAc, i.e. for the activity of the β4-specific GalNAc-transferases βGT3 and βGT4 (Miller et al., 2008). This determining cis-located peptide is characterized by a high content of basic amino acids and an α-helical structure. It is located C-terminally relative to the N-glycosylation site in carbonic anhydrase VI and could induce the modification of a normally unmodified protein (transferrin) after recombinant translocation of the peptide. However, the C- or N-terminal location of the cis-contolling peptide relative to the N-glycosylation site does not appear to play a significant role. By contrast, basic amino acids in the sequences PLRSKK or KRKKEK are essential for recognition by the βGT. Also partial structures of the 19-meric sequence conferred some substrate quality to the target protein, but the full 19-amino acid sequence was significantly more effective. Secondary structure prediction programs revealed that the 19-meric peptide adopts an α-helical structure, which may be necessary to place specific basic residues in the correct spatial relationship to interact with βGT3 and βGT4 (Miller et al., 2008). The glycosyltransferases βGT3 and βGT4 are large proteins comprizing about 1000 amino acids. Sequence homology is found primarily in the C-terminal part of the proteins, where the catalytic activity is encoded.

The functional implications of LacdiNAc-modified glycoproteins may be multifarious, but at least in a few selected cases experimental evidence could demonstrate the biological significance of this protein-specific modification. In particular, the human amniotic fluid-derived glycoprotein, glycodelin-A (GdA) was shown to be an effective inhibitor of gamete binding in an established sperm-egg binding system (Oehninger et al., 1995). Moreover, the same glycoprotein expresses immunosuppressive activities directed against a variety of different immune cell types (Okamoto et al., 1991). The inhibitory effect on gamete binding is likely related to specific N-glycosylation of GdA, since the same glycoprotein isolated from seminal fluid (GdS) does not inhibit human sperm-zona pellucida binding and exhibits a gender-specific N-glycosylation characterized by unusually fucose-rich complex N-glycans and absence of LacdiNAc modifications (Dell et al., 1995; Morris et al., 1996).

The finding that GdA interferes *via* its N-linked glycans with sperm-egg binding is directly related to the long established involvement of carbohydrates in the fertilization process. Murine sperm initiate fertilization by binding to the outer glycoprotein cover of the egg, the *zona pellucida.* Numerous studies have revealed that this interaction requires O-glycans linked to the *zona pellucida* protein 3 (ZP3) (Wassarman, 1999). Although the precise molecular basis of this interaction remains to be resolved, structural studies of mZP3 have revealed the presence of major core 2-based glycans with terminal LacNAc, and LacdiNAc modification (Dell et al., 2003). It is unlikely however that sialylated glycans participate in gamete binding, since on desialylation of mZP sperm binding increased *in vitro.*

A Cis-controlling Peptide in the Initiation of Mammalian O-mannosylation

As outlined above, most initial sugar transfer reactions exhibit some degree of control by the substrate position and local peptide environment. This holds true in particular for those initial transfer reactions that are under stringent control of a sequon. But there are types of initial O-glycosylation that lack such a control by the substrate site, while exhibiting a high degree of protein-specificity. One of these, the more rare modification of hydroxy amino acids with O-linked mannose, was initially regarded as restricted to yeast cells and other fungi (Sentandreu and Northcote, 1969). In yeast it is introduced into proteins with essential functions in the cell wall formation. The transfer reaction is catalysed by a family of protein mannosyltransferases residing in the endoplasmic reticulum (Lehle et al., 2006; Lengeler et al., 2008). In mammalian cells the modification appears to be restricted to a small number of proteins, primarily expressed in skeletal muscles, in the brain and in peripheral nerves (Krusius et al., 1986; Endo, 1999). The only structurally characterized protein with proven O-mannosylation is α-dystroglycan, an extracellular component of the dystrophin complex in membranes of skeletal muscle, nerves and brain (Michele and Campbell, 2003). While in yeast cells the core-mannose is elongated to linear oligomannose chains, the extensions on human dystroglycan can be quite complex by containing GlcNAc and Gal and by terminating with Fuc or NeuAc. The major glycan component of α-dystroglycan was characterized as a tetrasaccharide with the structure NeuAc2-3Gal1-4GlcNAc1-2Man and was claimed to represent the specific ligand for interaction with the laminin G domains (Chiba et al., 1997; Endo, 1999). Also structural variants were reported that express Lewis (Smalheiser et al., 1998) or HNK-1 epitopes (Yuen et al., 1997).

The initiation of O-mannosylation is catalyzed by two homologues of the yeast protein mannosyltransferases, the POMT1 and POMT2 (Manya et al., 2004). These are active only on co-expression and were shown to form a functional complex (Akasaka-Manya et al., 2006). The protein mannosyltransferases in yeast are able to transfer the sugar from a dolicholphosphate donor onto pentapeptide substrates indicating that a consensus sequence should not play a role in the initiation (Willer et al., 2005). The mammalian homologues were found to be inactive *in vitro* on short peptides with lengths up to 15-mers (Manya et al., 2004;

Breloy et al., 2008). However, when using 20-meric peptide substrates the *in vitro* transfer of mannose into two peptides with a spread consensus motif was observed (Manya et al., 2007). The 18-meric consensus sequence IXPT(P/X)TXPXXXXPTX(T/X)XX appears two times within regions adjacent to or upstream of the α-dystroglycan mucin domain, but BLAST searches for other proteins carrying this sequence failed.

The existence of a consensus motif as sufficient structural determinant for O-mannosylation is in conflict with a recent *in vivo* study demonstrating that O-mannosylation of recombinant α-dystroglycan probes was strictly dependent on upstream regions, N-terminal of the mucin domain (Breloy et al., 2008). A series of truncation probes comprizing sections of the mucin domain (p418-536) and more or less extended stretches of the upstream located regions (p286-417) was expressed in HEK293 cells and the affinity-isolated secretory fusion proteins were structurally characterized by mass spectrometry of the glycans. Unexpectedly, probes that contained the postulated consensus sequence (p400-417) were O-mannose-negative. Only if the peptide stretch p346-417 was linked to the mucin domain, a significant O-mannosylation was detectable (Breloy, 2008). Mannose was preferentially introduced into Ser/Thr clusters in the mucin domain that were flanked by diads of basic amino acids. A further specification of the peptide length necessary and sufficient for triggering O-mannosylation was possible by expression of deletion constructs that lacked parts of the upstream regions. While the extended strand forming peptide p368-377 could be deleted without affection of the O-mannosylation capacity, the deletion of p392-417 caused complete abrogation of O-mannose transfer. These findings allowed localizing the determining region to the 41-meric peptide p377-417: IQTPTLGPIQPTRVSEA GTTVPGQIRPTMTIPGYVEPTAVA (Breloy et al., 2008). A schematic view of the cis controlling peptide and the O-mannosylated mucin domain of human α-dystroglycan is shown in figure 1.

Different from the above referenced modification of N-glycans with LacdiNAc the determining peptide was necessary, but not sufficient for the induction of O-mannosylation in hybrid constructs. Fusion proteins consisting of the hDG determinant and the MUC1 repeat domain, which is normally not O-mannosylated, were O-mannose negative with respect to the MUC1 peptide (Breloy et al., 2008). This finding points to hDG-specific properties of the substrate sites in the hDG mucin domain not found in MUC1. *In vitro* transfer of GalNAc into Ser/Thr sites of the hDG mucin domain revealed only poor substrate qualities for the peptide GalNAc transferases ppGalNAc-T1 and –T2, which contrasts with the predictions by NetOGlyc 3.1 server (www.cbs.dtu.dk/services/NetOGlyc). The discrepancy between prediction and experimentally confirmed numbers of transferred GalNAc was most obvious for Ser/Thr clusters with flanking diads of basic amino acids. Interestingly, these sites were identified as the preferred targets for O-mannosylation *in vivo* (Breloy et al., 2008). In summary, the feature of a doubly controlled initiation process distinguishes mammalian O-mannosylation from other types of O-glycosylation, which are exclusively controlled by structural properties in the local environment of the substrate positions.

Although O-mannosylation represents a typical example of a protein-specific glycosylation, which is reflected in its rare occurrence, the knowledge on its functional implications is rather weak. Except for α-dystroglycan as an established interactor of the extracellular matrix protein laminin and the controversial discussion of O-mannose-based glycans contributing to the binding of G domains in α-laminin chains there is no hint to a specific function of this protein modification. The importance of O-mannose-based acidic glycans in laminin interaction however becomes evident from the existence of a series of congenital disorders of glycosylation with severe impact on skeletal and brain function. These forms of congenital muscle dystrophies or dystroglycanopathies are caused by mutation defects of glycosyltransferases involved in O-mannosylation of α-dystroglycan (Endo, 1999). Among these the Walker-Warburg syndrome represents an autosomal recessive form that is characterized by severe brain malfunction and eye anomalies. The patients suffer from a selective deficiency of glycosylated α-dystroglycan, which results from mutations in the POMT1/POMT2 genes and the concomitant loss of a functional O-mannosyltransferase. Lack of O-mannosylation causes a failure to firmly link α-dystroglycan to the "glia limitans", the basement membrane, which prevents neurons from migrating and escaping from the cortex during normal brain development (Van Reewijk et al., 2004). A further aspect of O-mannosyl glycans in mammals is their involvement in viral infections. α-Dystroglycan has been identified as the cellular receptor for arenaviruses including the Lassa fever virus, where binding of the virus occurs by the mucin domain and shows a dependency on the activity of LARGE (Kunz et al., 2005).

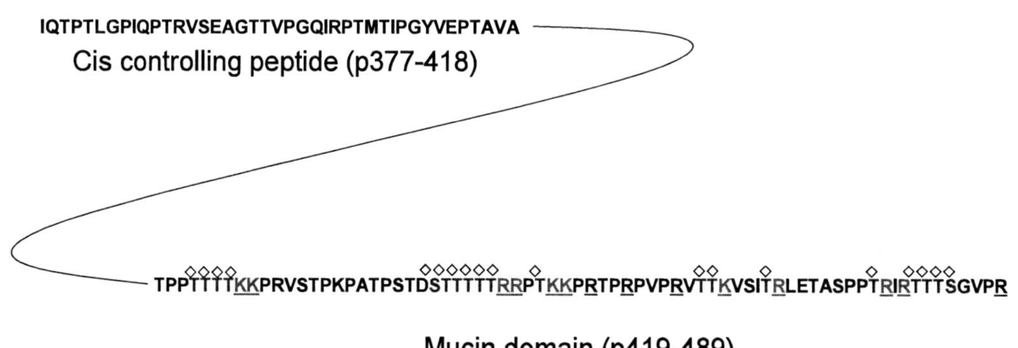

Figure 1. Cis control of protein specific O-mannosylation: An upstream located peptide (p377-418) serves as recognition determinant for the POMT1/POMT2 complex and controls in this way the human dystroglycan-specific transfer of mannose (◇) into Ser/Thr positions of the mucin domain (p419-489).

Conclusion

More than half of all human proteins have been claimed to be glycosylated in any way, although only a fraction of known proteins was actually characterized as glycoproteins. This number is steadily increasing with time, while our knowledge on the function of glycans modifying these proteins is often scarce. A series of specific functions attributed to glycoproteins were revealed in molecular and cellular interactions (as binding ligands or as anti-adhesive molecules), in the sorting and cellular trafficking of glycoproteins, in the stabilization of protein conformations or in the protection of proteins from proteolytic degradation. However, putative glycosylation sites, like those modified with O-GalNAc, may arise as a result of single point mutations, and in accordance with this the degree of specificity and hence of evolutionary conservation of O-glycosylation sites in mucin-like domains is expectedly low. This is not the case, if a protein-specific determinant, like a cis-controlling peptide sequence, which is independent of the putative glycosylation site, provides a higher level of specificity. We can propose that such independently controlling determinants must be associated with important and highly conserved functions. O-Mannosylation of proteins represents one of these posttranslational modifications with high evolutionary conservation from procaryotes to mammals and is still obscure with respect to its function. It will be challenging to identify other proteins than α-dystroglycan that are modified with O-linked mannose to learn on the potential involvement of this glycosylation variant in cellular differentiation and development.

REFERENCES

[1] Bause, E, Legler, G. (1981). The role of the hydroxy amino acid in the triplet sequence Asn-Xaa-Thr(Ser) for the N-glycosylation step during glycoprotein biosynthesis. *Biochem. J. 195*, 639-644.

[2] Harris, R.J., Spellman, M.W. (1993). O-Linked fucose and other post-translational modifications unique to EGF modules. *Glycobiology 3*, 219-224.

[3] Nishimura, H., Kawabata, S., Kisiel, W., Hase, S., Ikenaka, T., Takao, T., Shimonishi, Y., Iwanaga, S. (1989). Identification of a disaccharide (Xyl-Glc) and a trisaccharide (Xyl2-Glc) O-glycosidically linked to a serine residue in the first epidermal growth factor-like domain of human factors VII and IX and protein Z and bovine protein Z. *J. Biol. Chem. 264*, 20320-20325.

[4] Tarp, M.A., Clausen, H. (2008). O-Glycosylation and its potential use in drug and vaccine development. *Biochim. Biophys. Acta 1780*, 546-563.

[5] Ten Hagen, K.G., Fritz, T.A., Tabak, L.A. (2003). All in the family: the UDP-GalNAc: polypeptide N-acetylgalactosaminyltransferases. *Glycobiology 13*, 1R-16R.

[6] Hart, G.W., Housley, M.P., Slawson, C. (2007). Cycling of O-linked beta-N-acetylglucosamine on nucleocytoplasmic proteins. *Nature 446*, 1017-1022.

[7] Breloy, I.., Schwientek, T., Gries, B., Razawi, H., Macht, M., Albers, C., Hanisch, F.-G. (2008). Initiation of mammalian O-mannosylation in vivo is independent of a consensus sequence and controlled by peptide regions within and upstream of the a-dystroglycan mucin domain. *J. Biol. Chem. 283*, 18832-18840.

[8] Miller, E., Fiete, D., Blake, N.M.J., Beranek, M., Oates, E.I., Mi, Y., Roseman, D.S., Baenziger, J.U. (2008). A necessary and sufficient determinant for protein-selective glycosylation in vivo. *J. Biol. Chem. 283*, 1985-1991.

[9] Manzella, S.M., Hooper, L.V., Baenziger, J.U. (1996). Oligosaccharides containing beta 1-4-linked N-acetylgalactosamine, a paradigm for protein-specific glycosylation. *J. Biol. Chem. 271*, 12117-12120.

[10] Lowe, J.B., Marth, J.D. (2003). A genetic approach to mammalian glycan function. *Annu. Rev. Biochem. 72*, 643-691.

[11] Green, E.D., van Halbeek, H., Boime, I., Baenziger, J.U. (1985). Structural elucidation of the disulfated oligosaccharide from bovine lutropin. *J. Biol. Chem. 260*, 15623-15630.

[12] Dell, A., Morris, H.R., Easton, R.L, Panico, M., Patankar, M., Oehniger, S., Koistinen, R., Koistinen, H., Seppala, M., Clark, G.F. (1995). Structural analysis of the oligosaccharides derived from glycodelin, a human glycoprotein with potent immunosuppressive and contraceptive activities. *J. Biol. Chem. 270*, 24116-24126.

[13] Manzella, S.M., Dharmesh, S.M., Cohick, C.B., Soares, M.J., Baenziger, J.U. (1997). Developmental regulation of a pregnancy-specific oligosaccharide structure, NeuAcalpha2,6GalNAcbeta1,4GlcNAc, on select members of the rat placental prolactin family. *J. Biol. Chem. 272*, 4775-4782.

[14] Skelton, T.P., Kumar, S., Smith, P.L., Beranek, M., Baenziger, J.U. (1992). Pro-opiomelanocortin synthesized by corticotrophs bears asparagine-linked oligosaccharides terminating with SO4-4GalNAc beta 1,4GlcNAc beta 1,2 Man alpha. *J. Biol. Chem. 267*, 12998-13006.

[15] Fiete, D., Mi, Y., Oates, E.L., Beranek, M.C., Baenziger, J.U. (2007). N-linked oligosaccharides on the low density lipoprotein receptor homolog SorLA/LR11 are modified with terminal GalNAc-4-SO4 in kidney and brain. *J. Biol. Chem. 282*, 1873-1881.

[16] Martinez-Pomares, L., Crocker, P.R., DaSilva, R., Holmes, N., Colominas, C., Rudd, P., Dwek, R., Gordon, S. (1999). Cell-specific glycoforms of sialoadhesin and CD45 are counter-receptors for the cysteine-rich domain of the mannose receptor. *J. Biol. Chem. 274*, 35211-35218.

[17] Woodworth, A., Fiete, D., Baenziger, J.U. (2001). The man/GalNAc-4-SO4-receptor has multiple specificities and functions. *Glycobiology 11*, abstr. 45.

[18] Hooper, L.V., Beranek, M.C., Manzella, S.M., Baenziger, J.U. (1995). Differential expression of GalNAc-4-sulfotransferase and GalNAc-transferase results in distinct glycoforms of carbonic anhydrase VI in parotid and submaxillary glands. *J. Biol. Chem. 270*, 5985-5993 .

[19] Dell, A., Chalabi, S., Easton, R.L., Haslam, S.L., Sutton-Smith, M., Patankar, M.S., Lattanzio, F., Panico, M., Morris, H.R., Clark, G.F. (2003). Murine and human zona pellucida 3 derived from mouse eggs express identical O-glycans. *Proc. Natl. Acad. Sci. USA 100*, 15631-15636.

[20] Oehninger S., Coddington, C.C., Hodgen, G.D., Seppala, M. (1995). Factors affecting fertilization: endometrial placental protein 14 reduces the capacity of human spermatozoa to bind to the human zona pellucida. *Fertil. Steril. 63*, 377-383.

[21] Okamoto, N., Uchida, A., Takakura, K., Karyia, Y., Kanzaki, H., Riitinen, L., Koistinen, R., Seppala, M., Mori, T. (1991). Suppression by human placental protein 14 of natural killer cell activity. *Am. J. Reprod. Immunol. 26*, 137-142.

[22] Morris, H.R., Dell, A., Easton, R.L., Panico, M., Koistinen, H., Koistinen, R., Oehninger, S., Patankar, M.S., Seppala, M., Clark, G.F. (1996). Gender-specific glycosylation of human glycodelin affects its contraceptive activity. *J. Biol. Chem. 271*, 32159-32167.

[23] Wassarman, P.M. (1999). Mammalian fertilization: molecular aspects of gamete adhesion. *Cell 96*, 175-183.

[24] Sentandreu, R., Northcote, D.H. (1969). Yeast cell-wall synthesis. *Biochem. J. 115*, 231-240.

[25] Lehle, L., Strahl, S., Tanner, W. (2006). Protein glycosylation, conserved from yeast to man: a model organism helps elucidate congenital human diseases. *Angew. Chem. Int. Ed. Engl. 45*, 6802-6818.

[26] Lengeler, K.B., Tielker, D., Ernst, J.F. (2008). Protein-O-mannosyltransferases in virulence and development. *Cell Mol. Life Sci. 65*, 528-544.

[27] Krusius, T., Finne, J., Margolis, R.K., Margolis, R.U. (1986). Identification of an O-glycosidic mannose-linked sialylated tetrasaccharide and keratan sulfate oligosaccharides in the chondroitin sulfate proteoglycan of brain. *J. Biol. Chem. 261*, 8237-8242.

[28] Endo, T. (1999). O-Mannosyl glycans in mammals. *Biochim. Biophys. Acta 1473*, 237-246.

[29] Michele, D.E., Campbell, K.P. (2003). Dystrophin-glycoprotein complex: post-translational processing and dystroglycan function. *J. Biol. Chem. 278*, 15457-15460.

[30] Chiba, A., Matsumura, K., Yamada, H., Inazu, T., Shimizu, T., Kusunoki, S., Kanazawa, I., Kobata, A., Endo, T. (1997). Structures of sialylated O-linked oligosaccharides of bovine peripheral nerve alpha dystroglycan. The role of a novel O-mannosyl-type oligosaccharide in the binding of alpha-dystroglycan with laminin. *J. Biol. Chem. 272*, 2156-2162.

[31] Smalheiser, N.R., Haslam, S.M., Sutton-Smith, M., Morris, H.R., Dell, A. (1998). Structural analysis of sequences O-linked to mannose reveals a novel Lewis X structure in cranin (dystroglycan) purified from sheep brain. *J. Biol. Chem. 273*, 23698-23703.

[32] Yuen, C.T., Chai, W., Loveless, R.W., Lawson, A.M., Margolis, R.U., Feizi, T. (1997). Brain contains HNK-1 immunoreactive O-glycans of the sulfoglucuronyl lactosamine series that terminate in 2-linked or 2,6-linked hexose (mannose). *J. Biol. Chem. 272*, 8924-8931.

[33] Manya, H., Chiba, A., Yoshida, A., Wang, X., Chiba, Y., Jigami, Y., Margolis, R.U., Endo, T. (2004). Demonstration of mammalian protein O-mannosyltransferase activity: coexpression of POMT1 and POMT2 required for enzymatic activity. *Proc. Natl. Acad. Sci. USA 101*, 500-505.

[34] Akasaka-Manya, K., Manya, H., Nakajima, A., Kawakita, M., Endo, T. (2006). Physical and functional association of human protein-O-mannosyltransferases 1 and 2. *J. Biol. Chem. 281*, 19339-19345.

[35] Willer, T., Brandl, M., Sipiczki, M., Strahl, S. (2005). Protein-O-mannosylation is crucial for cell wall integrity, septation and viability in fission yeast. *Mol. Microbiol. 57*, 156-170.

[36] Manya, H., Suzuki, T., Akasaka-Manya, K., Ishida, H.K., Mizuno, M., Suzuki, Y., Inazu, T., Dohmae, N., Endo, T. (2007). Regulation of mammalian protein-O-mannosylation: preferential amino acid sequence for O-mannose modification. *J. Biol. Chem. 282*, 20200-20206.

[37] Van Reeuwijk, J., Brunner, H.G., Bokhoven, H. (2004). Glyc-O-genetics of Walker-Warburg syndrome. *Clin. Genet. 67*, 281-289.

[38] Kunz, S., Rojek, J.M., Kanagawa, M., Spiropoulou, C.F., Barresi, R., Campbell, K.P., Oldstone, M.B.A. (2005). Posttranslational modification of alpha-dystroglycan, the cellular receptor for arenaviruses, by the glycosyltransferase LARGE is critical for virus binding. *J. Virol. 79*, 1482-14296.

In: Glycobiology Research Trends
Editors: G. Powell and O. McCabe

ISBN: 978-1-60692-841-7

Short Communication B

ACTIVATION OF INVARIANT Vα19-Jα33 TCR α-BEARING CELLS BY STIMULATION WITH CERTAIN α-MANNOSYLATED GLYCOLIPIDS

Michio Shimamura*

Mitsubishi Kagaku Institute of Life Sciences, 11 Minamiooya, Macchida, Tokyo 194-8511, Japan

ABSTRACT

A novel invariant Vα19-Jα33 TCR α chain was first found from mammalian blood cells. This TCR α chain as well as an invariant Vα14-Jα18 TCR α chain was primarily expressed by NK1.1$^+$ T cell repertoire (Vα19 NKT cell). Vα19 NKT cells produce immunoregulatory cytokines in response to TCR engagement, thus are suggested to be a potent therapeutic target. Attempts have been made to find specific antigens for Vα19 NKT cells. A series of α- and β-glycosyl ceramides were synthesized and tested whether they had potential to stimulate the cells isolated from invariant Vα19-Jα33 TCR transgenic mice (where development of Vα19 NKT cells is facilitated). Following comprehensive examinations substantial antigenic activity was found in α-ManCer that was presented by MR1, one of the MHC class Ib molecules. To determine structural requirements for natural ligands for Vα19 NKT cells, naturally occurring and synthetic α-mannosyl glycolipids were further analyzed. As a result, α-mannosyl phosphatidyl inositols (PI) such as $(\alpha\text{-Man})_2$-PI and α-Man-α-$GlcNH_2$-PI (a partial structure of mycobacterial lipoarabinomannan and GPI-anchors) as well as α-ManCer derivatives with structural modification in their sphingosine portion were found to activate Vα19 NKT cells *in vivo* and *in vitro*. Taken together, it is strongly suggested that Vα19 NKT cells are responsive to certain α-mannosylated glycolipids in the context of MR1.

* E-mail: michio@ncnp.go.jp or michio@mitils.jp

Keywords: glycosphingolipid; glycosyl phosphatidylinositol; immunotherapy; MHC class Ib; NKT Cell

INTRODUCTION

Natural killer T (NKT) cells are a novel lymphocyte subset bearing both the common NK marker NK1.1, a product of a member of the NKR-P1 gene family, and TCR-CD3 complex [1]. Up to now, several types of NKT cells have been reported. They are classified according to MHC restriction, TCR structure, expression of NK and T cell markers such as NK1.1 CD4/8 and cytokine production potential [2]. The major component of NKT cells (Vα14 NKT cell) express invariant TCR α chains (mouse Vα14-Jα18, human Vα24-Jα18) [3, 4], and is positively selected by the non-polymorphic MHC class I-like CD1d molecule in association with β2-microglobulin (β2m) [5, 6]. Vα14 NKT cells are responsive to certain α-galactosyl glycolipids presented by CD1d, for example, α-galactosyl ceramide (α-GalCer) [7] isolated from marine sponge [8], α-glucuronosyl and α-galacturonosyl ceramides from α-proteobacteria [9, 10], and α-galactosyl diacylglycerol from *Borrelia* [11]. In addition, it has been proposed that Vα14 NKT cells are positively selected by intracellular lysosomal isoglobotriaosyl ceramide in the context of CD1d [12].

Recently, another invariant TCR α chain consisting of Vα19-Jα33 (conventionally Jα26) has been found in human, bovine, and TAP-deficient mouse peripheral blood cells by quantitative PCR analysis [13, 14]. We have demonstrated that cells expressing the Vα19-Jα33 invariant TCR α chain are mainly present as NKT cells in mouse [15]. These NKT cells (designated as Vα19 NKT cell in this review) are not assigned to any types of NKT cells according to the previously proposed criteria [2]. Invariant Vα19-Jα33 TCR$^+$ cells are absent in mice lacking the non-classical MHC class I molecule MR1, thus suggesting that they are positively selected by MR1 [16]. It is estimated that Vα19 NKT cells represent 1 % of mononuclear cells (MNCs) in the liver [15], thus they are a considerably large population as a lymphocyte clone. Localization of the invariant Vα19-Jα33 TCR$^+$ cells in gut lamina propria is also reported [16]. Similar to Vα14 NKT cells [17], Vα19 NKT cells immediately produce large amounts of both Th1 and Th2-promoting immunoregulatory cytokines in response to the engagement of the invariant TCR (Shimamura, M. *et al.*, Characterization of a novel NKT cell repertoire expressing an invariant Vα19-Jα26 TCRα chain using the invariant TCR transgenic mice, Abstract #4 for the 2nd International Workshop on CD1 Antigen Presentation and NKT Cells, Woods Hole, 2002) [18 - 20], thus are considered to have important roles in the regulation of the immune system. The finding that over-expression of Vα19 NKT cells suppressed the progress of experimental autoimmune encephalomyelitis [21], an animal model for multiple sclerosis, supports this notion. Therefore, the search for specific antigens for Vα19 NKT cells is quite important to develop new therapies for various immunoregulatory disorders based on the functional modulation of the lymphocyte repertoire.

The self-antigens presented by MR1 have not been identified [22, 23]. Judging from the primary structure, MR1 possibly has a three-dimensional structure similar to classical MHC class I molecules which have a groove for antigen-presentation. Although the several key amino acid residues located at the bottom of the antigen-presenting groove and interacting with the antigens are strictly conserved in classical MHC class I molecules, those in MR1 are considerably converted [22]. Thus it is strongly suggested that the molecular species presented by MR1 are different from those (peptides) presented by classical MHC class I molecules. The discovery of α-galactosyl glycolipids as stimulants for Vα14 NKT cells prompted us to investigate synthetic and natural glycolipids as agonists for Vα19 NKT cells.

Glycolipids with Non-reducing End α-mannosyl Residues as Ligands for Vα19 NKT Cells

In a continuous search for specific antigens for this novel NKT cell repertoire, we paid attention to a series of α- and β-glycosyl ceramides with a naturally occurring monosaccharide residue [18]. Total mononuclear cells prepared from livers of invariant Vα19-Jα33 TCR transgenic (Tg) mice with TCR $C\alpha^{-/-}$ genetic background (including invariant Vα19 TCR Tg^{+} cells as the sole component of TCR^{+} cells and antigen-presenting cells) were cultured in the presence of the synthetic glycosphingolipids and the immune responses were analyzed to screen any stimulants for Vα19 NKT cells (Figure 1). Vα19 Tg^{+} $TCR\alpha^{-/-}$ cells were most efficiently induced to proliferate and produce IL-4 and IFN-γ by α-ManCer. In contrast, C57BL/6 cells (including about 30 % of Vα14 NKT cells) were responsive to α-GalCer and to a certain extent to α-glucosyl ceramide and glucuronyl ceramide. Both C57BL/6 and Vα19 Tg^{+} cells showed no detectable responsiveness to β-glycosyl ceramides. β2-microglobulin-deficient cells (including neither Vα19 nor Vα14 NKT cells) showed no reactivity to the glycolipids. Thus, these results suggest that Vα19 NKT cells are responsive to the stimulation with α-ManCer. No significant immune responses of Vα19 NKT cells were induced in culture with either α- altrosyl ceramide or α-talosyl ceramide (both α-altrose and α-talose as well as mannose possess a 2-axial hydroxy group, but the 3- hydroxy group in α-altrose and the 4- hydroxy group in α-talose are reversed from those in α-mannose), thus suggesting a stringent recognition of the α-mannosyl residue by the invariant Vα19-Jα33 TCR. Since the natural occurrence of α-ManCer has not been reported yet, it is possible that this glycolipid mimics natural ligands for the NKT cells.

To determine structural requirements for ligands for Vα19 NKT cells, naturally occurring and synthetic glycolipids were further analyzed for potential to induce immune responses from Vα19 NKT cells. As well as α-ManCer [18] and its derivatives [24] (see below), 2, 6-di α-mannosyl phosphatidylinositol $(\alpha\text{-Man})_2$PI, a partial structure of mycobacterial lipoarabinomannan (LAM) [25]) and α-mannosyl 1-4 α-glucosamine1-6-phosphatidylinositol (α-Man-$GlcNH_2$-PI, a partial structure of GPI-anchor [26]) were found as a potent stimulator for Vα19 NKT cells [27]. The active glycolipids had α-mannosyl residue(s) at the non-reducing end in common. In contrast, glycolipids with α-mannosyl residue(s) such as yeast glycosyl phosphoinositol ceramide mixture (α-Man-Ino-PO_4^- Cer etc. [28], LAM and its partially degraded derivatives ($(\alpha\text{-Man})_n$-PI, 40kD) [29], bivalve α-mannosylated trihexosyl ceramides (α-Man-Man-Glc-Cer) [30] etc. did not stimulate Vα19 Tg^{+} cells up to 10 μg/ml. Taken together, it is strongly suggested that certain glycolipids with appropriately located non-reducing end α-mannosyl residue(s) have potential to stimulate Vα19 NKT cells.

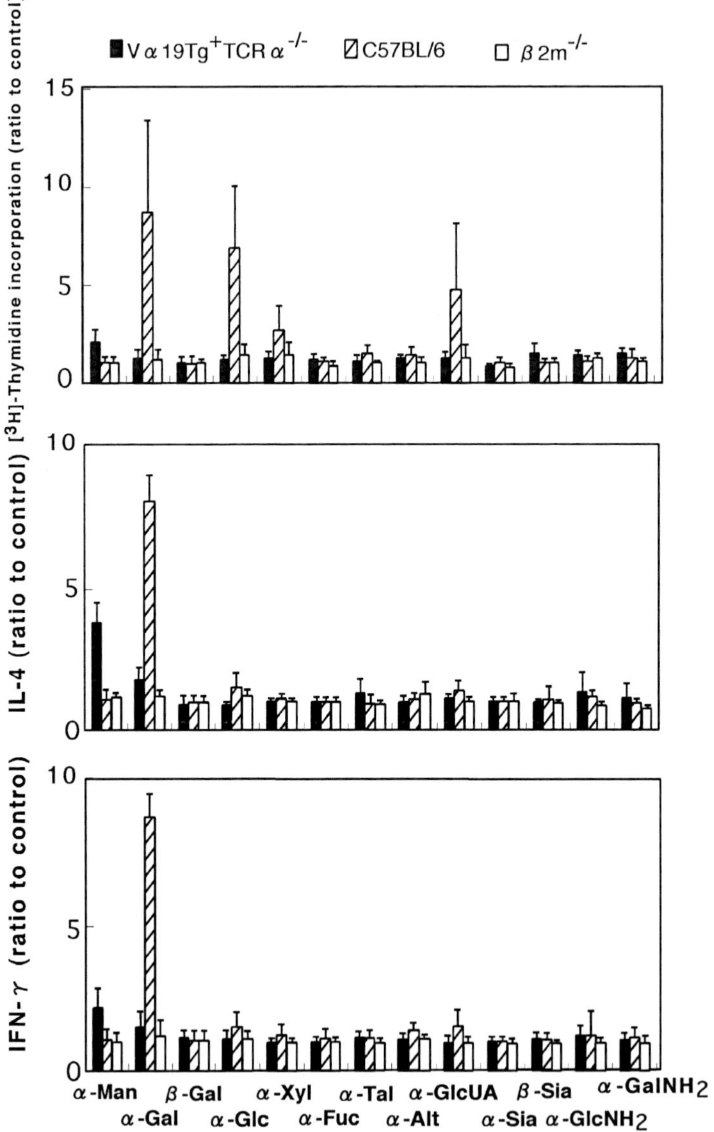

Figure 1. Activation of Vα19 Tg$^+$ cells with a-mannosyl ceramide in culture.
Liver MNCs prepared from Vα19Tg$^+$TCRα$^{-/-}$, C57BL/6 and β2m$^{-/-}$ mice were cultured in the presence (1 μg/ml) or absence of glycolipids. After 2 days, the immune-responses were monitored by measuring cell proliferation ([^{3}H]-thymidine incorporation for 5h), IL-4 and IFN-γ secretion in the culture supernatants. Results are shown as the fold increase relative to the control cultures with vehicle (1/200 v/v DMSO). Abbreviations: α-Gly, α-glycosyl ceramide; β-Gal, β-galactosyl ceramide; Sia, D-N-acetyl neuraminyl; Tal, D-talosyl, Alt, D-altrosyl.

The potential to respond to the α-mannosyl glycolipids was confined to the NK1.1$^+$ Vα19 Tg$^+$ (Vα19NKT) cells among the responders prepared from Vα19 Tg mice, because depletion of the NK1.1$^+$ or TCRαβ$^+$ population from the responders reduced the responsiveness to the glycolipids [24]. This finding suggests that the NK1.1$^+$ invariant Vα19 TCR Tg$^+$ cell population is distinguishable from the NK1.1$^-$ population in terms of antigen recognition although the use of invariant Vα19 TCR Tg is guaranteed in the TCR$^+$ cells of Vα19 Tg$^+$ TCR α$^{-/-}$ mice. Presumably, lymphoid precursors bearing invariant Vα19 TCR α chains paired with appropriate TCR β chains are capable of being positively selected by the MR1-α-mannosyl glycolipid complex and differentiated into NKT-lineage.

Presentation of α-mannosyl Glycolipids by MR1

Activation of Vα19 Tg$^+$ cells was induced by co-culture with cells of human B lymphoma line (Raji) transfected with the cDNA of MR1 [27], while their immune responses were not induced by the stimulation with the transfectants of any other MHC genes such as CD1, MR1, Qa2 and TL. Thus it is likely that invariant Vα19 TCR recognizes MR1 that is presenting certain endogenous antigens or chaperons. This result is in accord with the recent reports that invariant Vα19 TCR$^+$ cells are positively selected by MR1 [16, 19].

The immune responses of Vα19 Tg$^+$ cells toward MR1-transfectants were enhanced when the transfectants were previously loaded with the α-mannosyl glycolipids [27]. Presumably, putative intracellular ligands were replaced by these glycolipids at the antigen-presenting groove in MR1 molecules. Since the truncation of the N-acyl group length in α-glycosyl ceramides drastically reduced the activity toward Vα19 NKT cells [18], the lipid portion of antigenic glycolipids possibly binds to the antigen-presenting groove of MR1 leaving the sugar moiety available for the interaction with the invariant TCR. The immune responses were reduced in the presence of anti-MR1 antiserum but not pre-immune serum, thus also supporting the participation of MR1 in the antigen-presentation..

Collectively, it is strongly suggested that invariant Vα19 NKT cells recognize α-mannosyl glycolipids that are presented by MR1.

Induction of either Th1 or Th2 -dominant Immune Responses of Vα19 NKT Cells with α-ManCer Derivatives

It is possible that modification of the lipid moiety in α-ManCer will alter the interaction with the antigen-presenting molecule and improve the recognition of the α-mannose residue by the invariant TCR. An immunosuppressant FTY720, obtained by the structural modification of ISP-I, a product of *Isaria sincliairii* [31]. As expected by the structural homology between FTY720 and sphingosine, this drug targets sphingosine-1-phosphate receptors and acts as an agonist [32]. We synthesized a series of α-ManCer derivatives in which the sphingosine moiety was replaced with FTY720 or related aminoalcohols [33] (Figure 2). The modified α-ManCer induced promotive rather than suppressive immune responses from Vα19 NKT cells. They induced either Th1- or Th2-dominant immune responses [24]. The relative intensity of IL-4 to IFN-γ secretion by Vα19 NKT cells was dependent on the chemical structure of the stimulator. For instance, Man2HM4PhC16 (in which sphingosine moiety is replaced with FTY720) induced Th1-biased cytokine production by Vα19 Tg^+ cells. Presumably, manipulation of the sphingosine portion of α-ManCer alters the interaction between invariant Vα19 TCR and the α-mannosyl residue in the glycolipids, resulting in the modulation of the immune responses of Vα19 NKT cells. One of the modified α-ManCers, Man4PhC16, that has a phenyl group in the sphingosine hydrocarbon chain, induced the production of both Th1 (IFN-γ, IL-12, IL-17) and Th2 (IL-4, IL-5, IL-10) -promoting cytokines more intensively than the intact α-ManCer or any other derivatives [24].

CONCLUSION

The comparison between invariant Vα19 and Vα14 TCR-bearing cells is summarized in Figure 3. The structural requirements for natural ligands for invariant Vα19 TCR^+ cells were suggested by the comprehensive examination. Certain glycolipids, each possessing α-mannosyl residue(s) at the non-reducing end, have been shown to be stimulus for Vα19 NKT cells when they are presented by MR1. Recently, the structure of one of the MHC class Ib molecules Zn-α2-glycoprotein (ZAG) was determined by crystallography [34]. It is suggested that ZAG has a deep pocket in contrast to the two A' and F' pockets present in CD1d, and presents lipid-like antigens [35]. The T cell subsets recognizing ZAG have not been identified. Although the three-dimensional structure of MR1 has not been determined, it is possible that MR1 serve as the lipid-presenting MHC as well as CD1 and ZAG (Figure 3). The immune responses of Vα19 Tg^+ cells induced by the stimulation with α-mannosyl glycolipids were apparently less significant than those of Vα14 NKT cells stimulated with α-GalCer. Thus, it remains possible that glycolipids with non-reducing end α-mannosyl residue(s) are present that are more immunocompetent toward Vα19 NKT cells than the glycolipids so far characterized.

Specific activators or inhibitors for Vα19 NKT cells may be important for medical applications, since specific activators for Vα14 NKT cells such as α-GalCer and its homologues have been shown to be effective in a number of animal models of disease [36, 37]. The immune responses of Vα19 Tg^+ cells were induced not only in culture but also *in vivo* with the α-mannosyl glycolipids [24, 27]. In addition, structural modification of α-ManCer enhanced either Th1 or Th2 –dominant immune responses from Vα19 NKT cells [24]. Therefore, these glycolipids are prospective as lead compounds to develop new therapies for immunological disorders targeting Vα19 NKT cells.

Abbreviations	R^1	R^2	R^3	n	Hexose
Man4PhC16	H	H	4-octylphenyl	14	α-Man
GalC24	H	H	dodecanyl	22	α-Gal
ManC16	H	H	dodecanyl	14	α-Man
*Man2HM4PhC16	CH_2OH	H	4-octylphenyl	14	α-Man
*Gal2HM4PhC16	CH_2OH	H	4-octylphenyl	14	α-Gal
Man2HMC24	CH_2OH	H	dodecanyl	22	α-Man
Man2HMC16	CH_2OH	H	dodecanyl	14	α-Man
Man3OHC16	H	OH	dodecanyl	14	α-Man

*FTY720 derivatives

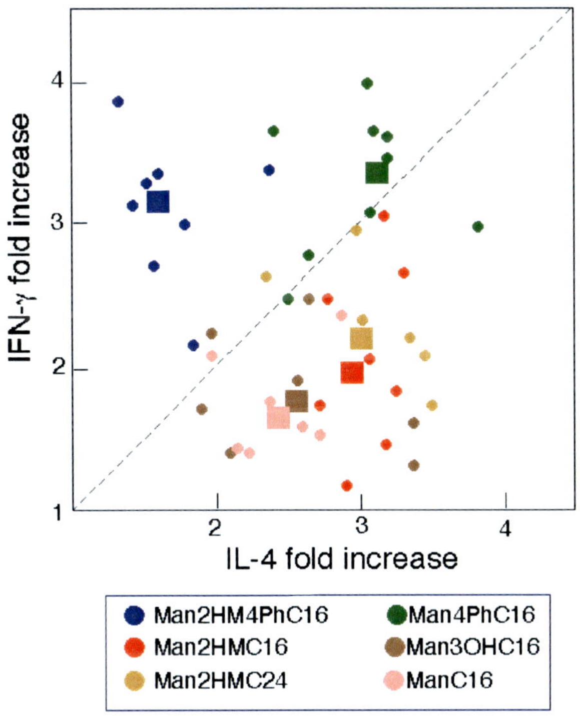

Figure 2. Immune responses of Vα19 NKT cells in culture elicited by α-ManCer and its derivatives. A list of the glycosyl ceramide derivatives with modification at the sphingosine portion is shown. Glycosphingolipids modified with a 2-hydroxymethyl, 3-hydroxyl, or 4-octylphenyl group are represented as 2HM, 3OH or 4Ph. Man3OHC16 is the original form of α-ManCer [18]. Man2HM4PhC16 and Gal2HM4PhC16 are the derivatives in which the sphingosine unit is replaced with FTY720 [31]. Liver MNCs from Vα19 Tg^+ $TCR\alpha^{-/-}$ and C57BL/6 mice were cultured with the addition of glycolipids dissolved in DMSO (final concentration, 2 μg/ml). After 1 and 2 days, the immune responses were monitored by measuring the concentrations of IL-4 and IFN-γ in the culture fluid. The concentrations of IL-4 on day 1 of culture and the IFN-γ on day 2 of culture are plotted. Results are shown as the fold-increase relative to the control cultures with the vehicle. Large squares represent the fold-increases in cytokine production on the average.

Invariant TCRα	Vα14-Jα18 (mouse) Vα24-JαQ (human)	Vα19-Jα33 (mouse) Vα7.4-Jα33 (human)
MHC restriction	CD1d	MR1
Antigen	α-Gal sphingolipids α-Gal PIs	α-Man glycolipids ?
Cytokine (upon TCR engagement)	IL-4, IFN-γ	IL-4, IFN-γ,
NK1.1 (CD161)	+	+
Distribution	lymphoid organs including liver and bone marrow	lymphoid organs including gut lamina propria and liver

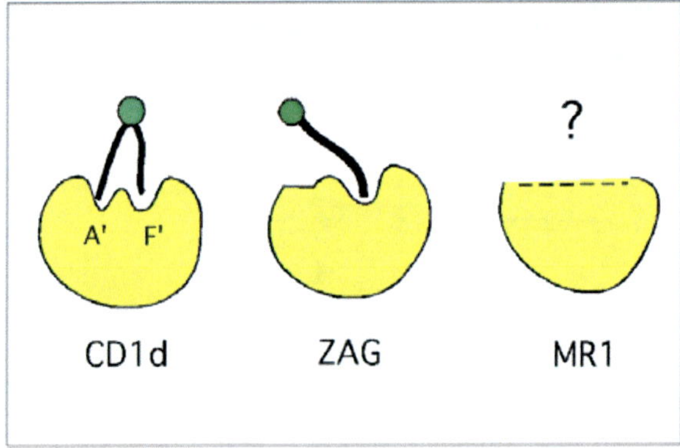

Figure 3. Summary of invariant Vα19-Jα33 and Vα14-Jα18-bearing cells.
Comparison between invariant Vα19-Jα33 and Vα14-Jα18-bearing cells in terms of MHC restriction, proposed antigens they recognize, cytokine profiles, phenotype and distribution are summarized.
In the lower panel, schematic drawings of MHC class Ib molecules CD1d, ZAG and MR1, that are supposed to present lipid antigens, are shown. The conserved amino acid residues commonly present at the antigen-presenting groove of classical MHC class I (class Ia) molecules are considerably converted in these MHC class Ib molecules. A' and F' represent antigen-presenting pockets in CD1d.

REFERENCES

[1] Bendelac, A., Rivera, M.N., Park, S.H., & Roark, J.H. (1997) Mouse CD1-specific NK1 T cells: development, specificity, and function. *Annu. Rev. Immunol.*, *15*, 535-562.

[2] Godfrey, D. I., MacDonald, H. R., Kronenberg, M., Smyth, M. I., & Van Kaer, L. (2004) NKT cells: what's in a name ? *Nat. Rev. Immunol. 4,* 231-237.

[3] Lantz, O., & Bendelac, A. (1999) An invariant T cell receptor α chain is used by a unique subset of major histocompatibility complex class I-specific $CD4^+$ and $CD4^-8^-$ T cells in mice and humans. *J. Exp. Med.*, *180*, 1097-1106.

[4] Makino, Y., Kanno, R., Ito, T., Higashino, K., & Taniguchi, M. (1995) Predominant expression of invariant $V\alpha14^+$ TCR α chain in $NK1.1^+$ T cell populations. *Int. Immunol.*, *7*, 1157-1161.

[5] Bendelac, A., Killeen, N., Littman, D.R., & Schwarz, R.H. (1994) A subset of $CD4^+$ thymocytes selected by MHC class I molecules. *Science*, *263,* 1774-1778.

[6] Bendelac, A., Lantz, O., Quimby, M.E., Yewdell, J.W.;,Bennink, J.R., & Brutkiewicz, R.R. (1995) CD1 recognition by mouse $NK1^+$ T lymphocytes. *Science*, *268*, 863-865.

[7] Kawano, T., Cui, J., Koezuka, Y., Toura, I., Kaneko, Y., Motoki, K., Ueno, H., Nakagawa, R., Sato, H., Kondo, E., & Taniguchi, M. (1997) CD1d-restricted and TCR-mediated activation of Vα14 NKT cells by glycosylceramides. *Science*, *278*, 1626-1629.

[8] Natori, T., Koezuka, Y., & Higa, T. (1993) Agelasphins, Novel α-galactosylceramides from the marine sponge *Agelas Mauritianus*. *Tetrahedron Lett.*, *34*, 5591-5592.

[9] Kinjo, Y., Wu, D., Kim, G., Xing, G.-W., Poles, M.A., Ho, D.D., Tsuji, M., Kawahara, K., Wong, C.-H., & Kronenberg, M. (2005) Recognition of bacterial glycosphingolipids by natural killer T cell. *Nature (London)*, *434*, 520-525.

[10] Mattner, J., DeBord, K.L., Ismail, N., Goff, R.D., Cantu III, C., Zhou, D., Saint-Mezard, P., Wang, V., Gao, Y., Yin, N., Hoebe, K., Schneewind, O., Walker, D., Beutner, B., Teyton, L., Savage, P.B., & Bendelac, A. (2005) Exogenous and endogenous glycolipid antigens activate NKT cells during microbial infections. *Nature (London)*, *434*, 525-529.

[11] Kinjo, Y., Tupin, E., Wu, D., Fujio, M., Garcia-Navarro, R., Benhnia, M. R.-E.-I., Zajonc, D. M., Ben-Menachem, G., Aihge, G. D., Painer, G. F., Khurana, A., Hoebe, K., Behar, S. M., Beurler, B., Wilson, I. A., Tsuji, M., Sellati, T. J., Wong, C.-H., & Kronenberg, M. (2006) Natural killer T cells recognize diacylglycerol antigens from pathogenic bacteria. *Nat. Immunol.,* *7*, 978-986.

[12] Zhou, D., Mattner, J., Cantu III, C., Schrantz, N., Yin, N., Gao, Y., Sagiv, Y., Hudspeth, K., Wu, Y.-P., Yamashita, T., Teneberg, S., Wang, D., Proia, R.L., Levery, S.T., Savage, P.B., Teyton, L., & Bendelac, A. (2004) Lysosomal glycosphingolipid recognition by NKT cells. *Science*, *306,* 1786-1789.

[13] Porcelli, S., Yockey, C. E., Brenner, M. B., and Balk, S. P. (1993) Analysis of T cell antigen receptor (TCR) expression by human peripheral blood $CD4^-8^-$ α/β T cells demonstrates preferential use of several Vb genes and an invariant a chain. *J. Exp. Med.* *178*, 1-16.

[14] Tilloy, F., Treiner, E., Park, S.-H., Garcia, G., Lemonnier, F., de la Salle, H., Bendelac, A., Bonneville, M., & Lantz, O. (1999) An invariant T cell receptor α chain defines a novel TAP-independent major hitocompatibility complex class Ib-restricted α/β T cell subpopulation in mammals. *J. Exp. Med.* *189*, 1907-1921.

[15] Shimamura, M. & Huang, Y. –Y. (2002) Presence of a novel subset of NKT cells bearing an invariant Vα19.1-Jα26 TCR α chain. *FEBS Lett.*, *516,* 97-100.

[16] Treiner, E., Duban, L., Bahram, S., Radosavijevic, M., Wanner, V., Tilloy, F., Affaticati, P., Gilfillan, S., & Lantz, O. (2003) Selection of evolutionarily conserved mucosal-associated invariant T cells by MR1. *Nature (London)*, *422*, 164-169.

[17] Yoshimoto, T., Bendelac, A., Watson, C., Hu-Li, J., & Paul, W.E. (1995) Role of $NK1.1^+$ T cells in a T_H2 response and in immunoglobulin E production. *Science*, *270,* 1845-1847.

[18] Okamoto, N., Kanie, O., Huang, Y.-Y., Fujii, R., Watanabe, H., & Shimamura, M. (2005) Synthetic α-mannosyl ceramide as a potent stimulant for a novel NKT cell repertoire bearing the invariant Vα19-Jα26 TCR α chain. *Chem. Biol., 12*, 677-683.

[19] Kawachi, I., Maldonado, J., Strader, C & Gillfillan, S. (2006) MR1-restricted Vα19*i* mucosal-associated invariant T cells are innate T cells in the gut lamina propria that provide a rapid and diverse cytokine response. *J. Immunol. 176*, 1618-1627.

[20] Shimamura, M., Huang, Y. –Y., Migishima, R., Yokoyama, M., Saitoh, T. and Yamamura, T. (2008) Localization of NK1.1+invariant Va19 TCR+ cells in the liver with potential to promptly respond to TCR stimulation. *Immunol. Lett. 121*, 38-44.

[21] Croxford, J. L.,Miyake, S., Huang, Y.-Y., Shimamura, M. & Yamamura, T. (2006) Invariant Vα19*i* T cells regulate autoimmune inflammation. *Nat. Immunol. 7*, 994.

[22] Hansen, T. H., Huang, S., Arnold, P. L., & Fremont, D. H. (2007) Patterns of nonclassical MHC antigen presentation. *Nat. Immunol. 8*, 563-568.

[23] Huang,S., Gilfillan, S., Cella, M., Miley, M. J., Lantz, O., Lybarger, L., Fremont, D. H., & Hansen, T. H. (2005) Evidence for MR1 antigen presentation to mucosal-associated invariant T cells. *J. Biol. Chem. 280*, 21183-21193.

[24] Shimamura, M., Okamoto, N., Huang, Y.-Y., Yasuoka, J., Morita, K., Nishiyama, A., Amano, Y., & Mishina, T. (2007) Modulation of Vα19 NKT cell immune responses by α-mannosyl ceramide derivatives consisting of a series of modified sphingosines. *Eur. J. Immunol. 37*, 1836-1844.

[25] Watanabe, Y., Yamamoto, T. & T. Okazaki, Synthesis of 2,6-di-O-α-D-mannopyranosylphosphatidyl-D-myo-inositol. (1997) Utilization of glycosylation and phosphorylation based on phosphite chemistry. *Tetrahedron*, *53*, 903-918.

[26] Murakata, C. & Ogawa, T. (1992) Stereoselective synthesis of glycobiosyl phosphatidylinositol, a part structure of the glycosyl-phosphatidylinositol (GPI) anchor of Trypanosoma brucei. *Carbohdr. Res. 234*, 75-91.

[27] Shimamura, M., Huang, Y.-Y., Okamoto, N., Watanabe, Y., Murakami, Y., Kinoshita, T., Hirabayashi, Y., Murakata, C., Ito, Y., & Ogawa, T. (2007) Glycolipids with non-reducing end α-mannosyl residues that have potentials to activate invariant Vα19 NKT cells. *FEBS J., 274*, 2921-2932.

[28] Dickson, R.C. & Lester, R.L. (1999) Yeast sphingolipids. *Biochim. Biophys. Acta, 14 26*, 347-357.

[29] Sumida, M., Hashimoto, M., Yasuoka, J., Okabe, E., Kusumoto, S., Takada, H., Hayashi, T. Tamura, T. & Kotani, S. (1996) Immuno-stimulating Macromolecular Glycolipids from Bacterial Cell Surfaces. *Polymer preprints 37*, 129-130.

[30] Mori, T., Ito, Y. & Ogawa, T. (1990) Total synthesis of mollu-series glycosyl ceramides. *Carbohydr. Res. 195,* 199-224.

[31] Fujita, T., Hirose, T., Hamamichi, N., Kitao, Y., Sasaki, S., Yoneta, M., & Chiba, K. (1995) 2-Substituted 2-aminoethanol: minimum essential structure for immunosuppressive activity of ISP-I (myocin). *Bioorg. Med Chem. Lett. 5*, 1857-1860.

[32] Mandala, S.; Hajdu, R., Bergstrom, J., Quackenbush, E., Xie, J., Millingan, J., Thornton, R., Shei, G.-J., Card, D., Keohane, C., Rosenbach, M., Hale, J., Lynch, C. L., Rupprecht, K., Parsons, W. & Rosen, H. (2002) Alteration of lymphocyte trafficking by sphingosine-1-phosphate receptor agonists. *Science*, *296*, 346-349.

[33] Shimamura, M., Okamoto, N., Huang, Y.-Y., Yasuoka, J., Morita, K., Nishiyama, A., Amano, Y., & Mishina, T. (2006) Induction of promotive rather than suppressive immune responses from a novel NKT cell repertoire Vα19 NKT cell with α-mannosyl ceramide analogues consisting of the immunosuppressant ISP-I as the sphingosine unit. *Eur. J. Med. Chem. 41*, 569-576.

[34] Sanchez, L. M., Chirino, A. J., & Bjorkman, P. J. (1999) Crystal structure of human ZAG, a fat-depleting factor related to MHC molecules. Science, *283*, 1914-1919.

[35] Delker, S. L., West Jr. A. P., McDermott, L., Kennedy, M. W., & Bjorkman, P. J. (2004) Crystallographic studies of ligand binding by Zn-α2-glycoprotein. *J. Struct. Biol. 148*, 205-213.

[36] Taniguchi, M., Harada, M., Kojo, S., Nakayama, T. & Wakao, H. (2003) The regulatory role of Vα14 NKT cells in innate and acquired immune response. *Annu. Rev. Immunol. 21,* 483-513.

[37] Miyamoto, K., Miyake, S. & Yamamura, T. (2001) A synthetic glycolipid prevents autoimmune encephalomyelitis by inducing TH2 bias of natural killer T cells. *Nature (London)*, *413*, 531-534.

INDEX

A

AAA, 39
ABC, vii, 1, 3, 7, 8, 21, 23, 25, 26
abiotic, 142
abnormalities, 19, 25, 52
absorption, 3, 46
academic, viii, 41
acarbose, 133
acceptor, ix, 85, 89, 90, 91, 93, 95, 96, 98, 103, 105, 111, 134, 138, 176, 180, 183, 186, 187, 190
acceptors, 96, 104, 111, 176
access, 144, 150
accessibility, 34
accounting, 46
acetate, 86
acetonitrile, 131, 148
acetylation, 62
acetylene, x, 97, 127, 130, 131, 132, 138, 139, 141, 142, 144, 145, 147, 148, 149, 150, 151, 152, 154
acid, viii, ix, 9, 12, 13, 17, 28, 32, 33, 46, 49, 51, 54, 56, 61, 62, 65, 73, 74, 75, 78, 85, 86, 87, 89, 90, 91, 93, 94, 95, 96, 97, 98, 99, 101, 102, 103, 104, 105, 109, 110, 111, 112, 113, 115, 130, 134, 136, 138, 139, 140, 142, 144, 148, 151, 154, 159, 163, 175, 176, 177, 181, 184, 185, 190, 191, 195, 197, 201, 206
acidic, 18, 95, 97, 194
acoustic, 56
activation, 6, 20, 102, 125, 130, 136, 150, 162, 163, 207
activators, 21, 164, 204
active site, 34, 92, 93, 102, 135, 136
Adams, 57, 170
adaptation, 145
adenocarcinoma, 18
adhesion, 42, 43, 48, 74, 77, 144, 166, 175, 184, 185, 186, 197
adhesion properties, 175
adhesives, 166
administration, 6, 67, 69, 72
ADP, 111
adrenal gland, 3
adriamycin, 20
adsorption, 156
adult, 79, 116, 117, 185
adulthood, 13
adverse event, 68, 72, 73
age, 13, 72
ageing, 66
agent, 9, 24, 46, 69, 135
agents, 5, 6, 55, 67, 72, 78, 79, 132, 134, 144, 145, 148, 151
aggrecan, 64, 67
aggregates, 64
aggregation, 33, 144
aging, 82
agonist, 203
agriculture, 46
aid, 124
alanine, 152
aldolase, 51
alkali, 65
alkaline, 47
alkaloids, 133
alkane, 156
alkynes, 133, 156
alpha, 23, 25, 37, 58, 91, 104, 105, 106, 184, 196, 197, 198
alternative, 9, 13, 14, 20, 51, 80, 89, 149
alters, 11, 23, 103, 123, 204

amide, 138, 140, 144, 145, 155
amine, 144, 145
amino, 4, 28, 32, 33, 42, 74, 75, 86, 92, 93, 94, 95, 96, 98, 105, 111, 112, 134, 136, 139, 140, 148, 181, 190, 191, 192, 193, 195, 197, 201, 206
amino acid, 4, 28, 32, 33, 42, 75, 92, 93, 94, 95, 96, 98, 105, 112, 136, 139, 140, 148, 181, 190, 191, 192, 193, 195, 197, 201, 206
amino acids, 4, 42, 75, 92, 96, 139, 190, 191, 192, 193
amino groups, 74
aminoglycosides, 129
amniotic, 191
amniotic fluid, 191
Amsterdam, 82
amyloid, 77
amylopectin, 147
anabolism, 9
anaerobic, 92
analgesics, 67
analog, 6, 12, 16, 69, 70, 92, 95, 105
analytical techniques, 156
analytical tools, viii, 42, 46, 52, 61, 144
angiogenesis, 62, 64, 76
anhydrase, xi, 130, 136, 137, 163, 164, 189, 191, 196
aniline, 146
animal models, 64, 204
animals, 45, 47, 80, 86, 99, 165
Anions, 162
Annexin, 76
Annexin V, 76
anode, 51
ANOVA, 70
antagonism, 17
antibacterial, 132, 141
antibiotic, 133, 141
antibiotics, 129, 133, 140, 141
antibody, 4, 28, 45, 46, 53, 55, 59, 100, 156, 184, 185
anticancer, 3, 9, 21, 26, 129, 133, 144, 147
anti-cancer, 44
anticancer drug, 3, 21, 26, 133
anticoagulant, vii, ix, 61, 75, 78, 79, 80, 84, 111, 112, 113, 121, 122, 129, 144
anticoagulants, 80
anticoagulation, 62, 75
anticonvulsant, 130
antidiabetic, 133
antidiuretic, 164
antigen, 3, 28, 43, 46, 47, 49, 50, 55, 68, 99, 100, 106, 134, 138, 169, 180, 184, 200, 201, 203, 206, 207, 208
antigen-presenting cell, 201
Antigens, 56
anti-HIV, 163
anti-inflammatory drugs, 67
antisense, 6, 14, 20
Antithrombin, 78, 113, 123, 130
antithrombin iii, 84
antiviral, 134, 162
antiviral agents, 134
AP, 56, 77
apoptosis, vii, 1, 5, 19, 20, 21, 23, 26, 152
apoptotic, 7, 20
apoptotic effect, 7
appendix, 78
application, viii, 37, 41, 46, 48, 51, 103, 157, 160, 170
aqueous humor, 81
aqueous solution, 138, 144, 150, 168
aqueous solutions, 168
arginine, 75, 96, 164
arrest, 5
arteriosclerosis, 136
arthritis, 66, 67, 69, 82, 125
articular cartilage, 66, 67, 82
aspartate, 31
assessment, 66, 69, 72, 73
asthma, 125
astrocyte, 185
astrocytoma, 26
atherosclerosis, 64, 66
atomic force, 16, 49
atomic force microscopy, 16, 49
atoms, 138
ATP, 3, 4, 7, 22, 23
ATPase, 2, 10, 11, 16, 22, 23, 39
attachment, 42, 77, 114, 157
attention, 129, 148, 154
Australia, 127
autoantibodies, 87
autocrine, 39
autoimmune, 43, 82, 86, 100, 200, 208, 209
autoimmune disease, 44, 86, 100
autoimmune diseases, 100
automation, 52
autoradiography, 8
autosomal recessive, 13, 194
availability, 48, 51

avian flu, 134
avian influenza, 87, 101, 134
axon, 116, 125
axon terminals, 116
axonal, 62, 100, 118, 120, 126
axons, 116, 118, 120, 126
azacrown ethers, 148

B

B lymphocytes, 3
bacteria, ix, 43, 44, 47, 50, 57, 59, 85, 86, 89, 91, 92, 94, 106, 153, 207
bacterial, ix, 15, 43, 45, 50, 56, 59, 85, 86, 87, 88, 91, 92, 93, 94, 95, 96, 97, 98, 99, 100, 104, 105, 106, 129, 144, 159, 166, 207
bacterial infection, 129
bacterium, 86, 92, 97, 104, 106
barrier, 7, 44, 130
barriers, 3
basement membrane, 115, 124, 176, 194
battery, 73
behavior, 59, 83, 120
bell, 32
bell-shaped, 32
benchmarks, 50, 59
benzene, 136, 137, 164
bias, 209
bicarbonate, 136
binding, ix, 2, 3, 4, 7, 10, 12, 16, 17, 18, 21, 22, 23, 24, 29, 31, 33, 34, 36, 37, 38, 39, 40, 43, 44, 45, 47, 48, 49, 50, 51, 53, 57, 58, 63, 64, 75, 76, 78, 79, 83, 84, 91, 92, 93, 94, 95, 96, 101, 103, 105, 109, 111, 112, 113, 114, 115, 116, 118, 120, 121, 122, 123, 124, 125, 128, 129, 135, 142, 143, 144, 147, 148, 151, 152, 153, 155, 156, 157, 166, 168, 179, 180, 181, 184, 191, 192, 194, 197, 198, 209
binding energy, 78, 135
bioavailability, 14, 78, 80, 129, 148
biochemistry, vii, 63
biocompatibility, x, 127, 131, 142, 146, 148
biocompatible, 144, 150, 157
biodegradable, 146
biological activity, 45, 80, 129, 132, 138, 140, 144, 160
biological interactions, 52, 123
biological processes, ix, 47, 85, 86, 99, 110
biological systems, 57
biomacromolecules, 146
biomarker, 57
biomarkers, 44, 54, 73, 142
biomaterial, 146
biomaterials, 80, 144, 147
biomedical applications, 54, 144, 146, 150, 168
biomimetic, 46, 166
biomolecular, 43, 154
biomolecule, 46, 51
biomolecules, viii, x, 37, 41, 44, 127, 131
bioreactor, 51
bioreactors, 45
Biosensor, 46, 56, 58
biosensors, viii, 41, 46, 47, 52, 54, 55, 56, 57, 59
biosynthesis, vii, ix, x, 1, 3, 5, 9, 12, 14, 15, 18, 22, 74, 85, 88, 89, 90, 91, 104, 109, 111, 113, 114, 116, 120, 122, 128, 140, 173, 175, 177, 178, 179, 180, 181, 182, 183, 184, 185, 186, 187, 195
biosynthetic pathways, 136
biotechnology, 45, 46, 55, 56, 157
biotin, 142, 143
bivalve, 79, 201
bleeding, 78
blocks, 129, 133, 138, 139, 140, 147, 148, 152
blood, xi, 3, 13, 17, 19, 23, 31, 56, 75, 79, 110, 113, 123, 153, 199, 200, 207
blood glucose, 56
blood group, 13
blot, 57, 116, 142
bonding, 75, 96, 144
bonds, 47, 58, 65, 138, 139, 146, 155, 158
bone marrow, 13, 19
bone remodeling, 64
Bose, 20
Boston, 82
bovine, 47, 50, 65, 78, 79, 115, 133, 186, 195, 196, 197, 200
brain, 115, 116, 117, 119, 120, 125, 130, 146, 175, 176, 177, 185, 186, 192, 194, 196, 197
brain development, 194
branching, 120, 129, 150, 154
breakdown, 14, 25, 66
breast cancer, 6, 20, 21, 26, 44, 52, 53, 54, 59, 155, 169
building blocks, 129, 133, 138, 139, 140, 147, 148, 152

C

C. thermocellum, 92
Ca^{2+}, 31, 92
Caenorhabditis elegans, 121

calcium, 32, 37, 38, 181
calixarenes, 148
calnexin, 28, 29, 37, 40
calreticulin, 28, 29, 37, 40
CAM, 175, 185
Campylobacter jejuni, 86, 94, 100, 104, 105, 106
Campylobacter jejuni infection, 100
Canada, 1
cancer, 2, 5, 6, 14, 15, 16, 20, 21, 23, 42, 43, 44, 45, 47, 52, 57, 58, 59, 62, 64, 66, 74, 80, 81, 128, 137, 142, 150, 152, 155, 159, 164, 165, 166, 169
cancer cells, 2, 5, 6, 15, 16, 19, 20, 21, 23, 26, 43, 44, 52, 58, 59
cancer progression, 52, 62, 64
cancer treatment, 14
cancerous cells, 43
Candida, 168
candidates, 44, 160
capacity, ix, 7, 11, 52, 64, 71, 85, 99, 125, 193, 196
capillary, 82, 152
caprolactone, 149, 168
capsule, 104
carbohydrate, vii, viii, ix, x, 12, 27, 28, 30, 33, 34, 37, 39, 40, 42, 45, 46, 47, 48, 49, 50, 51, 52, 53, 54, 56, 57, 58, 59, 62, 85, 86, 87, 93, 94, 96, 99, 100, 101, 105, 106, 127, 128, 129, 130, 132, 133, 135, 136, 138, 140, 141, 143, 144, 148, 150, 153, 154, 155, 156, 157, 159, 166, 168, 170, 173, 175, 176, 177, 178, 179, 180, 183, 184, 185, 186, 187
carbohydrates, 42, 44, 47, 48, 57, 129, 133, 144, 153, 157, 158, 159, 162, 166, 192
carbon, 8, 50, 91, 136, 138, 167
Carbon, 161, 164
carbon dioxide, 136
carbon nanotubes, 167
carbonic anhydrase inhibitors, 137
carboxyl, 74
carboxyl groups, 74
carcinoma, 5, 10, 12, 16, 48
cargo, 37, 38, 39
carrier, viii, 27, 29, 147, 175, 176, 178, 182, 186
cartilage, 64, 65, 66, 67, 73, 81, 82, 146, 166
caspase, 19
caspase-dependent, 19
catabolic, 39, 67
catabolism, 66, 129
catalysis, 94, 150
catalyst, x, 127, 130, 131, 148, 149, 151
catalysts, 136, 150
catalytic activity, 78, 191
catalytic system, 148
cathepsin G, 76
cation, 91
cDNA, 9, 15, 111, 112, 184, 185, 203
cell, vii, viii, ix, x, xi, 1, 2, 3, 5, 6, 7, 9, 10, 11, 12, 13, 14, 15, 16, 17, 18, 19, 20, 21, 22, 25, 28, 30, 42, 43, 44, 45, 46, 47, 48, 50, 52, 53, 57, 58, 59, 60, 61, 62, 63, 64, 74, 76, 79, 80, 81, 86, 87, 88, 91, 100, 109, 110, 111, 112, 114, 116, 117, 118, 120, 121, 123, 124, 125, 128, 129, 132, 141, 142, 143, 148, 152, 153, 157, 165, 166, 175, 176, 178, 184, 185, 186, 190, 191, 192, 197, 199, 200, 201, 202, 203, 207, 208, 209
cell adhesion, 48, 175, 184, 185, 186
cell culture, 45, 142
cell death, 5, 6, 20
cell differentiation, 5, 47
cell fusion, 112, 114, 124
cell growth, 80
cell line, 3, 5, 9, 10, 11, 12, 14, 18, 19, 20, 21, 44, 45, 57, 59, 80, 112, 142, 143, 176, 178
cell lines, 3, 5, 9, 10, 11, 14, 18, 19, 20, 21, 44, 45, 57, 59, 80, 142
cell membranes, viii, 61
cell signaling, 125
cell surface, vii, ix, x, 1, 11, 12, 14, 21, 25, 44, 48, 58, 64, 86, 87, 109, 110, 114, 116, 120, 123, 125, 128, 132, 141, 142, 143, 157, 165, 173, 175, 176
cellulose, 146, 166
Cellulose, 146, 166
central nervous system, 13
cerebellum, 116, 118, 119, 125
cerebral cortex, 116, 117
cerebrum, 119
channels, 19
chaos, 121
chaperones, 28, 29
charge density, 62, 74, 75
chemical approach, 43
chemical properties, 62
chemical reactions, 88
chemical structures, 100
chemosensitization, 16
chemotaxis, 67
chemotherapeutic drugs, 6
chemotherapy, 3, 20
Chemotherapy, 160
chiral, 129
chitin, 133
chloride, 5, 50, 97

CHO cells, 99, 178, 179
cholera, 19
cholesterol, vii, 1, 2, 5, 10, 11, 14, 16, 19, 22, 23
chondrocyte, 125
chondrocytes, 66, 67
chromatography, 29, 84, 111, 123
chromosome, 122
circadian, 115, 123
circadian rhythm, 115, 123
cis, vii, xi, 9, 189, 190, 191, 193, 195
cis-regulating, xi, 189
clams, 84
classes, 44, 62, 115, 123, 129
classical, 10, 23, 71, 130, 200, 206
classification, 95, 105
cleavage, 19, 30, 34, 78, 155, 174, 184
click chemistry, x, 127, 130, 131, 132, 133, 134, 135, 136, 139, 140, 141, 142, 143, 144, 145, 146, 147, 148, 149, 150, 151, 152, 153, 154, 155, 156, 157, 160, 161, 163, 165, 169, 170
clinical symptoms, 13, 64, 70
clinical syndrome, 66
clinical trial, 14, 58, 69, 72, 73, 137, 159
clinical trials, 14, 73, 137, 159
clone, 111, 112, 200
cloning, 56, 91, 102, 103, 104, 121, 184, 185, 186
clusters, 67, 113, 156, 193
CMV, 76
cnidaria, 79
CO2, 136
coagulation, 31, 38, 75, 79, 80, 110, 123
coagulation factor, 31, 38
coagulation factors, 31
coatings, 146
codes, 32, 35
coding, 102, 169
coenzyme, 112
co-existence, x, 173, 180
coil, 31
collaboration, 118
collagen, 65, 67
colon, 10, 18, 53, 165
colonization, 44, 59
colorectal cancer, 57
commercialization, 46
commissure, 118
communication, 28, 43, 48, 128, 138
compatibility, 75, 140, 146, 150
complement, 76, 79
complex carbohydrates, 158
complex interactions, 10
complexity, x, 51, 57, 173
compliance, 69
complications, 13
components, vii, 1, 34, 46, 62, 65, 67, 78, 110, 111, 134
composition, 3, 5, 10, 19, 44
compounds, 3, 4, 84, 104, 133, 136, 141, 154, 155, 161, 204
concentration, 3, 66, 67, 78, 79, 97, 134, 154, 155, 205
conception, 157
confinement, 59
conflict, 193
conjugation, x, 90, 127, 132, 138, 146, 154
connective tissue, 64, 73
consensus, 28, 75, 137, 190, 192, 193, 195
conservation, 75, 136, 195
constraints, 51
construction, 51, 55, 147, 150, 154
consumption, 70, 72
control, vii, x, 6, 7, 33, 49, 50, 69, 73, 80, 87, 117, 120, 128, 150, 154, 173, 174, 181, 182, 189, 190, 192, 194, 202, 205
control group, 69, 73
conversion, 38, 89, 90, 111, 148, 154
copper, 131, 150
corpus callosum, 118
correlation, 9, 79, 80
correlations, 44
cortex, 194
cortical neurons, 119, 120
cost-effective, 52
cotton, 67
coupling, 106, 138, 147, 165
covalent, 42, 106, 155
CPT-11, 20
CRC, 159, 163
CRD, 29, 30, 31, 32, 33, 34, 35
crosslinking, 146
cross-linking, 12
CRT, 28, 29, 30, 33, 36
crystal structure, 31, 32, 34, 37, 39, 92, 93, 95, 96, 102, 112
crystal structures, 31, 32, 34, 93, 95
crystallization, 104, 183
C-terminal, 4, 34, 39, 111, 176, 180, 183, 191
C-terminus, 38
culture, 12, 51, 57, 93, 111, 142, 201, 202, 203, 204, 205

culture media, 12
cycles, 31, 147
cyclodextrin, 11, 149, 168, 169
cyclodextrins, 168
cyclopentadiene, 157
Cyclosporin, 15, 26
cysteine, 103, 140, 196
cysteine residues, 103
cytokine, 76, 200, 204, 205, 206, 208
cytokine response, 208
cytokines, xi, 3, 5, 19, 66, 199, 200, 204
cytometry, 10, 142, 145
cytoplasm, 80
cytoplasmic membrane, 3, 97
cytoplasmic tail, 176, 181, 187
cytosine, 161
cytoskeleton, 10, 23
cytosol, 4, 28, 33, 34, 39, 181
cytosolic, vii, 1, 9, 10, 11, 14, 31, 39, 136, 152, 181
cytotoxic, 6, 7, 142
cytotoxicity, 19

D

database, 53, 94, 95, 96, 112
de novo, 6, 14, 20, 128, 138
death, 20, 175
deconvolution, 46
defects, 3, 65, 115, 118, 183, 184, 194
defense, 28, 54, 59, 80
defense mechanisms, 80
deficiency, 13, 31, 38, 74, 194
definition, 131
deformities, 72
degenerate, 111
degenerative joint disease, 67
degradation, viii, 13, 27, 28, 29, 30, 33, 34, 35, 36, 40, 129, 144, 181, 194
degradation pathway, viii, 27, 30, 33, 34, 35
dehydrogenase, 119
delivery, 136, 144, 145, 148, 150, 151, 156, 166, 168
denaturation, 92
dendrimers, 151, 168, 169
dendrites, 116, 117, 120
dendritic cell, 3, 17
density, 10, 16, 23, 32, 129, 142, 147, 154, 196
depolymerization, 78
depression, 136
derivatives, xi, 13, 47, 80, 86, 93, 97, 103, 130, 134, 143, 160, 161, 170, 199, 201, 203, 205, 208
destruction, 67
detection, viii, 41, 44, 46, 47, 48, 49, 50, 51, 55, 57, 59, 151, 155, 170
detergents, 10
developed nations, 66
developing brain, 116, 120
developmental process, 110, 115
dexamethasone, 19
DFT, 160
diabetes, viii, 41
diacylglycerol, 200, 207
dietary, 44, 59
diets, 58
differentiation, 5, 22, 42, 47, 62, 76, 124, 128, 174, 184, 195
dimer, 93, 125, 148, 184
dimeric, 131
dimerization, 93, 125, 165
dipole, x, 127, 130, 132, 133, 138
dipole moment, 138
direct action, 7
disability, 66
Discovery, 158, 159, 160, 161, 163, 165, 166
discrimination, 112
disease activity, 72
disease model, 14
disease progression, 174
diseases, 11, 13, 15, 25, 43, 47, 64, 65, 67
dispersion, 46
dispersity, 144
displacement, 50, 131, 140, 152
disposition, 3
dissociation, 49, 131
distribution, 3, 10, 11, 18, 23, 58, 78, 81, 87, 112, 158, 163, 181, 206
disulfide, 155, 184
diversity, 42, 51, 52, 99, 100, 110, 113, 129, 158, 183, 186
DMF, 146
DNA, 30, 43, 133, 144
donor, ix, 85, 88, 89, 91, 93, 94, 95, 96, 97, 98, 103, 105, 106, 134, 135, 138, 176, 181, 192
double-blind trial, 69
Drosophila, 115, 121, 175, 182, 185, 187
drug delivery, 144, 151, 166
drug design, 55, 57, 128, 129, 157
drug discovery, x, 53, 127, 131, 132, 153, 157, 159, 160, 163, 164, 168
drug efflux, vii, 1, 2, 7, 9, 11, 12, 19
drug interaction, 16

drug release, 144, 146
drug resistance, vii, 1, 2, 3, 12, 17, 21
drug targets, 129, 203
drug toxicity, 14
drug-induced, 19
drug-resistant, 18, 19, 20
drugs, vii, ix, 2, 3, 6, 12, 15, 17, 44, 45, 50, 61, 67, 69, 80, 129, 130, 135, 140, 145, 148, 159, 164
dual task, 39
duodenal ulcer, 44
duration, 72
dyes, 79
dysregulation, 136
dystrophin, 192

E

E. coli, 50, 92, 93, 94, 99, 106, 180
ecosystem, 53
edema, 67
effusion, 68, 70
egg, 20, 191, 192
elderly, 82
electrodes, 50
electrolyte, 136
electron, 32, 50, 130
electron density, 32
electrons, 50
electrophoresis, 82, 145, 152
ELISA, 47, 156
elongation, ix, 34, 109, 177
embryo, 116, 185
emergence, 134, 135, 140
emission, 148
employment, 150
encephalomyelitis, 200, 209
encoding, 9, 30, 92, 94, 111, 112, 121, 184, 185
endocytosis, 11, 12
endoplasmic reticulum, 10, 11, 17, 22, 28, 36, 37, 38, 39, 40, 43, 187, 192
endothelium, 13, 75, 84
endothermic, 131
endotoxins, 102
energy, 2, 7, 78, 128, 131, 135
cngagcmcnt, xi, 199, 200
enteritis, 87
environment, x, 2, 3, 10, 16, 22, 43, 51, 97, 174, 190, 192, 193
environmental factors, 45
enzymatic, ix, 6, 58, 65, 78, 85, 88, 89, 90, 91, 92, 93, 96, 97, 103, 104, 106, 110, 115, 129, 135, 158, 176, 180, 181, 197
enzymatic activity, 91, 181, 197
enzyme inhibitors, 128
enzyme interaction, 180, 181
enzymes, vii, viii, ix, x, 36, 41, 42, 47, 51, 56, 74, 79, 85, 92, 94, 95, 96, 97, 99, 105, 110, 111, 112, 114, 115, 116, 120, 133, 136, 173, 175, 176, 178, 180, 182, 186
epidemiology, 82
epidermal growth factor, 195
epithelia, 3
epithelial cell, 12, 19, 59
epithelial cells, 12, 59
epithelium, 15, 44, 60
epitope, x, 49, 143, 173, 175, 178, 184, 185, 186, 187
epitopes, xi, 47, 67, 82, 141, 156, 175, 185, 189, 192
equilibrium, 67
ERAD, 29, 33, 34
ER-associated degradation, 29, 33
erosion, 72
erythropoietin, 45
Escherichia coli, ix, 3, 9, 30, 85, 92, 94, 101, 102, 103, 104, 106, 133, 186
ESI, 154, 155
ester, 115, 123, 145, 146
esterase, 76
esterification, vii, 1, 2, 11, 14, 16, 146
esters, 65, 86
ethers, 148
eukaryotes, 99
evolution, 42, 52, 71
evolutional processes, 124
examinations, xi, 72, 199
excretion, 7
exfoliation, 64, 81
experimental autoimmune encephalomyelitis, 200
experimental condition, 11
expertise, vii, 43
exposure, 67
expressed sequence tag, 111, 112
externalization, 6
extracellular matrix, viii, ix, 61, 65, 67, 109, 110, 193
extrusion, 3

F

fabrication, 46, 156, 157
Fabry disease, 13, 14, 24, 25
factor VII, 38
factor VIII, 38
factor Xa, 75, 78, 123, 130
failure, 129, 194
family, ix, 7, 8, 9, 23, 64, 86, 94, 95, 96, 97, 105, 109, 112, 114, 115, 118, 121, 136, 152, 184, 187, 192, 195, 196, 200
family members, 187
fat, 209
F-box, 33, 39
FDA, 45
feedback, vii, 1
feedback inhibition, vii, 1
fertility, 73
fertilization, 42, 192, 196, 197
fetal, 79
FGF-2, 76, 115, 123, 125
FGFs, 115
fiber, 56
fibrillar, 67
fibrillation, 66
fibrils, 65
fibroblasts, 9, 23, 65
fidelity, 145, 152, 154, 155, 157
film, 49
filtration, 84
fission, 197
flexibility, 66, 74, 88, 103
floating, 23
flora, 58
flow, 10, 142, 145
flow cytometry analysis, 145
fluid, 62, 64, 65, 66, 136, 191, 205
fluorescence, 7, 46, 49, 57, 142, 148
fluorescent microscopy, 142
fluorogenic, 142, 143, 148
focusing, vii, ix, 86
folding, 28, 29, 35, 36, 37, 43, 128, 137
food, 46, 50
Ford, 24, 121, 161
forebrain, 118, 126
Fox, 19
fragmentation, 115
France, 73
free radical, 67
free radicals, 67
Friedmann, 185
functional analysis, 102
functional aspects, 187
fungi, 47, 192
fusion, 30, 53, 77, 103, 112, 114, 124, 193
fusion proteins, 30, 193

G

galectin-3, 35, 39
Galectins, 54, 152
gamete, 191, 192, 197
gangliosides, 2, 5, 8, 12, 14, 86, 99, 106
gastric, 44
gastritis, 44
gastrointestinal, 12, 44
gastrointestinal epithelial cells, 12
Gaucher disease, 13, 14, 25
GCS, 6, 9, 14
GDP, 106, 134, 135, 155, 163
gel, 44, 64, 70, 84, 123, 145
gender, 192
gene, vii, 2, 3, 7, 9, 15, 16, 17, 21, 22, 25, 39, 56, 92, 98, 99, 101, 102, 104, 116, 119, 120, 136, 144, 150, 166, 175, 176, 177, 184, 185, 200
gene targeting, vii, 56, 175
gene therapy, 25
gene transfer, 184
generation, 20, 23, 45, 66, 99, 111, 139, 152, 155
genes, 42, 60, 92, 103, 104, 112, 116, 117, 120, 125, 174, 194, 203, 207
genetic factors, 66
genetics, 198
genome, 43
genomic, 98, 122
genomics, 43, 46
genotoxic, 73
genotype, 15, 80
genotypes, 79
Germany, 158, 159, 160, 163, 164, 166, 189
gift, 15
glass, 157
glia, 194
glial, 175
Glucan, 57
glucoamylase, 130, 133
gluconeogenesis, 136
glucose, 13, 28, 29, 31, 36, 37, 42, 46, 56, 147, 190
glucosidases, 25, 36
glucoside, 152

Glucosylceramide, 10, 24, 26, 130
glutathione, 30
glycans, viii, x, xi, 28, 29, 31, 34, 35, 38, 39, 41, 42, 43, 44, 45, 47, 48, 49, 50, 53, 55, 57, 99, 128, 129, 132, 137, 142, 157, 165, 174, 175, 178, 182, 186, 189, 191, 192, 193, 194, 196, 197
glycerol, 163
glycine, 18
glycoconjugates, ix, x, 28, 42, 44, 47, 48, 49, 50, 53, 54, 55, 85, 86, 88, 89, 99, 128, 129, 133, 136, 158, 163, 164, 169, 174, 185
glycogen, 14, 25
glycol, 50, 59, 151, 156, 157, 164
glycolipid antigens, 207
glycolipids, ix, xi, 9, 44, 47, 59, 85, 86, 100, 101, 175, 176, 185, 186, 199, 200, 201, 202, 203, 204, 205
glycopeptides, 58, 99, 101, 138, 139, 141
glycopolymers, 143, 144, 145, 146, 150, 166
glycoprotein, vii, 1, 16, 17, 28, 29, 34, 35, 36, 37, 38, 39, 42, 45, 49, 54, 56, 57, 98, 99, 106, 107, 112, 113, 124, 128, 138, 141, 142, 160, 169, 176, 180, 186, 191, 192, 195, 196, 197, 204, 209
Glycoprotein, 17, 18, 23, 39, 45, 54, 99, 114, 137, 161, 165
glycoproteins, ix, 7, 16, 17, 18, 28, 29, 31, 32, 33, 34, 35, 36, 37, 38, 39, 40, 44, 45, 47, 48, 49, 50, 52, 58, 85, 86, 87, 96, 99, 100, 101, 106, 128, 138, 140, 141, 142, 175, 176, 184, 191, 194
glycosaminoglycans, ix, 79, 81, 84, 109
glycoside, 13, 143, 165
glycosides, 139, 144, 151, 152, 155, 156, 164
glycosyl, xi, 139, 140, 142, 144, 147, 150, 162, 199, 200, 201, 202, 203, 205, 208
glycosylated, 5, 33, 39, 42, 44, 140, 194
glycosylation, vii, x, 6, 20, 21, 40, 42, 43, 44, 45, 47, 48, 49, 53, 54, 55, 57, 58, 59, 60, 88, 99, 103, 106, 128, 138, 140, 141, 157, 158, 173, 174, 178, 183, 186, 189, 190, 191, 192, 193, 194, 195, 196, 197, 208
gold, 4, 49, 50, 156
gold nanoparticles, 50
Golgi complex, 29, 31, 43, 182, 187
GPD, 135
GPI, xi, 199, 201, 208
gram negative, 44
Gram-negative, 86, 91, 97
Gram-positive, 92
granule cells, 117, 125
granules, 80, 111
graph, 179
groups, ix, x, 62, 65, 69, 70, 72, 73, 74, 75, 78, 80, 86, 96, 109, 113, 123, 127, 129, 130, 136, 138, 140, 144, 145, 146, 149
growth, ix, 5, 22, 26, 43, 44, 45, 62, 64, 74, 80, 109, 110, 113, 115, 117, 120, 122, 124, 125, 153, 161, 166, 195
growth factor, ix, 64, 74, 109, 110, 113, 115, 122, 124, 125, 161, 166, 195
growth factors, ix, 64, 74, 109, 110, 113
growth inhibition, 153
GST, 30
guidance, 118, 120, 125, 126
Guillain-Barre syndrome, 44
Guillain-Barré syndrome, 86, 100
Guinea, 77
gut, 3, 57, 58, 200, 208

H

H. pylori, 44
H1, 101
H_2, 101
H3N2, 59
H5N1, 134
half-life, 25, 45, 191
hands, 68, 72, 83
healing, 80
health, 44, 46, 52, 128, 157, 183
heart, 13
heat, 92, 102
Helicobacter pylori, 44, 60, 106
helix, 75
hemagglutinin, 87
hemagglutinins, 58, 101
hematological, 13
hematopoietic, 3, 79
hematopoietic stem cell, 3, 79
hematopoietic stem cells, 3, 79
hemostasis, 113
Hendra, 44, 53
Heparin, 61, 74, 76, 81, 83, 111, 113, 115, 125, 129, 130
hepatitis, 44, 76
hepatitis B, 76
hepatocytes, 77
hepatoma, 21
hepatomegaly, 13
herpes, 112, 113, 122, 124
herpes simplex, 112, 113, 122, 124

herpes simplex virus type 1, 112, 113, 122, 124
heterocycles, x, 127, 130, 133
heterogeneity, 51
heterogeneous, 62, 66, 113
Higgs, 122
high resolution, 55
high-performance liquid chromatography, 82
hippocampal, 118
hippocampus, 176
hips, 69
histamine, 80
histidine, 32
histochemical, 47
HIV, 44, 49, 56, 58, 76, 77, 152, 156, 161, 163, 168
HIV infection, 77
HIV-1, 49, 56, 58, 76, 152, 156, 161, 168
homeostasis, vii, 1, 3
homogeneity, 92, 157
homolog, 22, 196
homology, 15, 28, 33, 112, 191, 203
homopolymers, 146
hormone, 18, 128, 191
hormones, 44, 191
horse, 142
host, 28, 42, 43, 44, 53, 54, 55, 56, 59, 76, 87, 101, 135, 136, 141, 148
HPLC, 111
HRP, 142
HSV-1, 112, 113, 114, 115, 124
human, x, xi, 5, 6, 9, 10, 16, 17, 18, 19, 20, 21, 23, 26, 30, 31, 37, 38, 39, 44, 45, 53, 54, 57, 59, 87, 93, 99, 100, 101, 103, 107, 111, 112, 115, 121, 122, 124, 134, 137, 142, 143, 165, 173, 175, 183, 184, 186, 187, 189, 191, 192, 193, 194, 195, 196, 197, 200, 203, 207, 209
humans, 7, 44, 47, 114, 124, 136, 206
Hungary, 53
hybrid, 49, 66, 74, 148, 162, 168, 181, 193
hybridization, 51, 116, 117, 118
hydration, 65, 136
hydrazine, 115
hydro, 148
hydrocarbon, 12, 204
hydrodynamic, 149
hydrogen, 51, 75, 78, 96, 138, 144
hydrogen peroxide, 51, 78
hydrolases, xi, 189
hydrolysates, 142
hydrolysis, 4, 7, 20, 133, 138
hydrolytic stability, 133
hydrophilic, 148
hydrophobic, 3, 7, 17, 135, 148, 154
Hydrophobic, 135
hydrophobicity, 48
hydroxyl, 65, 86, 95, 98, 129, 145, 149, 205
hydroxyl groups, 86, 129, 149
hypersensitive, 14
hypersensitivity, 17
hypothesis, 80, 115, 180

I

ibuprofen, 67
ice, 176, 177
ICE, 20
identification, viii, ix, 9, 41, 43, 47, 49, 50, 55, 57, 86, 104, 153, 154, 157, 170
identity, 34, 112
idiopathic, 70
IFN, 5, 201, 202, 203, 205
IgE, 44, 68, 79
IgG, 43, 54
IL-1, 5, 204
IL-10, 204
IL-17, 204
IL-2, 5, 17
IL-4, 5, 17, 201, 202, 203, 205
IL-8, 76
ileum, 78
imagination, 145
imaging, 25, 55, 165
immobilization, 51
immune cells, 64
immune function, 152
immune response, 43, 47, 56, 68, 138, 146, 201, 203, 204, 205, 208, 209
immune system, 43, 44, 53, 57, 80, 146, 200
immunity, 59
immunoglobulin, 207
immunohistochemical, 120
immunological, 82, 86, 166, 167, 204
immunology, 175
immunoprecipitation, 178, 179
immunosuppressive, 191, 196, 208
immunotherapy, 200
impedance spectroscopy, 50
implants, 67
implementation, 57
in situ, x, 116, 127, 131, 135, 151, 154, 155, 161
in situ hybridization, 116

in vitro, ix, xi, 19, 25, 26, 44, 46, 56, 59, 68, 73, 85, 97, 114, 145, 165, 176, 179, 180, 192, 199
in vivo, xi, 14, 19, 24, 25, 26, 44, 46, 73, 106, 129, 138, 141, 142, 153, 164, 165, 169, 176, 180, 193, 195, 199, 204
inactivation, 79
inactive, 123, 184, 192
inclusion, 148
incubation, 50, 92, 142
incubation time, 142
indicators, 22
indigenous, 59
indole, 133
indomethacin, 67
induction, 20, 165, 193
industrial, viii, 41, 42, 146
industry, 45, 46, 55, 146
inert, 138, 141
inertness, x, 127, 131, 141
infection, 28, 47, 60, 64, 77, 86, 100, 110, 114, 115, 135
infections, 44, 86, 87, 101, 129, 135, 194, 207
infectious, viii, 41, 80
infectious disease, viii, 41, 80
infectious diseases, viii, 41, 80
inflammation, 42, 56, 66, 67, 77, 80, 86, 100, 128, 134, 158, 159, 165, 208
inflammatory, 21, 44, 64, 67, 76, 79, 82, 161
inflammatory mediators, 64
inflammatory response, 21, 67
influenza, 44, 57, 59, 87, 101, 134, 163
inherited, 13, 65
inhibition, vii, 1, 7, 9, 12, 14, 15, 21, 25, 75, 78, 128, 130, 132, 133, 134, 135, 136, 153, 155, 161, 162, 163, 164, 170, 185
inhibitor, 6, 7, 8, 12, 14, 75, 79, 121, 133, 134, 135, 136, 137, 152, 155, 162, 191
inhibitors, viii, x, 2, 6, 7, 9, 11, 12, 13, 14, 20, 21, 24, 25, 67, 75, 128, 129, 132, 133, 134, 135, 136, 137, 143, 152, 155, 160, 161, 162, 163, 164, 170, 201
inhibitory, 7, 72, 125, 162, 191
inhibitory effect, 72, 191
initiation, 112, 190, 192, 193
injection, 25, 69
injections, 69
innate immunity, 55, 57
innovation, 46
inositol, 208
insecticides, 58
insertion, 32
insight, 113, 141
instruments, 46
insulin, 130
integration, 142
integrins, 74
integrity, 65, 197
intensity, 203
interaction, viii, 10, 11, 14, 22, 31, 38, 40, 41, 44, 46, 47, 48, 49, 50, 53, 55, 62, 64, 75, 78, 111, 113, 115, 116, 118, 121, 122, 123, 168, 178, 180, 181, 183, 187, 192, 194, 203
interactions, ix, 10, 12, 14, 16, 28, 29, 31, 33, 34, 37, 38, 42, 43, 50, 51, 52, 53, 54, 56, 57, 59, 62, 64, 65, 74, 75, 80, 81, 83, 93, 109, 113, 123, 125, 126, 144, 148, 150, 153, 166, 181, 182, 184, 194
interdisciplinary, viii, 41
interface, 4, 12
interference, 56
interferon, 17
interleukin, 17
interleukin-2, 17
internalization, 24
interpretation, 48
intervention, 125, 128, 136, 157
intestinal tract, 44
intestine, 78, 80
intracellular signaling, 115
intramuscular, 69
intramuscular injection, 69
intraocular, 136
intraocular pressure, 136
intrinsic, 3, 18
invertebrates, 78, 79, 84
Investigations, 123
investment, viii, 41
ionic, 10
ionization, 31
ions, 32
Ireland, 41
irradiation, 147, 149
Islam, 159
isoforms, 45, 110, 112, 115, 116, 121, 122, 136
isolation, 47, 112, 128
isomerization, 38
isozyme, 136, 137
isozymes, 130, 136, 164
Italy, 61, 73

J

Japan, 27, 36, 85, 100, 109, 173, 199
Japan Tobacco, 85
JI, 53
joint damage, 66
joint pain, 68, 69, 70, 71
joints, 64, 66, 71, 82
judgment, 72
Jung, 163

K

KB cells, 20
kidney, 3, 7, 13, 176, 178, 186, 196
killer cells, 165
killing, 20
kinase, 90, 102
kinases, 162
kinetic parameters, 49
kinetic studies, 130
kinetics, 103, 111
knees, 69, 72
knockout, 3, 9

L

L2, 185
label-free, 48, 52, 58
labeling, 4, 9, 53
lactose, 95, 96, 98, 105, 151, 152, 156, 167, 170
lamellae, 65
lamina, 200, 208
laminin, 192, 193, 197
Laminin, 186
large-scale, ix, 80, 86, 99
larva, 185
laser, 18
Lassa fever, 194
learning, 176, 177
lectin, x, 6, 20, 28, 31, 33, 36, 37, 38, 42, 45, 47, 48, 49, 50, 54, 55, 57, 58, 128, 132, 147, 153, 154, 155, 156, 157, 166, 167, 170, 184
leishmaniasis, 135
lentiviral, 25
lesions, 13, 71, 135
leukemia, 19, 23
leukemia cells, 19
leukemic, 7, 21
leukemic cells, 7, 21
leukocytes, 66
ligand, viii, 32, 34, 35, 40, 41, 47, 50, 121, 122, 129, 131, 143, 152, 155, 192, 209
ligands, ix, xi, 31, 47, 48, 49, 50, 51, 53, 56, 109, 112, 151, 152, 153, 155, 156, 194, 199, 201, 203, 204
Light-induced, 123
limitations, 43, 48, 49
linear, 51, 74, 138, 144, 149, 153, 192
linear polymers, 149
linkage, ix, 74, 85, 87, 95, 96, 98, 106, 129, 165, 177, 186
links, 2, 39
lipid, vii, 1, 2, 3, 6, 7, 10, 11, 12, 13, 14, 16, 17, 19, 20, 21, 23, 24, 128, 157, 203, 204, 206
Lipid, 7, 10, 12, 21, 22, 24, 26, 76
lipid metabolism, 10
lipid rafts, vii, 1, 2, 10, 14, 23
lipids, 2, 3, 5, 7, 10, 11, 16, 19, 47, 128
lipophilic, 148
lipopolysaccharide, 50, 105
lipopolysaccharides, 100
lipoprotein, 196
liquid chromatography, 43, 82
liver, 3, 13, 14, 15, 20, 36, 37, 64, 78, 96, 101, 133, 169, 200, 208
liver cells, 3
liver disease, 64
LMW, 78
localization, 2, 3, 5, 11, 12, 16, 17, 18, 33, 77, 124, 178, 181, 182, 184, 186
location, vii, 1, 4, 101, 191
London, 23, 82, 207, 209
longevity, 9, 22
low molecular weight, 78
low-density, 16
LSD, 11
LTA, 111
lumen, 9, 10, 22, 90
luminal, vii, 1, 9, 30, 31, 36
lung, 78, 125
lungs, 80
luteinizing hormone, 191
lymph, 78
lymph node, 78
lymphatic, 17
lymphocyte, 200, 208
lymphocytes, 3, 5, 59
lymphoid, 14, 80, 203

lymphoid organs, 80
lymphoma, 142, 143, 203
lysine, 32, 75
lysosomes, 13
lysozyme, 67

M

mAb, 49, 179
machinery, 22, 34, 43, 142, 183
macromolecules, viii, 43, 61, 62, 74, 81, 145
macrophage, 13, 25
macrophages, 13, 135
magnetic, 43
maintenance, 19, 65, 175, 186
Maintenance, 125
major histocompatibility complex, 206
malignant, 19, 55, 59, 64, 146
malignant mesothelioma, 64
maltose, 147
Mammalian, 15, 22, 53, 91, 93, 101, 125, 183, 192, 197
mammalian cell, 21, 22, 80, 183, 192
mammalian cells, 21, 22, 183, 192
mammalian tissues, 80, 84
mammals, 31, 36, 78, 79, 80, 90, 91, 101, 112, 175, 181, 194, 195, 197, 207
management, 14, 25, 44
Manganese, 32
manipulation, 204
mantle, 80
manufacturing, 45
mapping, 54, 57
market, 45, 46, 52, 56
market value, 45
mask, 45
mass spectrometry, 43, 54, 58, 186, 193
mast cell, 74, 79, 80, 84, 111
mast cells, 74, 79, 80, 84, 111
maternal, 73
matrix, viii, ix, 61, 62, 63, 64, 65, 67, 81, 82, 84, 109, 110, 123, 125, 193
matrix metalloproteinase, 125
matrix protein, 193
maturation, 117
MDR, 2, 3, 5, 6, 7, 10, 12, 15, 17, 18, 19
measurement, 46, 50, 71, 72
measures, 70, 73
mechanical properties, 144
media, 5, 12, 131
mediators, 64, 66
medical diagnostics, 59
medicine, 45, 166
melanoma, 7, 21, 48, 169
membranes, 3, 5, 10, 11, 16, 23, 192
memory, vii, x, 136, 173, 176, 178, 185
memory formation, x, 173, 178
memory loss, 136
men, 72
mergers, viii, 41
mesothelioma, 64
metabolic, 6, 128, 141
metabolic pathways, 128
metabolism, vii, 1, 7, 10, 13, 15, 16, 20, 22, 24, 25, 65, 66, 71, 73, 82, 99, 124, 128, 129
metabolite, 7
metabolites, 3, 5, 6
metalloproteinases, 125
metals, 50
metastasis, 42, 47, 56, 64, 77, 128, 134, 165
metazoan, 78
methylation, 113
Mg^{2+}, 92
MHC, xi, 59, 199, 200, 203, 204, 206, 207, 208, 209
MHC class II molecules, 59
mice, xi, 3, 7, 14, 15, 17, 21, 25, 26, 79, 84, 118, 121, 124, 175, 176, 177, 178, 199, 200, 201, 202, 205, 206
micelles, 48
microarray, 55
Microarrays, 57, 58, 153, 170
microbes, 44
microbial, 44, 55, 58, 62, 99, 103, 128, 150, 207
Microbial, 101
microbiota, 44
microenvironment, 23, 176
micro-environments, 166
microflora, 58
microheterogeneity, 74, 128
microorganisms, 86
microscopy, 9, 10, 16, 18, 49, 142
microsomes, vii, 1, 8
microwave, 147, 149, 150
middle-aged, 66
migraine, 130
migration, 17, 42, 77, 121, 175, 185
mimicking, 87
mimicry, 87, 88, 100, 133, 138
mineralized, 82
miniaturization, 52

Ministry of Education, 36
mirror, 49
misfolded, 30, 33, 39
mitochondria, 4, 18
mitochondrial, 3
mitochondrial DNA, 4
mitochondrial membrane, 3
mitogenic, 44
mitosis, 21
Mitsubishi, 199
mobility, 68, 69, 71
model system, vii, 1, 7
modeling, 83
models, 14, 25, 59, 64, 78, 153, 166, 204
modulation, 11, 18, 24, 136, 152, 200, 204
modules, 147, 195
moieties, 28, 34, 51, 86, 88, 99, 106
molecular biology, vii, 42
molecular dynamics, 113
molecular mass, 49, 62, 63, 64, 74, 78, 79
molecular mimicry, 86, 100, 101
molecular weight, 44, 66, 74, 78, 81, 84, 144
molecules, ix, x, xi, 7, 9, 10, 42, 43, 44, 45, 47, 49, 50, 51, 58, 59, 63, 64, 74, 109, 115, 120, 124, 125, 148, 151, 173, 175, 176, 178, 182, 185, 194, 199, 200, 203, 204, 206, 207, 209
monoclonal, 4, 100, 184
monoclonal antibodies, 100
monoclonal antibody, 4, 184
monolayer, 50
monolayers, 51, 154, 157
monomer, 152
monomeric, 146
monomers, 144
mononuclear cell, 19, 23, 200, 201
mononuclear cells, 19, 23, 200, 201
monosaccharide, 38, 91, 136, 137, 148, 201
monosaccharides, 65, 86, 98
Moon, 112, 122
morphogenesis, 64, 115, 120, 125
morphological, 115
mouse, 9, 25, 26, 30, 39, 59, 111, 116, 121, 125, 126, 176, 178, 185, 186, 196, 200, 207
mouse model, 25, 26
movement, 7, 22, 26, 69
mRNA, 116, 176
MRSA, 130
mucin, xi, 44, 47, 48, 60, 189, 190, 193, 194, 195
multidrug resistance, 12, 15, 16, 17, 18, 19, 20, 21, 22, 24
multiple sclerosis, 200
murine model, 25
muscle, 192, 194
musculoskeletal, 66
musculoskeletal pain, 66
mutagenesis, 92, 93
mutant, 12, 176, 177
mutants, 22
mutation, 92, 194
mutations, 18, 194
myelin, 186
myocardial infarction, 13
myo-inositol, 208

N

N-acety, v, ix, 29, 47, 58, 62, 65, 85, 86, 87, 89, 90, 93, 95, 96, 98, 100, 101, 102, 103, 109, 110, 130, 134, 140, 152, 153, 155, 159, 162, 169, 170, 176, 177, 184, 186, 187, 195, 202
NaCl, 97
nanoparticles, 50, 56
nanoparticulate, 169
nanostructures, 150
nanotechnology, 46
nanotubes, 167
naphthalene, 143
Nash, 16, 160
natural, vii, viii, x, xi, 3, 8, 46, 48, 51, 57, 61, 62, 63, 73, 78, 80, 97, 106, 114, 129, 133, 140, 145, 146, 160, 165, 166, 173, 175, 184, 196, 199, 201, 204, 207, 209
natural killer, x, 3, 165, 173, 175, 184, 196, 207, 209
natural killer cell, x, 3, 165, 173, 175, 184, 196
nausea, 69
necrosis, 20
nematode, 60, 112
neoglycoconjugates, x, 127, 132, 150, 152, 157, 166
Neomycin, 159
neonatal, 25
neoplasia, 165
neoplastic, 16, 52
neoplastic cells, 16, 52
nerve, 88, 185, 197
nerves, 192
nervous system, x, 116, 125, 173, 175, 178, 185
network, 36, 67, 175
neural crest, 175, 185
neural function, 175
neural network, 175

neural tissue, 86
neuraminidase, 98, 134
neuroblastoma, 7, 20
neurodegenerative, 13, 128
neurons, 116, 118, 120, 185, 194
neuropathic pain, 13
neuropathy, 86, 185
neuropeptides, 164
neuropsychopharmacology, 164
New York, 82, 99
NHS, 30, 50, 145
Nielsen, 84
nitric oxide, 67
nitrogen, 133
NMR, 31, 43, 75, 83, 102, 104, 148, 186
nodes, 72
non-native, 34
non-polymorphic, 200
normal, 5, 8, 11, 16, 17, 43, 67, 71, 80, 121, 194
novel materials, 145
NSAIDs, 67, 68, 69, 82
N-terminal, 4, 30, 34, 176, 181, 183, 191, 193
nuclear, 3, 17, 43, 101, 181
nuclear magnetic resonance, 43
nucleic acid, viii, 41, 42, 46, 51, 133, 144
nucleosides, 133, 161
nucleotides, 93, 133
nucleus, 3, 90, 181
nutrients, 44, 59

O

obesity, 136
observations, 5, 7, 10, 78, 190
oligomer, 64
oligomeric, 42
oligomerization, 125, 174
oligonucleotides, 150
oligosaccharide, 28, 66, 86, 90, 99, 100, 111, 114, 138, 141, 147, 148, 176, 190, 196, 197
oligosaccharides, viii, 14, 27, 29, 31, 32, 34, 35, 36, 37, 44, 47, 50, 54, 58, 59, 75, 82, 86, 88, 91, 99, 101, 103, 106, 107, 114, 115, 128, 134, 135, 146, 148, 153, 154, 165, 196, 197
Oligosaccharides, 54, 128, 159, 195
oncogenesis, 134, 136, 152
optic chiasm, 118
optical, 46, 54, 57
optimization, 57
oral, 68, 69, 70, 71, 72, 80, 83
organ, 13
organic, x, 127, 130, 131, 145, 146, 147, 150
organic solvent, 131, 146, 147
organic solvents, 131, 146, 147
organism, 197
organization, 14, 22, 23
orientation, 66, 96, 98
osmosis, 64
OST, ix, 109, 110, 111, 112, 113, 114, 115, 116, 117, 118, 119, 120
osteoarthritis, vii, ix, 61, 62, 64, 72, 73, 81, 82, 83
osteosarcoma, 17
ovarian cancer, 15, 19
ovary, 15, 22, 112, 124
oxide, 67, 82
oxygen, 67, 91, 96

P

paclitaxel, 7, 24
PAF, 5
pain, 13, 66, 68, 69, 70, 71, 72, 73, 82, 83
pain score, 70
paracrine, 67
paralysis, 86
parasite, 42, 59, 135
parasites, 80, 163
Paris, 82
parotid, 196
partial thromboplastin time, 78
passive, 69, 132
pathogenesis, 62, 87, 101, 115
pathogenic, 87, 153, 207
pathogens, 43, 44, 80, 104, 166
pathology, 13, 42, 63, 122, 128, 136, 157
pathophysiological, 129
pathophysiology, 24, 80
pathways, 9, 24, 30, 67, 88, 91, 126, 128, 136, 144
patients, 13, 25, 45, 66, 68, 69, 70, 71, 72, 73, 74, 83, 86, 100, 125, 194
patterning, 120
PCR, 111, 116, 119, 200
peptide, xi, 28, 29, 34, 39, 40, 51, 75, 97, 128, 133, 138, 139, 141, 164, 189, 190, 191, 192, 193, 194, 195
Peptide, 161, 167, 191, 192
peptides, 3, 17, 51, 75, 111, 137, 140, 145, 147, 148, 192, 201
performance, 30, 60, 81, 176
Peripheral, 60

peripheral blood, 19, 23, 200, 207
peripheral blood mononuclear cell, 19, 23
peripheral nerve, 13, 192, 197
peripheral nervous system, 122
periplasm, 97
peritoneal, 84
permeability, 66, 78, 129
permit, 43, 44, 51
perturbation, 128
P-glycoprotein, 2, 3, 7, 15, 16, 17, 18, 19, 20, 21, 22, 23, 24, 26
pH, x, 31, 32, 33, 35, 38, 47, 91, 92, 93, 97, 98, 104, 127, 131, 136, 157
phage, 51, 55
phagocytosis, 67
pharmaceutical, 75, 144, 146
pharmaceutical industry, 146
pharmaceuticals, 3, 45
pharmacodynamics, 80
pharmacokinetic, 3, 24, 45, 129
pharmacokinetics, 17, 78, 80, 129
pharmacological, 62, 75, 82, 136
pharmacology, 63
phenol, 154
phenotype, vii, viii, 2, 3, 10, 14, 15, 18, 65, 118, 184, 206
phenotypes, 177, 178
phenylalanine, 18
phosphate, xi, 33, 86, 89, 90, 91, 96, 112, 119, 145, 163, 169, 189, 203, 208
phosphatidylcholine, 7, 19, 21
phosphatidylethanolamine, 7, 17
phosphoenolpyruvate, 89, 90
phospholipids, 5, 7, 22, 176
phosphorescence, 46
phosphorylation, 57, 181, 208
photosynthesis, 136
phylogenetic, 78
phylogenetic tree, 78
physical sciences, viii, 41
physicians, 72
physicochemical, 129, 140, 167
physicochemical properties, 140, 167
physiological, viii, 3, 5, 16, 21, 23, 42, 43, 46, 61, 62, 81, 115, 116, 117, 128, 129, 136, 143, 157, 163, 183
physiology, 42, 80, 121
pilot study, 71, 83, 147
pineal, 113, 115, 123
pineal gland, 113, 115
PL, 19
placebo, 69, 70, 71, 72, 73, 82, 83
placenta, 112
placental, 196
plants, 45, 47, 48
plaque, 77
plasma, viii, 2, 3, 4, 5, 7, 10, 11, 12, 14, 16, 19, 21, 23, 24, 63, 66, 67, 78, 113
plasma membrane, viii, 2, 3, 4, 5, 7, 10, 11, 12, 14, 16, 19, 21, 23, 24, 63, 67
plasmid, 15, 30, 144
plasminogen, 76
plasticity, vii, x, 173, 176, 177, 178, 185
platelet, 5, 18, 78, 92
platelet-activating factor, 5, 18, 92
platelet-activating factor (PAF), 5
platforms, 48, 52, 54
platinum, 50, 51
play, ix, 2, 6, 11, 12, 42, 47, 48, 62, 85, 86, 92, 95, 99, 110, 114, 115, 120, 137, 191, 192
pleural, 64
pleurisy, 67
PNA, 47, 49, 50
point mutation, 194
polarized light, 49
polyacrylamide, 146
polyethylene, 50, 59, 157
polymer, 66, 74, 111, 144, 145, 146, 149, 152
polymer structure, 144
polymerase, 111, 116
polymerase chain reaction, 111, 116
polymeric macromolecules, 145
polymerization, 110, 181, 187
polymers, 106, 110, 143, 144, 145, 147, 149, 151, 152, 168
polypeptide, 28, 34, 35, 42, 146, 190, 195
polypeptides, 29, 146
polysaccharide, ix, 63, 65, 74, 80, 109, 110, 129, 147, 166
polysaccharides, vii, viii, 61, 62, 80, 91, 128, 144, 166, 167
Polysaccharides, v, 57, 61, 167, 168
polysialic acid, 104, 175, 184
polystyrene, 145, 154
pools, 9, 23
poor, 48, 98, 129, 193
population, 19, 73, 200, 202
pores, 148
portability, 46
post-translational, 42, 54, 58, 195, 197

post-translational modifications, 42, 54, 58, 195
post-translational modifications (PTM), 42
power, 3, 166
prediction, 191, 193
pregnancy, viii, 41, 46, 196
pregnancy test, viii, 41, 46
pressure, 69, 136
prevention, 24, 44, 78
primaquine, 169
primary tumor, 64
probe, 50, 98, 111, 117, 118, 142, 143, 186
procoagulant, 67
prodrugs, 129, 166
production, ix, 6, 42, 43, 45, 48, 51, 54, 60, 67, 84, 86, 87, 97, 99, 103, 106, 110, 200, 204, 205, 207
progenitors, 116
progeny, 80
prognosis, 54
prokaryotes, 56, 99
prolactin, 191, 196
proliferation, 5, 59, 62, 66, 76, 116, 125, 153, 202
proline peptide, 38
promote, 14, 66
prostate, 58
prostate cancer, 58
proteases, 67, 74, 75, 80
Proteasome, 33
protection, 4, 88, 115, 128, 134, 150, 194
protein, vii, ix, xi, 1, 3, 7, 10, 12, 13, 17, 18, 19, 21, 22, 23, 29, 30, 31, 33, 36, 37, 38, 39, 40, 42, 43, 45, 47, 49, 50, 53, 57, 58, 59, 62, 74, 75, 76, 77, 81, 83, 85, 91, 99, 101, 111, 112, 113, 115, 118, 120, 126, 128, 129, 137, 142, 143, 144, 150, 153, 154, 157, 164, 166, 174, 180, 181, 182, 186, 187, 189, 190, 191, 192, 193, 194, 195, 196, 197
protein binding, 129, 144, 155, 157
protein conformations, 194
protein engineering, 47
protein folding, 36, 43, 128, 137
protein function, 99
protein-protein interactions, 33
proteins, viii, ix, x, xi, 2, 3, 6, 8, 9, 10, 11, 14, 15, 22, 23, 24, 27, 28, 30, 31, 34, 36, 40, 41, 42, 43, 44, 45, 47, 48, 49, 51, 53, 54, 60, 62, 64, 65, 67, 74, 75, 76, 77, 80, 83, 99, 109, 110, 111, 116, 118, 120, 126, 128, 147, 152, 153, 154, 155, 156, 157, 158, 173, 174, 176, 180, 181, 182, 183, 187, 189, 190, 191, 192, 193, 194, 195
proteobacteria, 200
proteoglycans, viii, 61, 62, 81, 120, 121, 125, 177, 186
proteolysis, 34
proteome, 59
proteomics, 43, 46
protocol, 69
protocols, 155
prototype, 51
protozoa, 135
proximal, xi, 12, 189
PSA, 159, 175, 182
pseudo, 168
PTT, 78
PUB, 34
purification, 47, 52, 103
Purkinje, 116, 117, 118, 125
Purkinje cells, 117, 118, 125
pyridine ring, 133
pyrimidine, 162
pyrophosphate, 91
pyruvate, 51, 89

Q

QSAR, 165
quality control, viii, 28, 33, 36, 37, 47
quality of life, 73
quantum, 155, 170
quantum dots, 155, 170
quartz, 50, 56, 57, 156

R

radical, 144
radiological, 69, 70, 71, 72
Raman, 42, 46, 57, 81
range, viii, ix, x, 7, 52, 61, 74, 86, 98, 101, 127, 129, 131, 135, 136, 138, 143, 146, 150, 152, 157
rat, 31, 36, 37, 57, 60, 96, 101, 196
ratings, 70
rats, 58, 73, 84
reaction mechanism, 91, 93, 183
reaction rate, 131
reaction time, 150
reactivity, 129, 149, 201
reagents, 58
real time, 51, 154, 156
receptor agonist, 208

receptors, 44, 56, 57, 87, 115, 123, 148, 168, 196, 203
recognition, viii, xi, 27, 28, 35, 36, 37, 39, 43, 46, 48, 49, 50, 51, 53, 55, 57, 58, 59, 75, 86, 87, 105, 112, 128, 129, 136, 143, 151, 170, 175, 180, 186, 187, 189, 191, 194, 201, 203, 207
recovery, 15, 49, 78
redistribution, 2, 11, 16, 31
redox, 50, 147
reduction, 13, 14, 25, 50, 52, 64, 66, 68, 69, 70, 115
regional, 120, 125
regioselectivity, x, 127, 130, 147
regular, 65
regulation, vii, 1, 16, 20, 59, 81, 113, 116, 120, 123, 125, 136, 180, 183, 191, 196, 200
regulators, x, 5, 174
rejection, 165
relationship, vii, viii, ix, 2, 3, 5, 6, 10, 11, 14, 61, 75, 85, 95, 123, 124, 181, 191
relationships, 17, 23, 57, 81, 128, 164
relative size, 150
relevance, 157
reliability, 51
remodeling, 64, 66
remodelling, 165
renal, 13, 48, 136
renal failure, 13
repair, 67
replication, 43, 133
reproduction, 73
research, viii, ix, x, 2, 12, 13, 41, 42, 44, 46, 86, 110, 113, 120, 133, 141, 152, 153, 157, 173, 175
researchers, viii, 41, 42, 48, 113
residues, ix, 4, 18, 28, 29, 30, 31, 32, 33, 38, 62, 66, 74, 75, 87, 93, 95, 96, 100, 103, 109, 110, 111, 112, 113, 115, 123, 137, 142, 155, 176, 190, 191, 201, 206, 208
resilience, 67
resistance, vii, 1, 2, 3, 6, 12, 15, 16, 17, 18, 19, 20, 21, 22, 24, 44, 50
resolution, 39, 49, 55
resources, 43
respiration, 136
responsiveness, 201, 203
retention, 135, 180, 186
reticulum, 10, 11, 17, 22, 28, 36, 37, 38, 39, 40, 43, 187, 192
Reynolds, 20
rheological properties, 64
rheumatic, 64
rheumatic diseases, 64
rheumatoid arthritis, 64, 73
rigidity, 132, 133, 149
risk, 44, 76
risks, 66
RNA, 43, 56, 147, 167
RNAi, 115
robustness, 51, 155
room temperature, 139, 147
Royal Society, 151
Rutherford, 18

S

safety, 25, 72, 73
sample, 43, 46, 49, 50
sampling, 51
SAP, 154
saponins, 129
SARS, 44
SARS-CoV, 44
saturation, 11
scaffold, 134, 136, 144, 146, 152, 165, 166
scaffolding, 23
scaffolds, 129, 133, 136, 137, 140, 148, 150
scavenger, 25
Schmid, 103
scores, 70, 72
search, 98, 111, 112, 113, 200, 201
searches, 193
seawater, 97
secretion, 38, 39, 43, 136, 174, 202, 203
security, 46
selecting, 51
selectivity, 12, 14, 49, 50, 88, 136, 190
Self, 153, 154, 156, 170
self-antigens, 200
senescence, 5
sensing, 31, 32, 33, 34, 35, 42, 51
sensitivity, 3, 6, 9, 12, 15, 21, 46, 47, 49, 50, 51, 92, 98
sensor technology, 48, 51
sensors, 46, 47, 50, 51, 52, 53, 55, 56, 59
separation, 75, 82, 123, 153
sepsis, 125
sequencing, 47, 111
series, xi, 28, 34, 133, 140, 144, 145, 147, 148, 150, 151, 152, 191, 193, 194, 197, 199, 201, 203, 208
serine, 6, 22, 58, 65, 74, 75, 79, 181, 190, 195
serpin, 75, 76, 123

serum, 13, 14, 15, 26, 43, 45, 52, 57, 64, 65, 67, 111, 125, 203
services, 193
shape, 66, 148
shares, 33
sheep, 197
sialic acid, 51, 56, 86, 87, 89, 91, 96, 97, 99, 101, 102, 103, 104, 105, 142, 181, 184
side effects, 67, 69, 80, 135, 136
signal peptide, 97
signal transduction, 115, 181
signaling, 20, 23, 46, 59, 113, 115, 116, 117, 118, 121, 125
signaling pathway, 116
signaling pathways, 116
signals, 20, 56, 64, 116, 117, 118, 120
silver, 50
similarity, 28, 33, 34, 94, 95, 177, 180
Singapore, 82
singular, viii, 41
sites, xi, 10, 32, 34, 63, 66, 75, 95, 105, 112, 121, 142, 143, 148, 189, 193, 194
skeletal muscle, 192
skin, 13, 17, 78, 80
SLPI, 76
socioeconomic, 66
sodium, 17, 83, 97, 150
sodium butyrate, 17
solubility, 12, 129, 146, 148
solvents, 130, 146, 147
sorting, 32, 33, 37, 194
spatial, 66, 176, 185, 191
spatial learning, 176, 185
species, x, 4, 9, 55, 67, 78, 79, 112, 113, 123, 131, 152, 173, 175, 201
specificity, ix, 4, 17, 28, 31, 33, 38, 46, 47, 48, 51, 80, 85, 87, 92, 93, 94, 96, 98, 101, 102, 106, 135, 136, 138, 177, 178, 190, 192, 195, 206
spectroscopy, 50, 57, 148, 186
spectrum, 14
speed, 52
sperm, 191, 192
spermatozoa, 196
spheres, 32
sphingolipids, 6, 15, 16, 21, 22, 24, 208
sphingosine, xi, 6, 12, 199, 203, 205, 208, 209
spines, 116
spleen, 13
splenomegaly, 13
SPR, 49, 53, 156
SRT, 13, 14
stability, 43, 48, 64, 115, 129, 133, 137, 155
stabilization, 11, 71, 194
stabilize, 71
stages, 51, 52, 66, 69, 135
standard deviation, 119, 179
Staphylococcus aureus, 130
starch, 148, 168
statistical analysis, 70
stem cells, 19
steric, 31, 34, 130
steroid, 3
Steroid, 17
sterols, 8
stiffness, 144
stimulant, 208
stimulus, 204
stomach, 44
storage, 11, 13, 14, 15, 24, 25, 128, 130
strain, 30, 95, 98, 134, 142
strains, 86, 92, 99, 100, 134, 135, 140
strategies, viii, x, 13, 24, 41, 43, 57, 58, 107, 128, 132, 138, 146, 150, 154
strength, 47, 65, 143
Streptomyces, 133
stress, 33, 39, 66
stroma, 65
stromal, 65
stromal fibroblasts, 65
structural changes, 74, 105
structural characteristics, 43, 140
structural modifications, 66, 140
structuring, 14
Subcellular, 18
subjective, 71
substitution, 25, 93, 129, 138, 146, 149
substrates, 3, 10, 11, 12, 17, 34, 36, 37, 39, 47, 89, 90, 91, 92, 93, 94, 96, 98, 102, 103, 133, 136, 141, 142, 181, 192
suffering, 69, 71, 72
sugar, viii, x, 7, 13, 14, 27, 28, 29, 30, 31, 33, 34, 35, 37, 38, 39, 42, 43, 46, 48, 50, 54, 55, 89, 91, 93, 98, 99, 100, 105, 133, 137, 152, 156, 161, 169, 173, 181, 187, 190, 192, 203
sugars, viii, 14, 25, 50, 52, 58, 61, 98, 99, 181, 190
sulfamates, 136
sulfate, viii, ix, 61, 62, 65, 66, 81, 82, 83, 84, 86, 109, 110, 111, 112, 113, 115, 116, 120, 121, 122, 123, 124, 125, 126, 178, 180, 181, 182, 187, 197
sulfonamide, 136, 137

sulfonamides, 136, 164
sulfuric acid, 154
sulphate, 68, 71, 81, 82, 83, 121, 122, 124
Sun, 24, 54, 105, 106, 171
supply, 88, 91
suppression, 115
suppressors, 187
supramolecular, 167, 168
surgery, 53
surprise, 178
surveillance, 80, 128, 131
survival, 44, 135
swelling, 68, 70
Switzerland, 36, 70, 73
symptom, 72
symptoms, 13, 64, 66, 67, 68, 69, 71
synaptic plasticity, vii, x, 173, 176, 177, 178, 185
synaptogenesis, 120
syndrome, 64, 66, 81, 86, 100, 101, 194, 198
synovial fluid, 62, 64, 65, 66
synovitis, 66
synthesis, vii, ix, x, 1, 6, 7, 8, 9, 10, 12, 13, 14, 20, 22, 37, 42, 44, 51, 55, 64, 66, 67, 75, 80, 85, 89, 91, 93, 97, 99, 101, 103, 104, 106, 107, 113, 121, 123, 127, 128, 129, 130, 131, 133, 136, 138, 139, 140, 144, 145, 147, 149, 150, 151, 153, 158, 160, 161, 165, 166, 167, 168, 169, 178, 186, 187, 197, 208
systems, ix, 8, 42, 45, 53, 55, 57, 60, 85, 99, 136, 142, 152, 165

T

T and C, 181
T cell, xi, 54, 199, 200, 204, 206, 207, 208, 209
T cells, 206, 207, 208, 209
T lymphocyte, 17, 207
T lymphocytes, 17, 207
tamoxifen, 6, 20
targets, 29, 35, 45, 52, 54, 56, 120, 129, 157, 193, 203
task performance, 176
taxa, 79
Taxol, 7
taxonomy, 28
TCR, vi, xi, 199, 200, 201, 202, 203, 204, 205, 206, 207, 208
technology, 13, 47, 48, 49, 51, 52, 56, 154
temperature, 139, 147
tenascin, 76, 191
tensile, 67
tension, 67
teratogenic, 73
terminals, 151
ternary complex, 75, 105, 112
theory, 37, 168
therapeutics, 45, 55, 128, 129, 137, 158, 159, 165
therapy, 13, 14, 16, 25, 64, 69, 70, 73, 81, 82, 135, 164
Thomson, 163
three-dimensional, 28, 200, 204
threonine, 58, 65, 181, 190
threshold, 49
thrombin, 74, 75, 78
thromboembolism, 78
thymidine, 162, 202
thymocytes, 184, 207
thymus, 78
thyroid, 18
time, viii, 5, 41, 43, 44, 46, 47, 57, 65, 70, 72, 93, 115, 116, 131, 142, 148, 154, 156, 180, 194
time frame, 131
tin, 149
tissue, 3, 25, 26, 45, 52, 56, 64, 65, 67, 73, 74, 77, 78, 86, 112, 136, 144, 146, 166, 184
tissue engineering, 144, 146, 166
tissue plasminogen activator, 45
titration, 148
Tivela mactroides, 84
TM, 3, 12, 17, 24
TNF, 23, 161
TNF-alpha, 23
TNF-α, 161
Tokyo, 27, 36, 82, 103, 104, 105, 109, 199
tolerance, x, 69, 127, 131, 146, 150
toluene, 139
topology, 75
toxic, 73, 144, 160
toxic effect, 73
toxicities, 12
toxicity, 7, 14, 73, 135, 141
toxicological, 73
toxin, 17, 19
toxins, 44
TPA, 76
traffic, 6, 17, 24, 26, 37, 38
traits, 13
trans, 9, 95, 98
transcript, 118, 119, 176
transcription, 39, 43, 116

transcriptional, 120, 128
transcripts, 112, 116, 119
transduction, 76
transection, 185
transfection, vii, 1, 2, 6, 9, 14, 20, 144
transfer, ix, xi, 9, 22, 88, 89, 93, 95, 104, 106, 109, 111, 134, 176, 177, 184, 189, 190, 192, 193, 194
transferrin, 191
transformation, x, 59, 127, 150
transformations, 130, 138, 145, 146
transforming growth factor, 161
transgene, 25
transgenic, xi, 45, 115, 199, 200, 201
transgenic mice, xi, 199, 200
transition, 133, 135, 144, 163
translation, 43
translational, 158
translocation, 5, 6, 7, 8, 9, 10, 11, 14, 22, 28, 191
transmembrane, 3, 4, 12, 17, 18, 31, 136, 176, 181
transmembrane region, 31
transparency, 65, 81
transport, 2, 3, 4, 5, 7, 10, 11, 12, 14, 15, 16, 17, 19, 21, 24, 31, 32, 33, 37, 38, 76, 175
trial, 25, 58, 69, 70, 71, 72, 73
triggers, 38
triglycerides, 5
trimer, 148
tropism, 87
tuberculosis, 161
tumor, 2, 14, 16, 19, 47, 48, 64, 110, 164, 165, 187
tumor cells, 2, 14, 19, 48, 64
tumor growth, 64
tumor metastasis, 110, 165
tumor progression, 165
tumors, 167
tumour, 16, 20, 26, 128, 134, 136, 137, 152, 153
tumour growth, 153
tumours, 137
turnover, 39, 67, 81

U

ubiquitin, viii, 27, 28, 29, 33, 34, 35, 39
ubiquitous, viii, x, 47, 61, 78, 86, 115, 127, 130, 136, 142, 152
ulcerative colitis, 55
underlying mechanisms, 36
unfolded protein response, 39
uniform, 120
United States, 66
unmasking, 169
urinary, 3
urine, 65, 74, 115, 124

V

vaccination, 45
vaccine, 45, 58, 128, 144, 157, 167, 195
Valdez, 187
validity, 180
values, 30, 96, 146, 156
van der Waals, 48, 144
van der Waals forces, 48
vancomycin, 140
variability, 78
variable, 74, 120, 128, 129, 136, 152, 154
variation, 45, 78, 79, 101
vascular disease, 13, 64, 81
vasculature, 66
vasopressin, 164
vector, 30
vehicles, 129, 144, 148
ventricular zone, 116, 117
verapamil, 6, 7, 11, 12, 20
versatility, 157
vertebrates, 77, 80
vessels, 17
vinblastine, 12, 19
viral infection, 28, 86, 110, 194
virulence, 44, 59, 128, 163, 197
virus, 44, 46, 53, 59, 74, 77, 80, 87, 94, 101, 112, 113, 114, 122, 124, 134, 194, 198
virus infection, 87
viruses, 43, 44, 47, 57, 59, 87
vision, 167
visual system, 126
Vitronectin, 77

W

walking, 70, 72
water, 12, 50, 64, 67, 131, 146, 148, 152, 168
wild type, 9
withdrawal, 69
Wnt signaling, 117
women, 72
workers, 69, 136, 139, 140, 142, 144, 146, 147, 148, 150, 152, 153, 154, 155, 156, 157
wound healing, 64, 76

X

xenografts, 26
x-linked, 13
x-rays, 72

Y

yeast, 22, 36, 45, 99, 181, 192, 197, 201
yield, 88, 93, 152

Z

zebrafish, 175, 185